BUPRENORPHINE: COMBATTING DRUG ABUSE WITH A UNIQUE OPIOID

BUPRENORPHINE: COMBATTING DRUG ABUSE WITH A UNIQUE OPIOID

Editors

ALAN COWAN
Department of Pharmacology
Temple University School of Medicine
Philadelphia, Pennsylvania

JOHN W. LEWIS
School of Chemistry
University of Bristol
Bristol, England

WILEY-LISS

A JOHN WILEY & SONS, INC., PUBLICATION
New York • Chichester • Brisbane • Toronto • Singapore

Address All Inquiries to the Publisher
Wiley-Liss, Inc., 605 Third Avenue, New York, NY 10158-0012

Copyright © 1995 Wiley-Liss, Inc.

Printed in the United States of America.

Under the conditions stated below the owner of copyright for this book hereby grants permission to users to make photocopy reproductions of any part or all of its contents for personal or internal organizational use, or for personal or internal use of specific clients. This consent is given on the condition that the copier pay the stated per-copy fee through the Copyright Clearance Center, Incorporated, 27 Congress Street, Salem, MA 01970, as listed in the most current issue of "Permissions to Photocopy" (Publisher's Fee List, distributed by CCC, Inc.), for copying beyond that permitted by sections 107 or 108 of the US Copyright Law. This consent does not extend to other kinds of copying, such as copying for general distribution, for advertising or promotional purposes, for creating new collective works, or for resale.

While the authors, editors, and publisher believe that drug selection and dosage and the specifications and usage of equipment and devices, as set forth in this book, are in accord with current recommendations and practice at the time of publication, they accept no legal responsibility for any errors or omissions, and make no warranty, express or implied, with respect to material contained herein. In view of ongoing research, equipment modifications, changes in governmental regulations and the constant flow of information relating to drug therapy, drug reactions and the use of equipment and devices, the reader is urged to review and evaluate the information provided in the package insert or instructions for each drug, piece of equipment or device for, among other things, any changes in the instructions or indications of dosage or usage and for added warnings and precautions.

Library of Congress Cataloging-in-Publication Data

Buprenorphine : combatting drug abuse with a unique opioid / edited by
Alan Cowan and John W. Lewis.
    p.   cm.
  Includes bibliographical references and index.
  ISBN 0-471-56198-3
  1. Opioid habit—Chemotherapy. 2. Buprenorphine—Therapeutic use.
I. Cowan, Alan, 1942-   . II. Lewis, John W.
    [DNLM: 1. Buprenorphine—therapeutic use. 2. Narcotic Dependence—
  therapy. QV 92 B9443 1994]
RC568.O58B87   1994
616.86'32061 —dc20
DNLM/DLC
for Library of Congress   94-28470
                                                                                                    CIP

The text of this book is printed on acid-free paper.

CONTENTS

Contributors ix

Foreword
George E. Bigelow xi

Preface
Alan Cowan and John W. Lewis xv

CHEMISTRY

Buprenorphine—Medicinal Chemistry
John W. Lewis 3

PRECLINICAL PHARMACOLOGY

Buprenorphine: A Review of the Binding Literature
Richard B. Rothman, Qi Ni, and Heng Xu 19

Update on the General Pharmacology of Buprenorphine
Alan Cowan 31

Behavioral and Pharmacological Determinants of Buprenorphine's Effects on Schedule-Controlled Behavior
Linda A. Dykstra and S. Stevens Negus 49

Reinforcing Effects, Discriminative Stimulus Effects, and Physical Dependence Liability of Buprenorphine
S. Stevens Negus and James H. Woods 71

ASSAY, METABOLISM, AND PHARMACOKINETICS

Analysis
R. Andrew Moore 105

Absorption, Distribution, Metabolism, and Excretion of Buprenorphine in Animals and Humans
Donald S. Walter and Charles E. Inturrisi 113

Buprenorphine Kinetics in Humans
H. J. McQuay and R. A. Moore 137

CLINICAL PHARMACOLOGY AND EVALUATION

Clinical Pharmacology of Buprenorphine in Relation to Its Use as an Analgesic
John W. Lewis 151

Buprenorphine: Epidural and Intrathecal Use
Carl E. Rosow 165

Buprenorphine in Psychiatric Disorders
David Nutt, Simon Groves, Nick Coupland, and Paul Glue 175

STUDIES RELATING TO TREATMENT OF SUBSTANCE ABUSE

Laboratory Studies of Buprenorphine in Opioid Abusers
Donald R. Jasinski and Kenzie L. Preston 189

Clinical Efficacy Studies of Buprenorphine for the Treatment of Opiate Dependence
Paul J. Fudala and Rolley E. Johnson 213

Buprenorphine Treatment of Cocaine and Heroin Abuse
Nancy K. Mello and Jack H. Mendelson 241

Detoxification and Induction Onto Naltrexone
Marc Rosen and Thomas R. Kosten 289

PERSPECTIVE

Buprenorphine: What Interests the National Institute on Drug Abuse?
Doralie L. Segal and Charles R. Schuster 309

Index 321

CONTRIBUTORS

NICK COUPLAND, Psychopharmacology Unit, School of Medical Sciences, University of Bristol, Bristol BS8 1TD, UK [175]

ALAN COWAN, Department of Pharmacology, Temple University School of Medicine, Philadelphia, PA 19140 [31]

LINDA A. DYKSTRA, Departments of Psychology and Pharmacology and The Curriculum in Neurobiology, University of North Carolina, Chapel Hill, NC 27599 [49]

PAUL J. FUDALA, Department of Psychiatry, University of Pennsylvania School of Medicine, and Department of Veterans Affairs Medical Center, Philadelphia, PA 19104 [213]

PAUL GLUE, Psychopharmacology Unit, School of Medical Sciences, University of Bristol, Bristol BS8 1TD, UK; present address: Schering-Plough Research Institute, Kenilworth, NJ 07033 [175]

SIMON GROVES, Psychopharmacology Unit, School of Medical Sciences, University of Bristol, Bristol BS8 1TD, UK; present address: Ravenscraig Hospital, Greenock PA16 9HA, UK [175]

CHARLES E. INTURRISI, Department of Pharmacology, Cornell University Medical College, New York, NY 10021 [113]

DONALD R. JASINSKI, Department of Medicine, Francis Scott Key Medical Center, The Johns Hopkins University School of Medicine, Baltimore, MD 21224 [189]

ROLLEY E. JOHNSON, Department of Psychiatry, The Johns Hopkins University School of Medicine, Baltimore, MD 21224 [213]

THOMAS R. KOSTEN, Department of Psychiatry, Division of Substance Abuse, Yale University School of Medicine, New Haven, CT 06519 [289]

JOHN W. LEWIS, School of Chemistry, University of Bristol, Bristol BS8 1TS, UK [3,151]

The numbers in brackets are the opening page numbers of the contributors' articles.

H. J. McQuay, Oxford Regional Pain Relief Unit, Churchill Hospital, Oxford OX3 7LJ, and Nuffield Department of Anaesthetics, Radcliffe Infirmary, Oxford, UK [137]

Nancy K. Mello, Alcohol and Drug Abuse Research Center, McLean Hospital—Harvard Medical School, Belmont, MA 02178 [241]

Jack H. Mendelson, Alcohol and Drug Abuse Research Center, McLean Hospital—Harvard Medical School, Belmont, MA 02178 [241]

R. Andrew Moore, Oxford Regional Pain Relief Unit, Churchill Hospital, Oxford OX3 7LJ, Nuffield Department of Anaesthetics, Radcliffe Infirmary, Oxford, and Euro/DPC Ltd., Glyn Rhonwy, Llanberis, Caernarfon, Wales, UK [105,137]

S. Stevens Negus, Department of Pharmacology, University of Michigan Medical School, Ann Arbor, MI 48109; present address: Alcohol and Drug Abuse Research Center, McLean Hospital, Belmont, MA 02178 [49,71]

Qi Ni, Clinical Psychopharmacology Section, Intramural Research Program, National Institute on Drug Abuse, National Institutes of Health, Baltimore, MD 21224 [19]

David Nutt, Psychopharmacology Unit, School of Medical Sciences, University of Bristol, Bristol BS8 1TD, UK [175]

Kenzie L. Preston, Addiction Research Center, National Institute on Drug Abuse, Baltimore, MD 21224 [189]

Marc Rosen, Department of Psychiatry, Division of Substance Abuse, Yale University School of Medicine, New Haven, CT 06519 [289]

Carl E. Rosow, Department of Anesthesia, Massachusetts General Hospital, Boston, MA 02114 [165]

Richard B. Rothman, Clinical Psychopharmacology Section, Intramural Research Program, National Institute on Drug Abuse, National Institutes of Health, Baltimore, MD 21224 [19]

Charles R. Schuster, National Institute on Drug Abuse, Addiction Research Center, Baltimore, MD 21224 [309]

Doralie L. Segal, National Institute on Drug Abuse, Medications Development Division, Rockville, MD 20857 [309]

Donald S. Walter, Reckitt & Colman Products, Hull HU8 7DS, UK [113]

James H. Woods, Department of Pharmacology, University of Michigan Medical School, Ann Arbor, MI 48109 [71]

Heng Xu, Clinical Psychopharmacology Section, Intramural Research Program, National Institute on Drug Abuse, National Institutes of Health, Baltimore, MD 21224 [19]

FOREWORD

It is a pleasure to introduce the present volume on buprenorphine. For many years, scientists involved in studies of opioid pharmacology and of drug abuse have included among their goals the development of effective analgesics with reduced potential for abuse and dependence, and the development of effective pharmacologic agents for treatment of opioid abuse and dependence. Buprenorphine appears to make an important scientific and clinical contribution on both of these fronts. Thus it is with considerable scientific and professional enthusiasm that the drug abuse research community welcomes this volume summarizing the pharmacology and clinical effects of this interesting and promising opioid. The present comments focus on the potential of buprenorphine as a new pharmacotherapy for opioid abuse.

The partial agonist character of buprenorphine and its tenacious binding to the opioid receptor confer upon this analgesic a profile of pharmacological activity that appears to offer significant advantages over currently available opioid abuse/dependence pharmacotherapies. These advantages include a ceiling on the magnitude of opioid effects that results in much greater safety; a long duration of receptor binding and opioid activity that results in a slow onset and relatively mild opioid abstinence syndrome; a reduced abuse liability relative to that of full opioid agonists; sufficient opioid agonist activity to be reinforcing and to sustain treatment acceptance/participation by drug abuser patients; opioid antagonist activity that, under certain circumstances, may offer therapeutic advantages and reduce abuse liability; and the possibility of attracting into treatment populations of patients for whom the previously existing treatment modalities have been unacceptable. The hope is that these features will contribute both to improving treatment and to increasing the availability of treatment for opioid abuse and dependence.

The other primary pharmacotherapies available for opioid abuse are methadone and naltrexone. Methadone is an orally active opioid agonist used as a substitute opioid maintenance agent. It is a very effective treatment when administered properly, but it is not ideal. Methadone sustains a considerable degree of physical dependence such that patients experience withdrawal if a daily dose is missed; also, it is relatively unacceptable to some communities. Naltrexone is an orally active opioid antagonist that specifically blocks opioid receptors and prevents opioid effects or opioid dependence even if opioids are used. Naltrexone is in many ways a pharmacological wonder drug, but as a pharmacotherapy it is not ideal. It is useful in selected circumstances, but, in general, it has had relatively little clinical impact on drug abuse treatment largely because patients choose not to take the medication. It has been announced that levo-alpha-acetylmethadol (LAAM) has been approved for clinical use by the U.S. Food and Drug Administration and will therefore soon be available as yet another opioid abuse pharmacotherapy. LAAM, like methadone, is a substitute opioid agonist maintenance agent. Its main advantage over methadone is a longer duration of activity, permitting less than daily dosing and reducing the need to provide take-home medication that might be diverted to illicit or inappropriate use. The major disadvantage with LAAM is a slow onset of agonist activity at the outset of therapy as active metabolites accumulate over several days, such that patients often experience withdrawal discomfort. Thus, LAAM also does not represent the ideal pharmacotherapy.

While buprenorphine has many positive and attractive features, it is not the ideal pharmacotherapy for opioid abuse and dependence either. While praising the benefits and promise of buprenorphine, we should also acknowledge its limitations. Its sublingual route of administration is somewhat cumbersome and inconvenient. Also, this route of administration requires that buprenorphine be prepared in a highly concentrated and water-soluble dosage form, and such dosage forms may have increased potential for illicit diversion to abuse by parenteral injection. Buprenorphine's μ-agonist (i.e., morphine-like) profile of pharmacological activity is one that in other opioid drugs has been associated with risk for diversion and abuse, and buprenorphine itself, based on both laboratory data and epidemiological experience, appears to have potential for diversion and abuse. Another potential liability relates to buprenorphine's duration of activity, or frequency of required dosing. At present, buprenorphine is being developed as a daily-dosing pharmacotherapy. In the U.S. this will likely be associated with a requirement for daily visits to a treatment clinic; such visits can be inconvenient and unattractive to patients and may decrease the acceptability of buprenorphine maintenance therapy. Adequate studies evaluating the feasibility of providing buprenorphine on a less-than-daily basis have not yet been conducted.

Thus, while buprenorphine holds great promise and appears likely to enter the clinical arena shortly as a recognized and approved pharmacotherapy with acceptable safety and efficacy for the treatment of opioid abuse and dependence, the end of the story is not yet written. The research presented in this volume outlines much important information about buprenorphine as an opioid abuse pharmacotherapy. Much more is yet to be learned as we proceed with investigating and documenting how to use this new pharmacotherapy most effectively. The most important point is that clinicians will likely soon have available to them a valuable new pharmacological option in dealing with problems of opioid abuse. Buprenorphine will be a significant and important addition to the therapeutic menu available to drug abuse therapists. Perhaps other medications will follow and will offer additional options.

It is gratifying that new treatments for drug abuse are being developed. The many scientists who have contributed to this progress should be proud of their roles in this accomplishment. The present volume provides a valuable summary of both the methods and the results of drug abuse pharmacotherapy development. In addition to documenting what is known about buprenorphine, the volume offers useful methodological guidance relevant to the development of other pharmacotherapies.

GEORGE E. BIGELOW, PHD

The Johns Hopkins University School of Medicine
Baltimore, Maryland

PREFACE

Over two decades ago, we crouched uncomfortably in a small observation box and peered through a hazy one-way mirror at three patas monkeys playing in a large cage. The animals had been injected three times daily for a month with an oripavine analgesic designated RX 6029-M and had just been challenged with naloxone, the recently introduced opioid antagonist. During the following hour we kept our fingers crossed; signs of abstinence were not precipitated . . . and buprenorphine came of age. This enigmatic compound has gradually emerged into the foreground of opioid research and development—first, as a clinically useful analgesic; second, as a unique pharmacological tool; and third, as a new pharmacotherapy for substance abuse. Indeed, the U.S. Food and Drug Administration has recently granted orphan status to buprenorphine for the treatment of opioid dependence.

This book provides a summary of current knowledge about buprenorphine—from receptor to clinic—and should be of interest to students and seasoned investigators from the biomedical community. The authors come from both sides of the Atlantic and represent academic medicine, the pharmaceutical industry, and the U.S. National Institute on Drug Abuse; they played major roles in the buprenorphine story, and we are very grateful for their enthusiasm and cooperation.

We also acknowledge the dedication of our talented colleagues in Hull whose contributions in the early days laid the foundation for the eventual development of buprenorphine. In this context, Drs. Kenneth W. Bentley, Alan L. A. Boura, and Michael J. Rance deserve an extra special mention.

Finally, it is a pleasure to thank Kelly Franklin, Ginger Berman, John Hanley, and Dr. Brian Crawford of Wiley-Liss for their advice, encouragement, and friendship along the way.

ALAN COWAN
JOHN W. LEWIS

CHEMISTRY

BUPRENORPHINE—MEDICINAL CHEMISTRY

JOHN W. LEWIS

School of Chemistry, University of Bristol, Bristol BS8 ITS, UK

SYNTHESIS AND STRUCTURE

Thebaine is chemically the most reactive of the morphine alkaloids and contains a dienol ether system that enables it to undergo Diels–Alder reactions to produce a range of adducts in very high yields. The first of these adducts (1) was prepared by the reaction of p-benzoquinone with thebaine [Schöpf et al., 1938]. Comparison of its structure with that of morphine (2) shows how the added portion transforms one side of the adduct. Bentley recognized the potential of such adducts as starting materials for chemical elaboration in the search for new analgesics but rejected the benzoquinone adduct because the dihydro derivative (3), to which the adduct had to be converted to avoid aromatization, had very low solubility. Attention was then directed to the adduct (4a) of thebaine with methyl vinyl ketone [Bentley and Hardy, 1967a], and this was the starting point for most of the subsequent work that ultimately led to the synthesis of buprenorphine.

The Diels-Alder reaction is stereo- and regiospecific. With unsymmetrical dienophiles such as methyl vinyl ketone, thebaine reacts only at the exposed face of the diene system to give the endostereochemistry for the etheno bridge. The acetyl substituent in the methyl vinyl ketone adduct is overwhelmingly in the 7α position (4a); only a very small quantity of the $C_{7\beta}$ epimer is produced. The cycloaddition must be under electronic control because no adduct with a C_8-acetyl substituent is formed.

The methyl vinyl ketone adduct (4a), by reaction with Grignard reagents (R^IMgBr), was converted into a series of tertiary alcohols (5) [Bentley et al., 1967b]. The reaction is stereospecific, being controlled by an intramolecular complex formed between the Grignard reagent and the oxygen atoms of the C_6-methoxy group and the carbonyl group, which can deliver the alkyl group (R^I) only to the less-hindered face of the carbonyl group (part structure 6). It follows that di-

---

Buprenorphine: Combatting Drug Abuse With a Unique Opioid, pages 3–16
© 1995 Wiley-Liss, Inc.

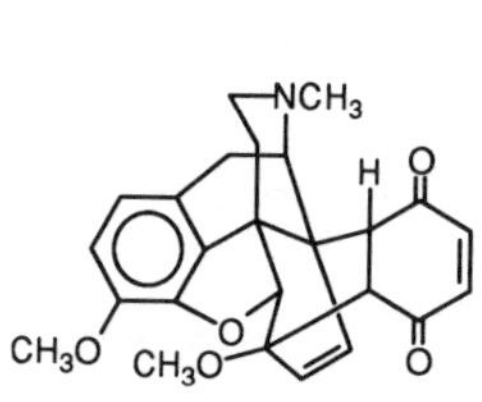

4   LEWIS

1

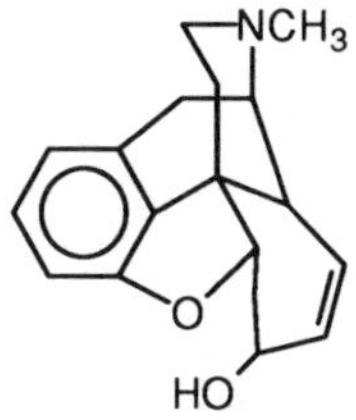

2

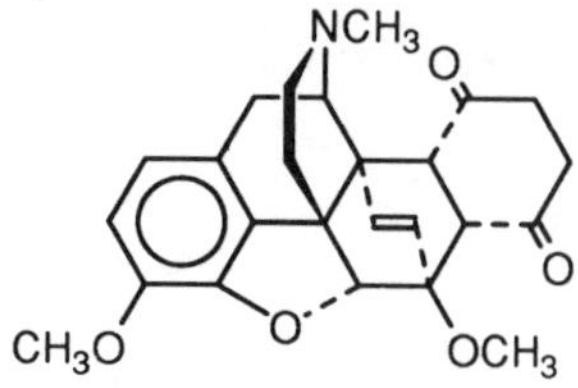

3

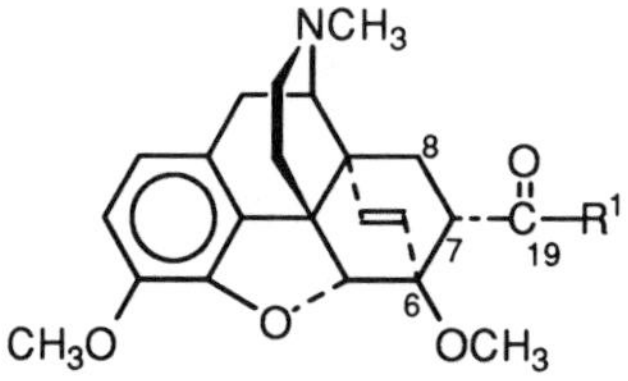

4

(a) R¹=CH₃

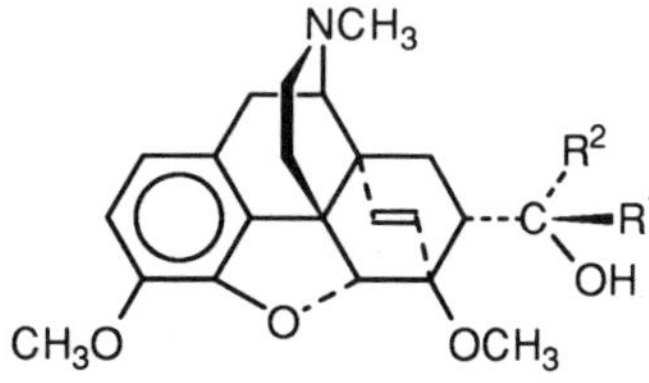

5

(a) R¹=CH₂CH₂CH₃ ; R²=CH₃

(b) R¹=CH₃ ; R²=CH₂CH₂CH₃

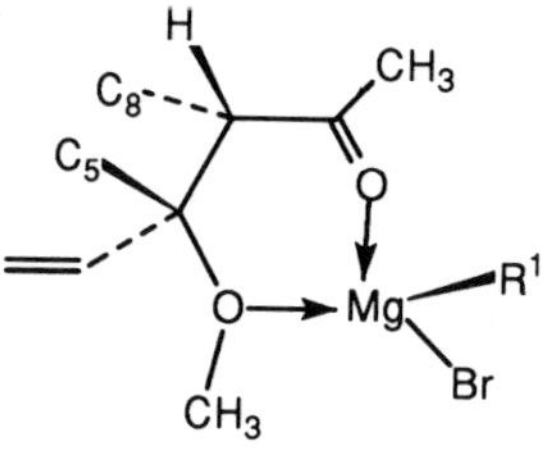

6

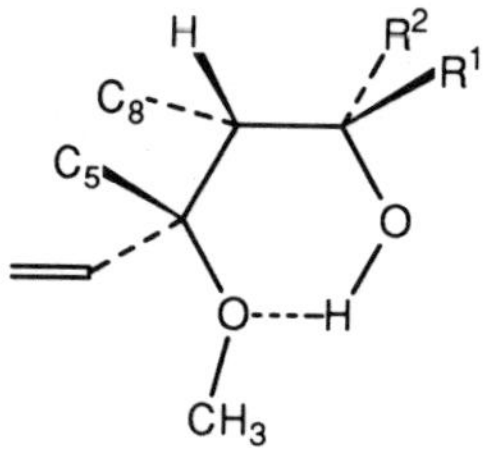

7

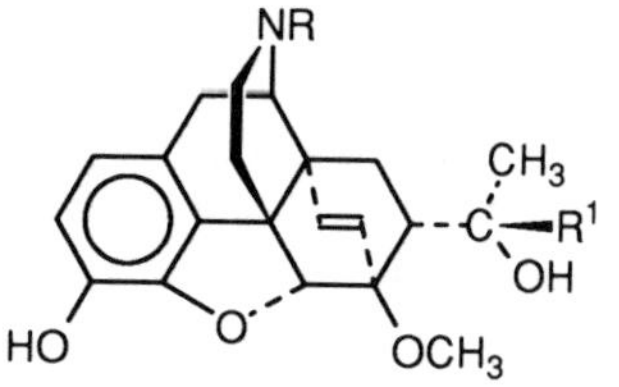

8

(a) R=CH₃

(b) R=CH₂─◁

(c) R=H

astereoisomers of structure 5 in which the groups R¹ and CH₃ attached to C_{19} are interchanged can be prepared by reaction of methyl magnesium bromide with the appropriate ketones of structure 4 [Bentley et al., 1967c].

Demethylation of the phenolic methyl ether group in the tertiary alcohols (5) was effected by sodium hydroxide in boiling diethylene glycol [Bentley and Hardy, 1967b]. The more usual acidic conditions could not be used, since the tertiary

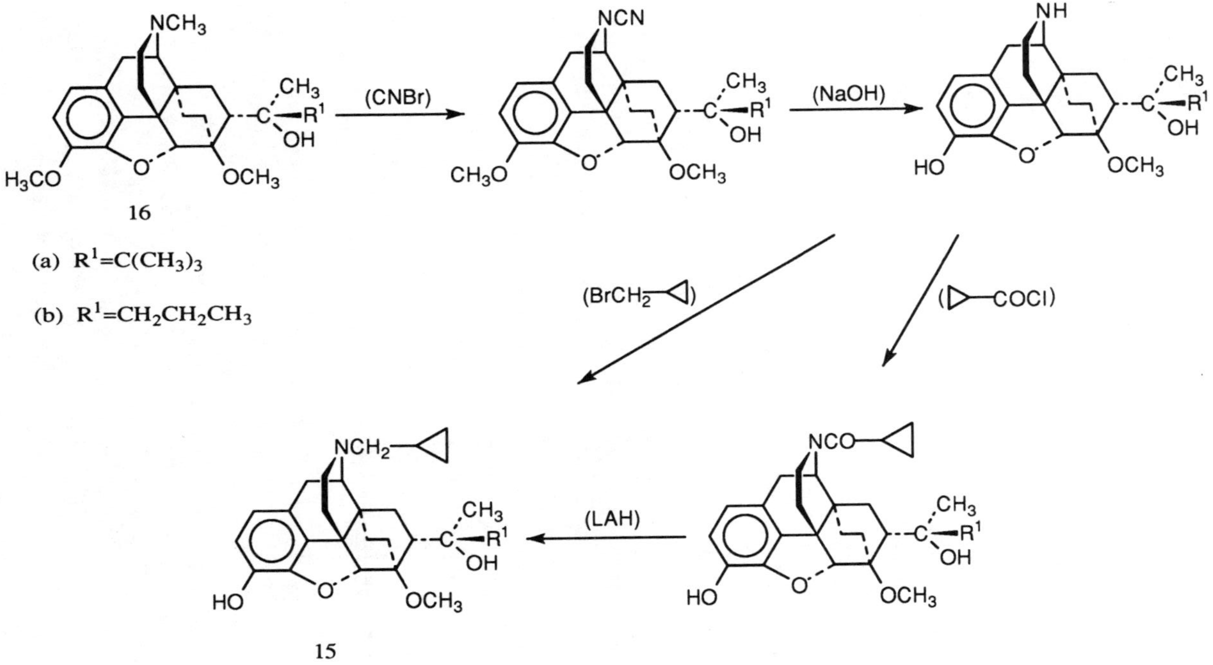

(a) $R^1 = C(CH_3)_3$

(b) $R^1 = CH_2CH_2CH_3$

Scheme 1. Synthesis of buprenorphine.

alcohols undergo acid-catalyzed rearrangements [Bentley et al., 1967a,c). The demethylated products (8a) are derivatives of oripavine (3-O-demethylthebaine) and were given the generic name orvinols; the ethers (5) from which they were prepared were designated thevinols [Bentley et al., 1967b]. Orvinols in which the N-methyl group was replaced by another alkyl, alkenyl, or cyclopropylmethyl group (e.g., 8b) were prepared from thevinols. The procedure involved conversion of the N-methyl group to N-cyano and thence to the nor-O-demethylated product (8c) before appropriate N-alkylation (or N-acylation followed by lithium aluminum hydride reduction). This procedure is illustrated in Scheme 1 for the synthesis of buprenorphine (22d) from the dihydrothevinol (16a) [Lewis, 1974].

The conformation of the C_7 side chain in the thevinols (5) and orvinols (8a) has been studied by nuclear magnetic resonance and infrared spectroscopy [Fulmor et al., 1967], X-ray crystallography [Van den Hende and Nelson, 1967] and semiempirical quantum mechanical methods [Loew and Berkowitz, 1979].

Generally, the conformation (7) with the intramolecular hydrogen bond is of significantly lower energy than other rotamers. This does not apply when both R^1 and R^2 are hydrogen or methyl individually or together. In these compounds there is no preference for the hydrogen-bonded conformer. When either R^1 or R^2 is tertiary butyl, this space-demanding alkyl group occupies the space furthest from the C_6-methoxyl group and the C_8 hydrogen atoms. This was calculated for the orvinol (8a; R^1 = tBu) and its diastereoisomer (structure 8a; R^1 = tBu, in which the methyl and hydroxyl groups are interchanged) [Loew and Berkowitz, 1979]. In the lowest-energy conformers the S-diastereoisomer[1] (8a; R^1 = tBu) having the same configuration at C_{19} as buprenorphine has the intramolecular hydrogen bond; but for the R-diastereoisomer, placing the t-butyl group in its preferred position requires that the hydroxyl group occupy space below C_8, which seems likely to correspond to a lipophilic site on the receptor (see below).

STRUCTURE–ACTIVITY RELATIONSHIPS

In the thevinols (5; $R^2 = CH_3$) antinociceptive potency in the tail pressure test in rats increased as the size of R^1 increased. When R^1 was a straight-chain alkyl group, peak of activity was reached at propyl to butyl at the level of 50–60 times levels for morphine. When a phenyl group was attached to the end of the alkyl chain, the effect of lengthening the chain was more pronounced. There was a 2,000-fold increase in changing phenyl to benzyl, with a small further increase to phenylethyl (300 times morphine) followed by nearly a 300 fold decrease to phenylpropyl [Bentley et al., 1967b]. When the C_{19}-methyl group (R^2) in the thevinols (5) was replaced by ethyl, with R^1 being kept constant as methyl, the effect generally was to decrease potency by about tenfold. Changing R^2 from ethyl to n-propyl resulted in a

---

[1]The orvinols of structure 8 and thevinols of structure 5 in which R^1 is an alkyl group larger than methyl have the C_{19} R-configuration except when R^1 = tBu, when application of the Kahn, Ingold, and Prelog rules (IUPAC, 1970) results in assignment of S-configuration.

further small decrease. As a result of the opposite effects of increasing the size of the two C_{19}-alkyl groups, the activity of diastereoisomeric pairs showed very pronounced differences. For example, the R-n-propyl thevinol (5a) was over 80 times more potent than the S-diastereoisomer (5b) [Bentley and Lewis, 1972]. O-demethylation of the thevinols (5; $R^2 = CH_3$) to the orvinols (8a) resulted in potency increases of 10- to 100-fold. Thus, some of the orvinols are among the most potent of opioids, reaching 10,000 times the potency of morphine [Bentley and Hardy, 1967b].

The magnitude of the effects on analgesic activity of very small changes of structure in the region of C_{19} in the orvinols strongly suggested that receptor binding of these opioids must involve a stereospecific component that was missing in morphine and its derivatives, and in structurally simpler analogs. The C_{19}-hydroxyl group appeared to exert a very specific influence on the interaction. This could be due to hydrogen-bonding to the receptor protein [Bentley and Lewis, 1972], but the discovery that extremely high-potency oripavine analgesics, e.g., 9, could be prepared without a C_{19}-hydroxyl group suggested that an alternative influence of the hydroxyl group was to fix the conformation about C_{19} (by intramolecular hydrogen-bonding to the C_6-methoxyl group) [Loew and Berkowitz, 1979]. This would explain activity differences between diastereoisomers, since the important alkyl group (R^1 in structure 8) would be in different positions and unequally able to interact with a lipophilic site on the receptor. This hypothesis would place the lipophilic site distant from C_7 and C_8 so as to allow favorable interaction with the R- but not the S-diastereoisomer.

Rapoport's group explored further the effects of structural modification of the orvinols; they prepared analogs in which the opportunity for intramolecular hydrogen-bonding between C_{19} and the C_6-methoxyl group was denied. Thus, the methoxyl group was replaced by hydrogen and methyl in structures 10 and 11, retaining the tertiary alcohol function attached at $C_{7\alpha}$ [Hutchins et al., 1981; Knipmeyer and Rapoport, 1985]. In these analogs, as in the orvinols (8a), the R-configuration of C_{19} (10) was consistently associated with very much higher analgesic potency than the S-configuration (11). This showed that intramolecular hydrogen-bonding was not a dominant factor determining the preferred conformation for agonist binding to opiate receptors. Confirmation was obtained from the activity of the orvinans (12), in which the C_6-methoxyl group of the orvinols was retained but the C_{19}-hydroxyl group was replaced by hydrogen [Hutchins and Rapoport, 1984]. The R-orvinan (12a) was nearly 50 times more potent than the S-diastereoisomer (12b).

The authors also prepared the epimeric tetrahydrofuran derivatives (13) by ring closure from the C_{20} primary alcohols (14). In structure 13 the position of the longer alkyl chain (R^1 or R^2) is fixed, whereas in the orvinols (8a) and analogs 10, 11, and 14, relatively free rotation about the $C_{7\alpha}$–C_{19} bond allows it to occupy several positions in conformers of different energy requirements. The diastereoisomer of 14 with the R-configuration of C_{19}, and the diastereoisomer of 13 with the S-configuration of C_{19} into which it was converted, were both very much more potent than the S-(14) and R-(13) diastereoisomers. From these results Hutchins and Rapoport [1984] were able to conclude that the lipophilic site on the receptor is located below

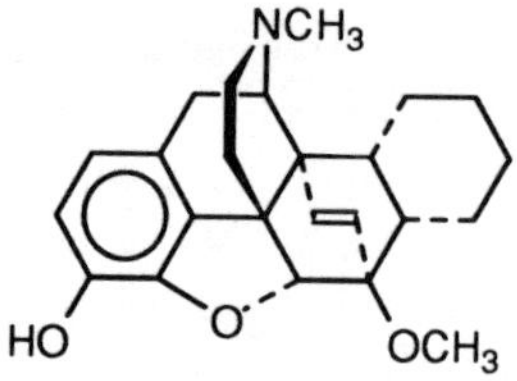

9

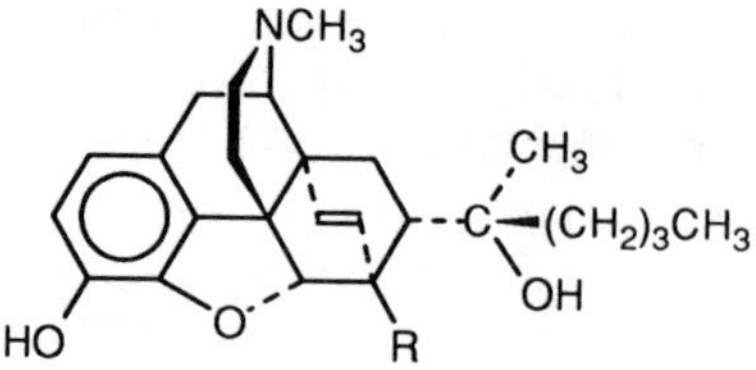

10
(a) R=H
(b) R=CH₃

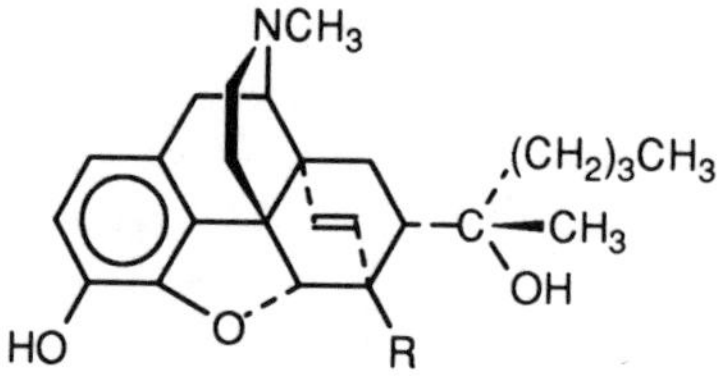

11
(a) R=H (b) R=CH₃

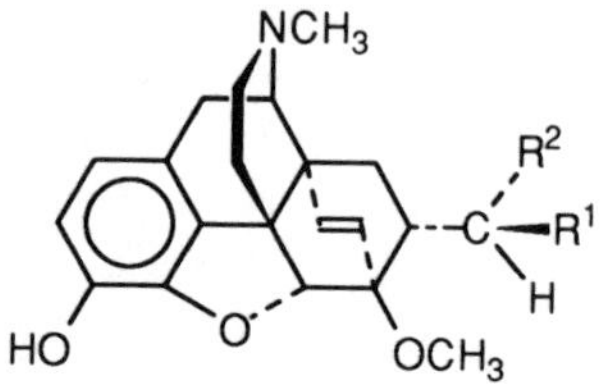

12
(a) R¹=CH₃(CH₂)₃ ; R²=CH₃
(b) R¹=CH₃ ; R²=CH₃(CH₂)₃

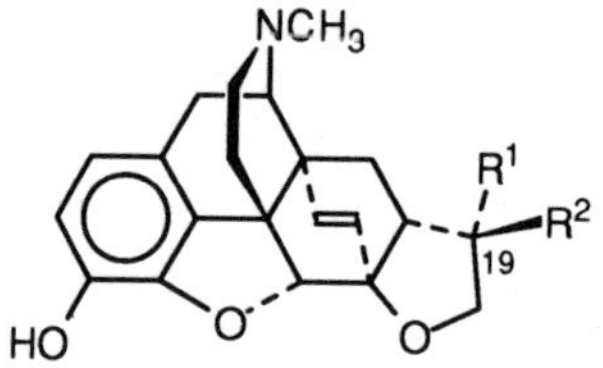

13
(a) R¹=CH₃(CH₂)₃ ; R²=H
(b) R¹=H ; R²=CH₃(CH₂)₃

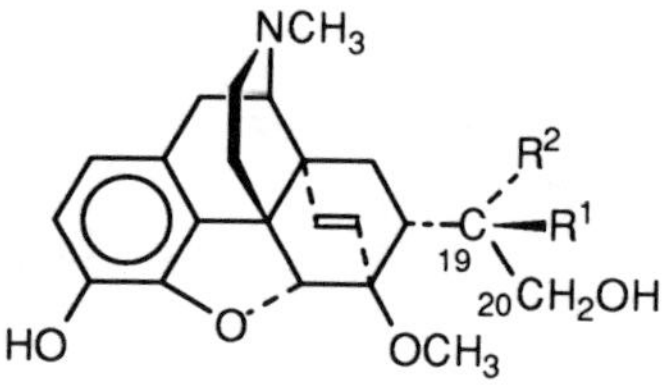

14
(a) R¹=CH₃(CH₂)₃ ; R²=H
(b) R¹=H ; R²=CH₃(CH₂)₃

C_8 and close to the etheno bridge so as to permit optimum agonist interaction with an alkyl chain in these orvinols and close analogs. This allows the C_{19}-hydroxyl group in the more potent diastereoisomers of the orvinols (8a) and analogs (10, 12a, and 14a) to occupy space above the bicyclic ring system. That these derivatives had greater potency than the orvinans (12) and tetrahydrofuran derivatives (13), which lacked such a hydroxyl group, could be explained by a favorable interaction with a secondary hydrophilic site through intermolecular hydrogen-bonding. Rapoport postulated that it is "synergism and competition" of binding at the lipophilic and hydrophilic sites that determines the activity of the bridged oripavine derivatives. If binding to both sites occurs, the analgesic response is significantly greater than when binding only to the lipophilic site is possible [Hutchins and Rapoport, 1984].

This model explains the substantial differences in activity between the C_{19}-R and C_{19}-S diastereoisomers when a C_{19}- or C_{20}-hydroxyl group is present. In these structures the R-diastereoisomer can have synergistic lipophilic/hydrophilic binding, whereas the S-diastereoisomer can have only lipophilic binding. The model offers less convincing rationalization of the potency difference between C_{19}-R and C_{19}-S diastereoisomers of the orvinans (11). In this case, with the n-butyl group binding to the proposed lipophilic site, the hydrophilic site will be occupied by the methyl group in the R-diastereoisomer and the hydrogen atom in the S-diastereoisomer. This difference in binding conformation does not easily explain the nearly 50-fold greater potency of the R-diastereoisomer.

Loew and Berkowitz [1979] showed that the preferred conformation for both the t-butyl orvinol (8a; R = tBu) and its diastereoisomer has the t-butyl group similarly located in the least-hindered space. In 8a (R = tBu), but not in the diastereoisomer, this conformation has the intramolecular H-bond. Since the t-butyl group appears to occupy the same space in the lowest-energy conformer of both diastereoisomers, the authors predicted that they would be more similar in potency than other diastereoisomeric pairs in the series. This contrasts with the prediction from Rapoport's model; assuming that conformations in which the t-butyl group is forced into the space below C_8 are attainable for binding to the receptor, the diastereoisomers would show the normal differentiation. A comparative investigation of these diastereoisomers is needed to answer this point.

The location of the lipophilic site proposed by Hutchins and Rapoport [1984] was different from that implied in the intramolecular hydrogen-bonding hypothesis of Loew and Berkowitz [1979]. Rapoport proposed that it is below C_8, whereas Loew's hypothesis required that it be distant from C_7 and C_8 to correspond to the position of R^1 (in structure 8) in the intramolecularly H-bonded conformation. The high potency of the oripavine (9), as it lacks the C_{19}-hydroxyl group but having the cyclohexano ring linking $C_{7\alpha}$ $C_{8\alpha}$, which can bind to a lipophilic site below C_8, lends support to Rapoport's proposal.

N-CYCLOPROPYLMETHYLNORORVINOLS

From the outset of the exploration of the orvinol series in the search for improved opioid analgesics the importance of replacing the N-methyl group (N-Me) in opioid structures with unsaturated groups such as allyl and cyclopropylmethyl (CPM) was recognized. The expected effect of this change was to convert opioids having typically morphine-like profiles into opiate antagonists or, more usually, mixed agonist–antagonists. In the series of orvinols (8) and dihydroorvinols (e.g., 15), a wide range of N-alkenyl, N-n-propyl, N-CPM, and other analogous derivatives were prepared and evaluated. The N-CPM derivatives provided the greatest interest.

Structure–activity relationships based on rodent antinociceptive and morphine antagonism tests were established [Lewis et al., 1971; Bentley and Lewis, 1972]. It was shown that antagonists were extremely rare and few members of the series had profiles analogous to nalorphine, the prototype antagonist–analgesic. At the time it seemed surprising that many of the N-CPM-nororvinols (8b, 15b) were analgesics and not morphine antagonists (as was nalorphine) in the tail pressure test in rats. It was later shown that some members of this subgroup were antagonists in tail flick assays and thus were mixed agonist–antagonists [Lewis, 1974], though different from nalorphine and cyclazocine, another early mixed agonist–antagonist analgesic.

We can now relate these differences to intrinsic activity at μ and κ receptors. The orvinols that showed antinociceptive activity in the tail pressure test had higher intrinsic activity at μ receptors than nalorphine and cyclazocine, but since they were ineffective or only partially effective in the tail flick assay, which requires a higher receptor output, they can be classified as partial agonists at μ receptors. The antagonism of morphine in the tail flick test confirmed this profile. Some members of the series were also very efficacious κ agonists and in particular the isopentylorvinol (8b; R^I = i-Pent; M320) was a full k agonist. The profile of M320 is similar to that of the prototype κ agonist, ethylketazocine [Katz et al., 1982].

The general effect of replacing an N-Me group by CPM in the orvinols is to change predominant intrinsic activity from μ to κ [Katz et al., 1982], as was also shown to be the case in the benzomorphan, morphinan, and benzomorphan series [Magnan et al., 1982]. It appears that the primary effect that brings about this change is reduction in μ intrinsic activity, with κ intrinsic activity being very little affected. Affinity at both μ and κ receptors is enhanced by changing N-Me to N CPM. A short series of N-CPM-nororvinols (8b) was evaluated in the rat vas deferens, which effectively has only μ opioid receptors, and the rabbit vas deferens, which has only κ receptors. The methyl- and ethyl-orvinols (8b; R^I = Me, Et) were antagonists in both preparations, whereas the n-propyl analog (8b; R^I = n-Pr) was an antagonist in the rat but a partial agonist in the rabbit. The potency of all three orvinols in each of the preparations was similar. However, there was generally tenfold greater potency in the rat, indicating higher affinity at μ than at κ receptors [Lewis, 1985]. Since both the rat and rabbit vas deferens require high receptor output to produce their response, only ligands of high intrinsic activity show up as agonists or partial agonists in these preparations. It was therefore not surprising that in whole-animal studies these orvinols showed substantial evidence of μ and/or κ agonist effects. From the nature of the abstinence syndromes produced in direct dependence studies in patas monkeys [Cowan, 1974], in the N-CPM series (8b) it was concluded that the predominant intrinsic activity was κ for R^I = methyl and n-propyl and μ for R^I = ethyl. Thus, increasing chain length of the alkyl group R^I in 8b generally had little effect on μ/κ selectivity and affinity but resulted in higher intrinsic activity at receptors of both types.

The effect of replacing the C_{19}-methyl group in series 8b by larger alkyl groups was investigated. As in the corresponding N-Me series (8a) the change from methyl to ethyl reduced potency, but in the N-CPM series there was also a loss of intrinsic activity to give a more antagonist profile. This is well exemplified by the difference between the R-propyl orvinol (8b, R^I = nPr) and its S-diastereoisomer, in which the methyl and n-propyl groups are interchanged. In the rat tail pressure test the R-diastereoisomer was an agonist 1,000 times more potent than morphine, whereas the S-diastereoisomer was a morphine antagonist of nalorphine-like potency [Bentley and Lewis, 1972].

The N-CPM-nordihydroorvinols (15) showed a pattern of activity similar to that observed for the series 8b but with the indication of some attenuation of μ intrinsic activity. No systematic investigations of receptor binding or effects in isolated tissue preparations for this series have been reported, but structure–activity relationships were established in rodent antinociceptive tests for agonist and antagonist effects [Lewis, 1974]. Intrinsic activity increased with the length of the alkyl chain R^I and also, to a much smaller extent, with chain branching. Buprenorphine (15a) was selected from a short list of candidates that was eventually reduced to two, the other being its close analog (15b), which has the n-propyl group instead of t-butyl in the C_{19} tertiary alcohol function. This compound had much higher apparent intrinsic activity but was rejected in favor of buprenorphine because it showed substantial κ agonist behavioral effects in studies in monkeys, in which buprenorphine failed to produce any such effects or significant morphine-like physical dependence [Cowan, 1974].

There is now considerable evidence to show that buprenorphine is a μ partial agonist and a κ antagonist and that it has very high affinity for both μ and κ receptors. It antagonized the diuretic effect of the κ agonist bremazocine in rats [Richards and Sadée, 1985; Leander, 1987] and the rate-suppressing effects of the κ agonist U50488 in pigeons that responded under a multiple fixed-interval/fixed-ratio schedule of food presentation [Leander, 1988]. Buprenorphine suppressed the effects of U50488 in the shock titration procedure in squirrel monkeys [Negus and Dykstra, 1988]. In all of these situations the κ antagonist effect of buprenorphine was observed at doses similar to or sometimes lower than doses producing μ partial agonist effects. These results were consistent with the results from studies in the mouse vas deferens [Miller et al., 1986; Smith, C.F.C., private communication], though Kajiwara et al. [1986] interpreted their data from mouse vas deferens and guinea pig ileum as indicating that buprenorphine was a partial agonist at κ as well as μ receptors.

The μ partial agonist/κ antagonist profile of buprenorphine is unique among those opioids that have been subjected to full receptor binding, isolated tissue, and in vivo evaluation. However, there may be very similar profiles among the close structural analogs of buprenorphine. In view of the significance of buprenorphine as a potential treatment for opiate abuse, in which its lack of κ agonist dysphoric effects plays an important part, it may be justified to reevaluate already synthesized close analogs of buprenorphine and to synthesize new related structures.

CHEMICAL REACTIVITY OF THE ORVINOLS

The thevinols and orvinols have good thermal stability but are subject to a range of transformations under acidic conditions. The structure of the products depends on

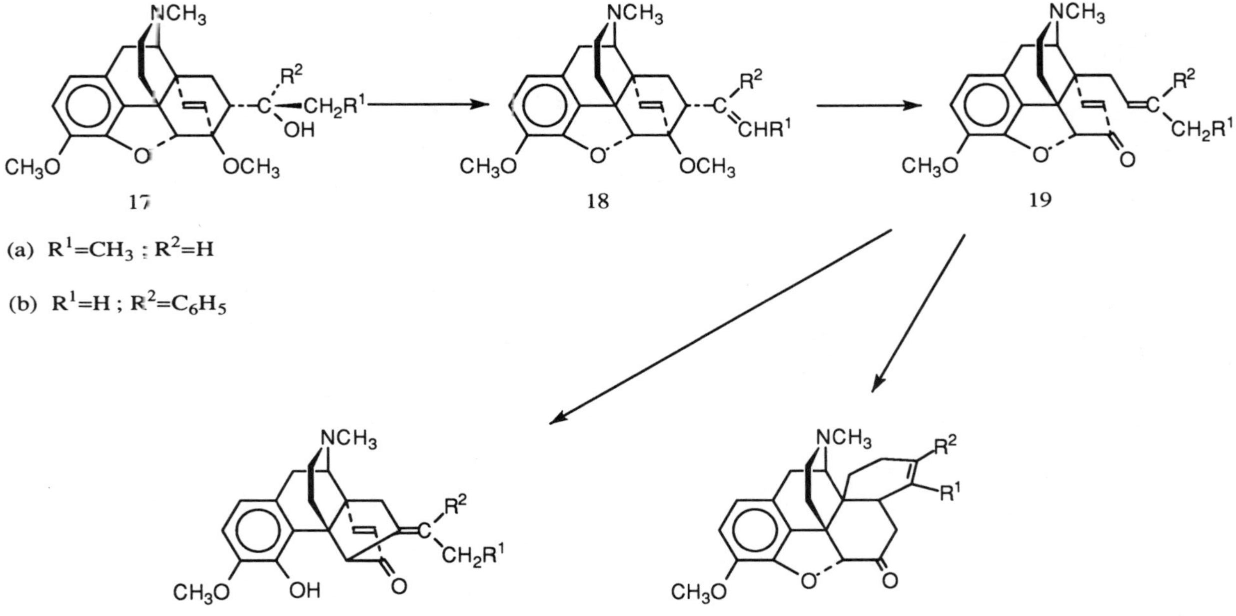

(a) $R^1=CH_3$; $R^2=H$
(b) $R^1=H$; $R^2=C_6H_5$

Scheme 2. Acid-catalyzed rearrangements of thevinols (and orvinols).

the nature of R' (in structures 5 and 8) and the conditions. Transformation of the thevinols (17) is shown in Scheme 2. Dehydration to the olefins (18) is the primary outcome of refluxing the thevinols and orvinols with 98%–100% formic acid [Bentley et al., 1967c]. In the case of the olefins (18a,b) longer treatment with formic acid results in opening of the bridged ring system with the formation of the alkenylcodeinones (19). The α,β-unsaturated ketones are further transformed by treatment with concentrated hydrochloric acid to give new ketonic products. The isopentenylcodeinone (19a) was converted to a mixture of a phenolic unsaturated ketone (20a) by opening of the C_4–C_5 oxide bridge and a saturated ketone (21a) through cyclization of the alkenyl group on to the α,β-unsaturated ketone system [Bentley et al., 1967c]. Saturated ketones of structure 21 were the predominant products of hydrochloric acid treatment of thevinols in which R' is aryl or an unbranched alkyl group larger than methyl, though in the latter cases the intermediate alkenylcodeinones could not be isolated from milder acid conditions [Bentley et al., 1967a]. The usual products from hydrochloric acid treatment of the dihydrothevinols (16) are saturated phenolic ketones (2), since the alternative cyclization involving the α,β-unsaturated ketone system cannot take place [Bentley et al., 1967c].

Bentley et al. [1969] found that dihydrothevinols with branching at the α position in the alkyl group R' (22) were cyclized to tetrahydrofuranocodides (23) (Scheme 3) by formic acid treatment. Thus the isopropylthevinol (22a) and cyclohexylthevinol (22b) were converted into 23a and 23b. The t-butylthevinol (22c) was also transformed into a furanocodide (23c) in which migration of a methyl group from C_{20} to

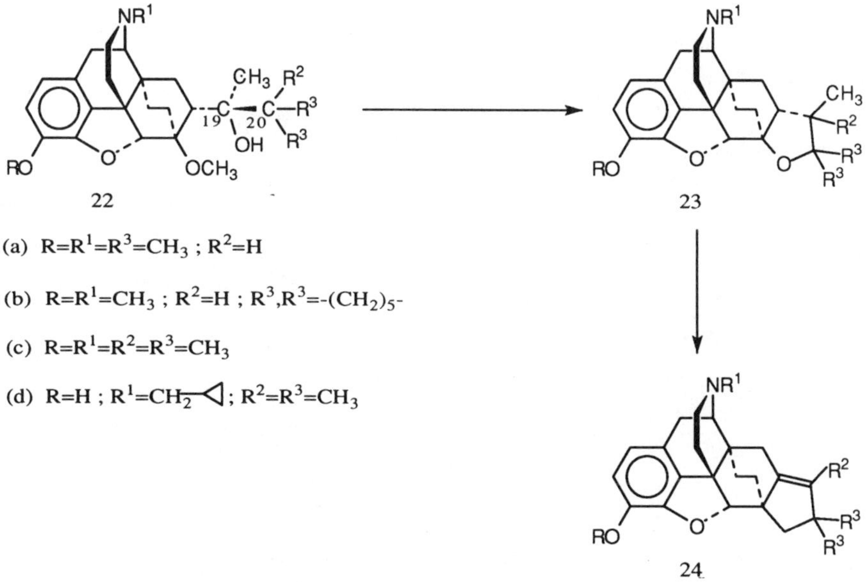

(a) $R=R^1=R^3=CH_3$; $R^2=H$

(b) $R=R^1=CH_3$; $R^2=H$; $R^3,R^3=-(CH_2)_5-$

(c) $R=R^1=R^2=R^3=CH_3$

(d) $R=H$; $R^1=CH_2\text{-}\triangleleft$; $R^2=R^3=CH_3$

Scheme 3. Acid-catalyzed rearrangements of branched thevinols and orvinols.

C_{19} had also occurred. With hydrochloric acid the furanocodides underwent further transformation to give cyclopentenocodides (24) [Bentley et al., 1969]. In similar procedures buprenorphine (22d) was converted to the furanomorphide (23d) and the cyclopentenormorphide (24d) (Lewis et al., unpublished). It has also been shown that buprenorphine is converted to 23d when sterile solutions are autoclaved at pH < 3 [Cone et al., 1984]. Orvinols of structures 8a and 8b in which R' is a straight-chain alkyl group were included in this study and in all cases there was evidence for the formation of a furanomorphide (26a,b) usually as the single identified product. This was surprising, since Bentley et al. [1967a] did not find the equivalent furanocodides (26c) among the transformation products in Scheme 2. However, the furanocodides could be intermediates in the conversion of 18 to 19 and would not be isolated if, under conditions used by Bentley, they were transformed faster than they were produced.

CONCLUSIONS

Intensive investigations of structure–activity relationships in the series of tertiary alcohols derived from the thebaine–methyl vinyl ketone adducts demonstrated the significance of the structure and stereochemistry of the C_{19} tertiary alcohol function. Understanding of these relationships was combined with a knowledge of the effect of the piperidine N-cyclopropylmethyl group to produce a range of candidates with antagonist–analgesic profiles. Many of these orvinols had profiles that we can now interpret as indicating very high affinity and intrinsic activity at both μ and κ receptors. Buprenorphine, which was eventually selected for development, was unusual in that it was an antagonist at κ receptors so that it was characterized *in vivo* as a μ partial agonist. Its very high affinity for κ receptors without significant intrinsic activity has subsequently been demonstrated in a number of animal models. The extent to which this characterization influences its clinical profile, other than to avoid κ agonist unwanted effects, is uncertain.

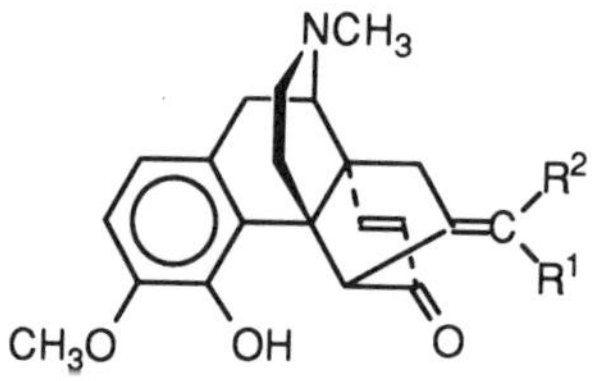

25

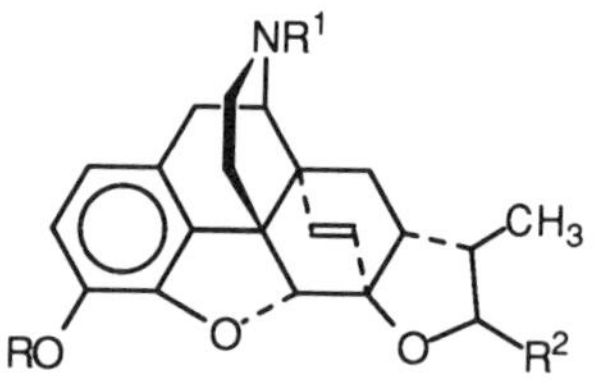

26

(a) R=H ; R^1=CH$_3$

(b) R=H ; R^1=CH$_2$–◁

(c) R=R^1=CH$_3$

REFERENCES

Bentley KW, Hardy DG (1967a): Novel analgesics and molecular rearrangements in the morphine–thebaine group. I. Ketones derived from 6,14-endo-ethenotetrahydrothebaine. J Am Chem Soc 89:3267–3273.

Bentley KW, Hardy DG (1967b): Novel analgesics and molecular rearrangements in the morphine–thebaine group. III. Alcohols of the 6,14-endo-ethenotetrahydro-oripavine series and derived analogs of N-allylnormorphine and -norcodeine. J Am Chem Soc 89:3281–3292.

Bentley KW, Hardy DG, Howell CF, Fulmor W, Lancaster JE, Brown JJ, Morton GO, Hardy RA Jr (1967a): Novel analgesics and molecular rearrangements in the morphine–thebaine group. V. Derivatives of 7,8-dihydrocyclohexeno-[1′,2′:8,14]-codeinone. J Am Chem Soc 89:3303–3311.

Bentley KW, Hardy DG, Meek B (1967b): Novel analgesics and molecular rearrangements in the morphine–thebaine group. II. Alcohols derived from 6,14-endoetheno- and 6,14-endoethanotetrahydrothebaine. J Am Chem Soc 89:3273–3280.

Bentley KW, Hardy DG, Meek B (1967c): Novel analgesics and molecular rearrangements in the morphine–thebaine group. IV. Acid-catalyzed rearrangements of alcohols of the 6,14-endoethenotetrahydrothebaine series. J Am Chem Soc 89:3293–3303.

Bentley KW, Hardy DG, Meek B, Taylor JB, Brown JJ, Morton GO (1969): Novel analgesics and molecular rearrangements in the morphine–thebaine group. X. Further acid-catalyzed rearrangements of alcohols in the 6,14-endoethenotetrahydrothebaine series. J Chem Soc (C): 2229–2232.

Bentley KW, Lewis JW (1972): The relationship between structure and activity in the 6,14-endoethenotetrahydrothebaine series of analgesics. In Kosterlitz HW, Collier HOJ, Villarreal JE (eds): "Agonist and Antagonist Actions of Narcotic Analgesic Drugs" London: Macmillan, pp. 7–16.

Cone EJ, Gorodetsky CW, Darwin WD, Buchwald WF (1984): Stability of 6,14-endoethenotetrahydro-oripavine analgesics; acid catalyzed rearrangement of buprenorphine. J Pharm Sci 73:243–246.

Cowan A (1974): Evaluation in nonhuman primates: Evaluation of the physical dependence capacities of oripavine–thebaine partial agonists in patas monkeys. Adv Biochem Psychopharmacol 8:427–438.

Fulmor W, Lancaster JE, Morton GO, Brown JJ, Howell CF, Nora CT, Hardy RA Jr (1967): Nuclear magnetic resonance studies in the 6,14-endoethenotetrahydrothebaine series. J Am Chem Soc 89:3322–3330.

Hutchins CW, Cooper GK, Pürro S, Rapoport H (1981): 6-Demethoxythebaine and its conversion to analgesics of the 6,14-ethenomorphinan type. J Med Chem 24:773–777.

Hutchins CW, Rapoport H (1984): Analgesics of the orvinol type. J Med Chem 27:521–527.

IUPAC (1970): Tentative rules for the nomenclature of organic compounds. Section E. Fundamental stereochemistry. J Org Chem 35:2849–2867.

Kajiwara M, Aoki K, Ishii K, Numata H, Matsumiya T, Oka T (1986): Agonist and antagonist actions of buprenorphine on three types of opioid receptor in isolated preparations. Jpn J Pharmacol 40:95–101.

Katz JL, Woods JH, Winger GD, Jacobson AE (1982): Compounds of novel structure having kappa-agonist behavioral effects in rhesus monkeys. Life Sci 31:2375–2378.

Knipmeyer L, Rapoport H (1985): Analgesics of the 6,14-ethenomorphinan type. 6-Deoxy-7α-orvinols and 6-deoxy-8α-orvinols. J Med Chem 28:461–466.

Leander JD (1987): Buprenorphine has potent kappa opioid receptor antagonist activity. Neuropharmacology 26:1445–1447.

Leander JD (1988): Buprenorphine is a potent κ-opioid receptor antagonist in pigeons and mice. Eur J Pharmacol 151:457–461.

Lewis JW (1974): Ring C-bridged derivatives of thebaine and oripavine. Adv Biochem Psychopharmacol 8:123–136.

Lewis JW (1985): Buprenorphine. Drug Alcohol Depend 14:363–372.

Lewis JW, Bentley KW, Cowan A (1971): Narcotic analgesics and antagonists. Annu Rev Pharmacol 11:241–270.

Loew GH, Berkowitz DS (1979): Intramolecular hydrogen bonding and conformational studies of bridged thebaine and orpavine narcotic agonists and antagonists. J Med Chem 22:603–607.

Magnan J, Paterson SJ, Tavani A, Kosterlitz HW (1982): The binding spectrum of narcotic analgesic drugs with different agonist and antagonist properties. Naunyn-Schmiedeberg's Arch Pharmacol 319:197–205.

Miller L, Shaw JS, Whiting EM (1986): The contribution of intrinsic activity to the actions of opioids in vitro. Br J Pharmacol 87:595–601.

Negus SS, Dykstra LA (1988): κ-Antagonist properties of buprenorphine in the shock titration procedure. Eur J Pharmacol 156:77–86.

Richards ML, Sadée W (1985): Buprenorphine is an antagonist at the κ receptor. Pharm Res 2:178–181.

Schöpf C, Gottberg KV, Petri W (1938): Thebaine–maleic anhydride, thebainequinone, thebainehydroquinone and the acid rearrangement product flavothebaone. Ann 536:216–257.

Van den Hende JH, Nelson NR (1967): The crystal and molecular structure of 7α-(1-(R)-hydroxy-1-methylbutyl)-6, 14-endoethenotetrahydrothebaine hydrobromide (19-propylthevinol hydrobromide). J Am Chem Soc 89:2901–2905.

PRECLINICAL PHARMACOLOGY

BUPRENORPHINE: A REVIEW OF THE BINDING LITERATURE

RICHARD B. ROTHMAN, QI NI, and HENG XU
Clinical Psychopharmacology Section, Intramural Research Program, National Institute on Drug Abuse, National Institutes of Health, Baltimore, MD 21224

INTRODUCTION

Buprenorphine is an oripavine analgesic structurally related to etorphine and diprenorphine. Several aspects of its pharmacology set it apart from most other opioid analgesics: (1) Buprenorphine is one of the most lipophilic opioid analgesics known [Hambrook and Rance, 1976]. (2) Its antinociceptive effect is readily blocked by narcotic antagonists when they are administered prior to or simultaneously with buprenorphine, but not after the antinociceptive effect is already established [Cowan et al., 1977b; Orwin, 1977; Gibbs et al., 1982]. (3) It apparently dissociates very slowly from the μ receptor in vivo [Hambrook and Rance, 1976; Wüster and Herz, 1976; Tallarida and Cowan, 1982]. (4) Buprenorphine characteristically produces a bell-shaped dose-response curve that is shifted to the right in a symmetrical manner by naloxone [Lewis, 1974; Cowan et al., 1977b; Rance, 1979; Dum and Herz, 1981]. (5) Buprenorphine is often classified as a member of the mixed agonist–antagonist analgesics. However, in contrast to nalorphine, which is a κ agonist and μ antagonist [Martin et al., 1976], buprenorphine is a partial or full μ agonist, and a κ antagonist in an assay that measures primarily the activity of the κ_1 receptor [Leander, 1983a,b].

Of the various pharmacological properties of buprenorphine listed above, pharmacologists have sought to explain the most puzzling: the characteristic bell-shaped dose response curve that is shifted to the right by naloxone [Cowan et al., 1977a; Rance et al., 1980; Dum and Herz, 1981]. The purpose of this chapter is to review the binding data on buprenorphine with the goal of providing an explanation for, or at least a testable hypothesis of, the bell-shaped dose-response curve.

REVIEW OF THE BINDING DATA

Direct Binding Studies

[³H]Buprenorphine binding has been studied by relatively few groups. In a preliminary study, Hambrook and Rance [1976] reported the stereospecific binding of [³H]buprenorphine to rat brain membranes. Although they did not report K_d and B_{max} values, Scatchard plots of [³H]buprenorphine binding were apparently linear. The addition of 100 mM Na⁺ increased the B_{max}, but did not alter the dissociation rate. This finding is characteristic of antagonists, since Na⁺ typically increases the dissociation rate of agonists [Blume, 1978]. Only the first phase of the [³H]buprenorphine dissociation curve, which is biphasic, was reported, with a $t_{1/2}$ of 45 min.

Villiger and Taylor [1981] compared [³H]buprenorphine binding with the binding of [³H]dihydromorphine, [³H]naloxone, and [³H][D-Ala²,Met⁵]enkephalin. In contrast to [³H]buprenorphine, which labeled an apparent single class of binding sites (K_d = 0.9 nM, B_{max} = 30 pmol/g tissue), [³H]dihydromorphine and [³H]naloxone had curvilinear Scatchard plots, with a total B_{max} value of about 15 pmol/g tissue. [³H][D-Ala²,Met⁵]enkephalin labeled an apparent single class of sites (K_d = 2 nM, B_{max} = 10.7 pmol/g tissue). As observed by Hambrook and Rance [1976], [³H]buprenorphine binding was increased by sodium. Unfortunately, it was not determined if the increase was due to an increase in the B_{max}, or a decrease in the K_d. Villiger and Taylor [1981] also reported a slow initial dissociation rate with a $t_{1/2}$ of about 40 min.

To explain the large [³H]buprenorphine B_{max} value, Villiger and Taylor [1981] suggested that [³H]buprenorphine labeled both μ and δ receptors. Although this hypothesis was supported by the apparent high affinity of buprenorphine for μ and δ sites determined by competition studies (IC_{50} values about 0.5–1 nM), it was not supported by the relatively weak interaction of morphine, naloxone, and [D-Ala²,Met⁵]enkephalin with [³H]buprenorphine binding (IC_{50} values greater than 50 nM). These inhibition curves were shallow, supporting the hypothesis that [³H]buprenorphine bound to more than one site. Similar data were reported for [³H]buprenorphine binding to rat dorsal spinal cord [Villiger and Taylor, 1982].

A subsequent study [Villiger, 1984] examined the effect of sodium chloride and guanosine triphosphate (GTP) on [³H]buprenorphine binding. Unlike previous studies (see above), sodium chloride did not alter either the K_d or the B_{max} of [³H]buprenorphine binding sites. GTP (0.1 mM) also had no effect on [³H]buprenorphine binding, although 1 mM $MnCl_2$ decreased [³H]buprenorphine binding. In competition assays with [³H]naloxone (μ sites), [³H][D-Ala²,D-Leu⁵]enkephalin (δ sites), and [³H]diprenorphine blocked with 10 μM morphiceptin and 100 nM [D-Ala²,D-Leu⁵]enkephalin (benzomorphan sites), buprenorphine inhibition curves had Hill coefficients approximating to unity. Both GTP and NaCl decreased the K_i of buprenorphine for μ and δ sites, but not benzomorphan sites. Generally, GTP increases the K_i of agonist ligands and does not alter the K_i of antagonist drugs [Childers and Snyder, 1980]. GTP produces moderate increases in the K_i of benzomorphan mixed agonist–antagonist analgesics [Chang et al., 1980]. Thus buprenorphine exhibited a unique profile in these GTP-shift experiments.

In a related study, Sadée et al. [1982] presented evidence that although buprenorphine can interact in vitro with high affinity at μ, δ, and κ binding sites, it apparently labels primarily μ binding sites in vivo; and that the dissociation of [³H]buprenorphine from the μ site is increased by 100 mM NaCl. These authors concluded that "there was no marked difference between the three oripavine alkaloids [etorphine, diprenorphine, buprenorphine] in their reactions towards Na⁺ and Mn-II that could account for their markedly different pharmacological behavior."

In retrospect, the major contribution of these early [³H]buprenorphine binding studies was that they demonstrated that [³H]buprenorphine has a slow dissociation rate. These studies did not comment on the fact that the dissociation of [³H]buprenorphine is actually biphasic, with a relatively rapid first phase of dissociation, after which about 50% of bound [³H]buprenorphine remains persistently bound [Boas and Villiger, 1985]. Similar data are observed for [³H]etorphine binding [Tolkovsky, 1982]. The slow dissociation rate provides a potential mechanism to explain the long-lasting antinociceptive effects of buprenorphine. A secondary contribution of these early studies is that they indicated that [³H]buprenorphine must be labeling more than one binding site in vitro. Unfortunately, the experimental designs used in these early studies were not entirely adequate for determining the identity of the binding sites.

Competition Binding Studies

The availability of [³H]ligands selective for the various types and subtypes of opioid receptors simplifies the task of determining the affinity of buprenorphine for each of these sites. Villiger [1984] reported K_i values of 0.31 nM, 0.38 nM, and 4.16 nM for μ, δ, and benzomorphan binding sites. Su (1985) reported an IC_{50} of about 1 nM at κ binding sites. Sadée et al. [1982] reported an IC_{50} value of 0.1 (K_i about 0.05 nM) measured with [³H]diprenorphine. The finding that buprenorphine is relatively nonselective among opioid receptor subtypes is representative of the in vitro binding literature. There exist, it should be noted, some discrepancies regarding the actual potency of buprenorphine (see below). In addition, the exact relationship between κ binding sites and benzomorphan binding sites remains to be established.

Our laboratory performed perhaps the most thorough examination of the interaction of buprenorphine with opioid receptor subtypes in vitro. These data were initially presented at the 1990 annual meeting of the Committee on Problems of Drug Dependence Inc. [Grayson et al., 1991]. This study determined the K_i values of (−)-buprenorphine, (+)-buprenorphine, (−)-diprenorphine, (+)-diprenorphine, (−)-etorphine and (+)-etorphine for μ, δ, $κ_1$, $κ_{2a}$, and $κ_{2b}$ binding sites [Rothman et al., 1990]. As reported in Table I, both (−)-etorphine and (−)-buprenorphine had very high affinity for the μ binding site. (−)-Etorphine was relatively selective for the μ binding site, having 53-fold, 23-fold, 271-fold, and 277-fold lower affinity for the δ, $κ_1$, $κ_{2a}$, and $κ_{2b}$ binding sites, respectively. (−)-Buprenorphine showed a

TABLE I. Apparent K_i (M) Values and Slope Factors of Test Agents at Opioid Receptors of Various Subtypes

	μ[a]	δ[b]	κ_1[c]	κ_{2a}[d]	κ_{2b}[d]
RTI-4614-4[e]	5.50×10^{-12}	148×10^{-9}	84.8×10^{-9}	2275×10^{-9}	22.3×10^{-9}
(−)-Buprenorphine	6.0×10^{-12}	480×10^{-12}	4.0×10^{-12}	0.78×10^{-9}	54×10^{-9}
(−)-Etorphine	49.3×10^{-12}	2.62×10^{-9}	1.12×10^{-9}	13.4×10^{-9}	13.7×10^{-9}
(−)-Diprenorphine	4.68×10^{-12}	27.9×10^{-12}	14.2×10^{-12}	210×10^{-12}	240×10^{-12}
(−)-Pentazocine	6.57×10^{-9}	107×10^{-9}	8.92×10^{-9}	60.6×10^{-9}	146×10^{-9}
(+)-Buprenorphine	3.69×10^{-6}	251×10^{-6}	17×10^{-6}	71.2×10^{-6}	$>0.125 \times 10^{-3}$
(+)-Etorphine	17.2×10^{-6}	785×10^{-6}	$>0.5 \times 10^{-3}$	0.29×10^{-3}	$>0.5 \times 10^{-3}$
(+)-Diprenorphine	26.9×10^{-6}	1.95×10^{-3}	28.9×10^{-6}	57.5×10^{-6}	$>0.25 \times 10^{-3}$
(+)-Pentazocine	1.10×10^{-6}	36.3×10^{-6}	0.25×10^{-6}	0.52×10^{-6}	14.4×10^{-6}

[a]Mu binding sites were labeled with [^{3}H]DAMGO and rat brain membranes treated as previously described [Rothman et al., 1988]. Briefly, incubations proceeded for 4–6 hr at 25° C in 50 mM Tris-HCl, pH 7.4, containing a protease inhibitor cocktail. Nonspecific binding was determined with 20 μM levallorphan.

[b]δ binding sites were labeled with [^{3}H][D-Ala^2,D-Leu^5]enkephalin and rat brain membranes treated as previously described [Rothman and McLean, 1988]. Briefly, incubations proceeded for 4–6 hr at 25° C in 50 mM Tris-HCL, pH 7.4, containing 100 mM choline chloride, 3 mM MnCl$_2$, 100 nM of the highly μ-selective peptide LY164929 (to block binding to μ sites), and a protease inhibitor cocktail. Nonspecific binding was determined with 20 μM levallorphan.

[c]κ_1 binding sites were labeled with [^{3}H]U-69,593 and guinea pig brain membranes depleted of μ and δ binding sites by pretreatment with BIT and FIT as previously described [Rothman et al., 1990], except that the incubation temperature was at 25° C. Briefly, incubations proceeded for 4–6 hr at 25° C in 50 mM Tris-HCl, pH 7.4, containing a protease inhibitor cocktail. Nonspecific binding was determined with 1 μM U-69,593.

[d]κ_2 binding sites were assayed with [^{3}H]bremazocine using 5 μM [Leu^5]enkephalin to block binding to the κ_{2b} binding site and 5 μM (−)-(1S,2S)-U-50,488 to block binding to the κ_{2a} binding site [Rothman et al., 1990]. Each [^{3}H]ligand was displaced by 9 {(−)-opiates} or 5 {(+)-opiates} concentrations of test drug, two times, each time with a different stock of test drug. The data were combined and fit to the two-parameter logistic equation for the best-fit estimates of the IC$_{50}$. The K_i values were calculated by standard equations. The SD values were between 3% and 10% of the mean.

[e]RTI-4614-4 is (±)-cis-N-[1-(2-hydroxy-2-phenylethyl)-3-methyl-4-piperidyl]-N-phenylpropanamide (data from Rothman et al. [1991]).

different profile, having 80-fold, 0.7-fold, 130-fold, and 257-fold lower affinity for the δ, κ_1, κ_{2a}, and κ_{2b} binding sites, respectively, than the μ binding site. (−)-Diprenorphine was relatively nonselective among the μ, δ, and κ_1 binding sites but had 44-fold and 51-fold lower affinity for the κ_{2a} and κ_{2b} binding sites, respectively. As reported previously [Rothman et al., 1991], the methylfentanyl analog (±)-cis-N-[1-(2-hydroxy-2-phenylethyl)-3-methyl-4-piperidyl]-N-phenylpropanamide (RTI-4614-4) is highly selective and potent at the μ binding site. The (+)-enantiomers were all considerably less potent than their corresponding (−)-enantiomers. The enantiomers of pentazocine were included for the sake of comparison.

The in vitro data show that (−)-buprenorphine has very high affinity for the μ and κ_1 binding sites. The order of potency of (−)-buprenorphine for the binding sites is $\kappa_1 = \mu > \delta > \kappa_{2a} > \kappa_{2b}$. The affinities of (−)-buprenorphine for opioid receptor subtypes in vitro agree reasonably well with their relative affinities determined in vivo by Richards and Sadée [1985]. As observed in our study, they reported that (−)-buprenorphine had about the same affinity for μ and κ binding sites but considerably lower affinity for the δ site. Unfortunately, it is not known with certainty which κ receptor subtypes were actually being measured in that study [Richards and Sadée, 1985]. Similarly, as observed in our study, Sadée and associates reported that (−)-etorphine is highly μ-selective in vivo [Rosenbaum et al., 1984; Richards and Sadée, 1985]. The moderate preference of (−)-diprenorphine for μ and κ_1 sites was also observed in vivo [Richards and Sadée, 1985].

The highly potent interactions of (−)-buprenorphine (K_i values of about 6 pM) were an unexpected finding. We therefore tested a different batch of (−)-buprenorphine and achieved similar results. Why the K_i values observed in our study are lower than reported by others is not clear. The most likely explanation is that the longer incubation times used in our laboratory (4–6 hr at 25° C versus 20–30 min commonly used in other studies) obviously permit a more complete equilibration of buprenorphine. Indeed, unpublished studies show that the IC_{50} value of buprenorphine for inhibiting [^{3}H][D-Ala^2-MePhe^4,Gly-ol^5]enkephalin ([^{3}H]DAMGO) binding decreases as a function of incubation time, an observation that is characteristic of ligands that bind irreversibly to receptors [Bylund and Yamamura, 1990]. The observation that about 50% of [^{3}H]buprenorphine remains persistently bound in dissociation experiments supports the idea that buprenorphine behaves in part as a pseudoirreversible ligand, and predicts that buprenorphine would have very high affinity for its binding sites. Interestingly, the approximately 1,000-fold difference between the K_i values of buprenorphine and morphine for the μ receptor in vitro is similar to the greater than 2,000-fold difference observed for binding to the μ receptor in vivo [Rosenbaum et al., 1984].

CONCLUSIONS

As mentioned in the Introduction, several aspects of buprenorphine's pharmacology set it apart from other opioid analgesics. (1) Its antinociceptive effect is readily

TABLE II. Comparison of Selected Pharmacological Properties of Buprenorphine, Methylfentanyl, and Etorphine

Pharmacological property	Buprenorphine	Methylfentanyl	Etorphine
Antagonized by naloxone when administered after the agonist	No	Yes	Yes
Slow dissociation	Yes	Yes	Yes
Long-lasting effects	Yes	Yes	Yes
Lipophilic	Yes	Yes	Yes
Full agonist in antinociception	Yes	Yes	Yes
Bell-shaped dose-response curve	Yes	No	No
Receptor subtypes acted on in vivo	μ (agonist) κ (antagonist) ? δ	μ (agonist)	μ (agonist)

blocked by narcotic antagonists when they are administered prior to or simultaneously with buprenorphine, but not after the antinociceptive effect is already established [Cowan et al., 1977b; Orwin, 1977; Gibbs et al., 1982]. (2) Buprenorphine characteristically produces a bell-shaped dose-response curve that is shifted to the right in a symmetrical manner by naloxone [Lewis, 1974; Cowan et al., 1977b; Rance, 1979; Dum and Herz, 1981].

Buprenorphine, (+)-cis-3-methylfentanyl, and etorphine share several pharmacological properties. Comparison of the similarities and differences among these agents might facilitate identification of what pharmacological properties of buprenorphine contribute to its unique features. As summarized in Table II, all three agents have very slow dissociation rates. For example, (+)-cis-3-methylfentanyl acts as a pseudoirreversible inhibitor of μ receptor binding in vitro [Xu et al., 1991] and in vivo [Band et al., 1990]. Using etorphine and buprenorphine we have observed similar in vitro data (unpublished data). Consistent with these data are the findings in dissociation binding studies, reviewed above, that showed that [^{3}H]buprenorphine and [^{3}H]etorphine have a component of essentially irreversible binding. In that all three agents are very lipophilic, and lipophilic drugs are generally rapidly eliminated from the brain [Herz and Teschemacher, 1971], the simplest explanation for the relatively long duration of action of these agents is a slow dissociation rate. In support of this idea, fentanyl, which does not produce pseudoirreversible binding [Xu et al., 1991], does not have an irreversible component of dissociation [Boas and Villiger, 1985], and also has a very short duration of action.

All three agents act as μ receptor agonists. Administration of naloxone, after the establishment of full agonist effects with methylfentanyl, results in a prompt and potent reversal of the agonist effects [Van Bever et al., 1974; Band et al., 1990]. In that methylfentanyl acts as a pseudoirreversible inhibitor of the μ receptor, these data demonstrate the occurrence of apparent negative cooperative interactions between naloxone and μ receptors already occupied with (+)-cis-3-methylfentanyl [Band et al., 1990]. Naloxone can also promptly reverse the agonist effects of etorphine (A. Cowan, personal communication), suggesting that similar negative cooperative actions also occur between naloxone and μ receptors already occupied with etorphine. These negative cooperative interactions are not observed in vitro [Band et al., 1990].

In contrast, administration of naloxone after the establishment of agonist effects with buprenorphine fails to potently and promptly reverse its agonist effects [Cowan et al., 1977b; Orwin, 1977; Gibbs et al., 1982]. The binding data provide no direct explanation for this. It is possible that buprenorphine binds to a domain of the μ receptor binding site whose conformation does not change when a neighboring domain is occupied by naloxone. In this case, naloxone cannot reverse buprenorphine because the buprenorphine remains persistently bound to the μ receptor. That buprenorphine binds differently to the μ receptor than do other agonists is supported by the observation that GTP and sodium chloride increase its affinity for the μ receptor [Villiger, 1984] rather than decrease it, as is generally observed with most μ agonists [Childers and Snyder, 1978, 1980]. Thus, the naloxone-resistant property of buprenorphine is most likely related to its slow dissociation from the μ receptor, and the fact that naloxone cannot apparently "kick" buprenorphine off the μ receptor once it is bound.

Regarding bell-shaped dose-response curves, it is the properties not shared among the three agents (Table II) that are most likely to gives clues as to possible mechanisms. The major difference between buprenorphine and etorphine and (+)-cis-3-methylfentanyl is that the latter two interact mostly with μ receptors. This suggests that interactions with κ receptors might be the underlying mechanism of the bell-shaped curves. Indeed, Sadée et al. [1982] presented a compelling case that the data are most simply explained by a two-receptor model in which agonist effects are mediated by one receptor (at the lower dose range) and a second inhibitory receptor (at the higher dose range) counteracts the effects of the first. This model was termed "noncompetitive autoinhibition," which has also been invoked to explain bell-shaped morphine dose-response curves [Zhu and Szeto, 1989]. The ability of naloxone to produce a symmetrical shift to the right therefore implies that naloxone must have about equal affinity for both binding sites.

The identity of the second inhibitory receptor is not known. Unlike etorphine and methylfentanyl, buprenorphine acts as a κ antagonist in vivo. Thus, it is tempting to speculate that the second receptor might be a κ receptor. Experimentally testing this hypothesis may be difficult, since several laboratories now report the existence of κ receptor subtypes [Nock et al., 1988; Zukin et al., 1988; Clark et al., 1989; Rothman et al., 1990; Tiberi and Magnan, 1990]. It is possible, however, to reach some tentative conclusions.

Ligand binding and pharmacological studies distinguish two classes of κ receptors termed κ_1 and κ_2 [Wood and Iyengar, 1986; Horan et al., 1991]. The κ_1 receptors are selectively activated by U-50, 488H and similar drugs such as U-69,593; κ_1 agonists produce a brisk diuresis [Leander, 1983b] and antagonize the antinociceptive [Porreca and Tortella, 1987], anticonvulsant [Porreca and Tortella, 1987], and micturition-inhibiting [Sheldon et al., 1987] effects of morphine. One simple way to test the hypothesis that the second inhibitory receptor is the κ_1

receptor is to generate a dose-response curve by using a constant ratio of etorphine and U69,593. If the hypothesis is correct, a bell-shaped dose-response curve should be observed.

Two sets of data suggest that the κ_1 receptor is not the second inhibitory receptor. (1) Buprenorphine acts as a κ_1 antagonist in the diuresis model [Leander, 1983a,b], suggesting that it should potentiate, not antagonize, the agonist effects of morphine. (2) Naloxone is about 10 times weaker in reversing the effects of κ_1 agonists than μ agonists, whereas its ability to shift the bell-shaped curve symmetrically to the right predicts that it should have about equal affinity at both receptors. (3) Buprenorphine has similar affinity at both κ_1 and μ binding receptors. Thus, the second inhibitory receptor is unlikely to be the κ_1 receptor.

Little is known of the physiological functions of the κ_2 and κ_3 [Clark et al., 1989] receptors. Thus, it is not known if buprenorphine acts as an agonist or antagonist at these κ receptors, and how these receptors might alter μ receptor-mediated antinociception. This is assuming, of course, that these binding sites do function as receptors. As mentioned above, the ability of naloxone to shift the bell-shaped curve symmetrically to the right predicts that it should have about equal affinity at both inhibitory and stimulatory receptors. Although naloxone has lower affinity for κ_2 receptors than for μ receptors, this is true only in the absence of sodium chloride. In the presence of sodium chloride, the K_i values of naloxone for κ_2 and μ sites are quite comparable [Rothman et al., 1990]. The fact that buprenorphine is considerably weaker at the κ_2 binding site than at the μ binding site is consistent with the notion that the κ_2 site might be the second inhibitory receptor. Testing this hypothesis will require a selective κ_2 antagonist, which is not yet available.

In that [Met[5]]enkephalin and its analogs attenuate morphine-induced antinociception [Lee et al., 1980; Vaught et al., 1982; Heyman et al., 1989], it is tempting to speculate that the δ receptor might be the second inhibitory receptor. This is unlikely, since coadministration of [Met[5]]enkephalin (or its analogs) shifts the morphine dose-response curve to the right, not into a bell-shaped curve [Vaught and Takemori, 1979; Heyman et al., 1989]. The δ receptor hypothesis is potentially testable, with newly developed δ receptor antagonists such as naltrindole [Portoghese et al., 1988], naltrindole 5'-isothiocyanate [Portoghese et al., 1990], and [D-Ala[2],Leu[5],Cys[6]]enkephalin [Jiang et al., 1990].

In summary, the pharmacology of buprenorphine is most simply explained by the hypothesis of noncompetitive autoinhibition proposed by Sadée and colleagues [Sadée et al., 1982]. The comparisons reported in Table II suggest that the ability of buprenorphine to interact with κ receptors might be the underlying mechanism of the bell-shaped curve. Further studies will be needed to determine which κ receptor is the inhibitory receptor postulated by Sadée et al. [1982].

REFERENCES

Band L, Xu H, Bykov V, Greig N, Kim C-H, Newman A, Jacobson AE, Rice KC, Rothman RB (1990): The potent opiate agonist (+)-cis-3-methylfentanyl binds pseudoirreversibly to the opioid receptor complex in vitro and in vivo: Evidence for a novel mechanism of action. Life Sci 47:2231–2240.

Blume AJ (1978): Interaction of ligands with the opiate receptors of brain membranes: Regulation by ions and nucleotides. Proc Natl Acad Sci USA 75:1713–1717.

Boas RA, Villiger JW (1985): Clinical actions of fentanyl and buprenorphine. Br J Anaesth 57:192–196.

Bylund DB, Yamamura HI (1990): Methods for receptor binding. In Yamamura HI (ed): "Methods in Neurotransmitter Receptor Analysis." New York: Raven Press, pp 1–35.

Chang KJ, Hazum E, Cuatrecasas P (1980): Possible role of distinct morphine and enkephalin receptors in mediating actions of benzomorphan drugs (putative κ and σ agonists). Proc Natl Acad Sci USA 77:4469–4473.

Childers SR, Snyder SH (1978): Guanine nucleotides differentiate agonist and antagonist interactions with opiate receptors. Life Sci 23:759–761.

Childers SR, Snyder SH (1980): Differential regulation by guanine nucleotides of opiate agonist and antagonist receptor interactions. J Neurochem 34:583–593.

Clark JA, Liu L, Price M, Hersh B, Edelson M, Pasternak GW (1989): Kappa opiate receptor multiplicity: Evidence for two U50,488-sensitive κ_1 subtypes and a novel κ_3 subtype. J Pharmacol Exp Ther 251:461–468.

Cowan A, Boardman S, Robinson T (1977a): Buprenorphine and intestinal motility: A pharmacological analysis of the biphasic dose-response curve. Fed Proc 36:994.

Cowan A, Lewis JW, Macfarlane IR (1977b): Agonist and antagonist properties of buprenorphine, a new antinociceptive agent. Br J Pharmacol 60:537–545.

Dum JE, Herz A (1981): *In vivo* receptor binding of the opiate partial agonist, buprenorphine, correlated with its agonist and antagonist actions. Br J Pharmacol 74:627–633.

Gibbs JM, Johnson HD, Davis FM (1982): Patient administration of i.v. buprenorphine for postoperative pain relief using the "Cardiff" demand analgesia apparatus. Br J Anaesth 54:279–284.

Grayson NA, Rothman RB, Xu H, Rice KC (1991): Pharmacological properties of (+)-buprenorphine and (+)-diprenorphine. NIDA Res Monogr 105:380–381.

Hambrook JM, Rance MJ (1976): The interaction of buprenorphine with the opiate receptor: Lipophilicity as a determining factor in drug-receptor kinetics. In Kosterlitz HW (ed): "Opiates and Endogenous Opioid Peptides." Amsterdam: North-Holland, pp 295–301.

Herz A, Teschemacher HJ (1971): Activities and sites of antinociceptive action of morphine-like analgesics and kinetics of distribution following intravenous, intracerebral and intraventricular application. Adv Drug Res 6:79–119.

Heyman JS, Jiang Q, Rothman RB, Mosberg HI, Porreca F (1989): Modulation of mu-mediated antinociception by delta agonists: Characterization with antagonists. Eur J Pharmacol 169:43–52.

Horan P, De Costa BR, Rice KC, Porreca F (1991): Differential antagonism of U69,593- and bremazocine-induced antinociception by (−)-UPHIT. Evidence of kappa opioid receptor multiplicity in mice. J Pharmacol Exp Ther 257:1154–1161.

Jiang Q, Bowen WD, Mosberg HI, Rothman RB, Porreca F (1990): Opioid agonist and antagonist antinociceptive properties of enkephalin: Selective actions at the $\delta_{\text{non-complexed}}$ site. J Pharmacol Exp Ther 255:636–641.

Leander JD (1983a): Further study of kappa opioids on increased urination. J Pharmacol Exp Ther 227:35–41.

Leander JD (1983b): A kappa opioid effect: Increased urination in the rat. J Pharmacol Exp Ther 224:89–94.

Lee NM, Leybin L, Chang J-K, Loh HH (1980): Opiate and peptide interaction: Effect of enkephalins on morphine analgesia. Eur J Pharmacol 68:181–185.

Lewis JW (1974): Ring C-bridged derivatives of thebaine and oripavine. Adv Biochem Psychopharmacol 8:123–136.

Martin WR, Eades CG, Thompson JA, Huppler RE, Gilbert PE (1976): The effect of morphine and nalorphine-like drugs in the nondependent chronic spinal dog. J Pharmacol Exp Ther 197:517–532.

Nock B, Rajpara A, O'Connor LH, Cicero TJ (1988): [^{3}H]U-69593 labels a subtype of kappa opiate receptor with characteristics different from that labeled by [^{3}H]ethylketocyclazocine. Life Sci 42:2403–2412.

Orwin JM (1977): Pharmacological aspects in man. In Harcus AW, Smith R, Whittle B (eds): "Pain—New Perspectives in Measurement and Management." London: Churchill Livingstone, pp 141–159.

Porreca F, Tortella FC (1987): Differential antagonism of mu agonists by U50,488H in the rat. Life Sci 41:2511–2516.

Portoghese PS, Sultana M, Takemori AE (1988): Naltrindole, a highly selective and potent non-peptide delta opioid receptor antagonist. Eur J Pharmacol 146:185–186.

Portoghese PS, Sultana M, Takemori AE (1990): Naltrindole 5'-isothiocyanate: A nonequilibrium, highly selective delta opioid receptor antagonist. J Med Chem 33:1547–1548.

Rance MJ (1979): Animal and molecular pharmacology of mixed agonist–antagonist analgesic drugs. Br J Clin Pharmacol 7:281S–286S.

Rance MJ, Lord JAH, Robinson T (1980): Biphasic dose response to buprenorphine in the rat tail flick assay: Effect of naloxone pretreatment. In Way EL (ed): "Endogenous and Exogenous Opiate Agonists and Antagonists." New York: Pergamon Press, pp 387–390.

Richards ML, Sadée W (1985): In vivo opiate receptor binding of oripavines to μ, δ, and κ sites in rat brain as determined by an ex vivo labeling method. Eur J Pharmacol 114:343–353.

Rosenbaum JS, Holford NH, Sadée W (1984): Opiate receptor binding-effect relationship: Sufentanil and etorphine produce analgesia at the mu-site with low fractional receptor occupancy. Brain Res 291:317–324.

Rothman RB, Bykov V, Ofri D, Rice KC (1988): LY164929: A highly selective ligand for the lower affinity [^{3}H]D-Ala^2-D-Leu^5-enkephalin binding site. Neuropeptides 11:13–16.

Rothman RB, McLean S (1988): An examination of the opiate receptor subtypes labeled by [^{3}H]cycloFOXY: An opiate antagonist suitable for positron emission tomography. Biol Psychiat 23:435–458.

Rothman RB, Bykov V, De Costa BR, Jacobson AE, Rice KC, Brady LS (1990): Interaction of endogenous opioid peptides and other drugs with four kappa opioid binding sites in guinea pig brain. Peptides 11:311–331.

Rothman RB, Xu H, Seggel M, Jacobson AE, Rice KC, Brine GA, Carroll FI (1991): RTI-4614-4: An analog of (+)-cis-3-methylfentanyl with a 27,000-fold binding selectivity for mu versus delta opioid binding sites. Life Sci 48:PL111–PL116.

Sadée W, Rosenbaum JS, Herz A (1982): Buprenorphine: Differential interaction with opiate receptor subtypes in vivo. J Pharmacol Exp Ther 223:157–162.

Sheldon RJ, Nunan L, Porreca F (1987): Mu antagonist properties of kappa agonists in a model of rat urinary bladder motility in vivo. J Pharmacol Exp Ther 243:234–240.

Su TP (1985): Further demonstration of kappa opioid binding sites in the brain: Evidence for heterogeneity. J Pharmacol Exp Ther 232:144–148.

Tallarida RJ, Cowan A (1982): The affinity of morphine for its pharmacological receptor *in vivo*. J Pharmacol Exp Ther 222:189–201.

Tiberi M, Magnan J (1990): Demonstration of the heterogeneity of the kappa-opioid receptors in guinea-pig cerebellum using selective and nonselective drugs. Eur J Pharmacol 188:379–389.

Tolkovsky AM (1982): Etorphine binds to multiple opiate receptors of the caudate nucleus with equal affinity but with different kinetics. Mol Pharmacol 22:648–656.

Van Bever WFM, Niemegeers CJE, Janssen PAJ (1974): Synthetic analgesics. Synthesis and pharmacology of the diastereoisomers of N-[3-methyl-1-(2-phenylethyl)-4-piperidyl]-N-phenylpropanamide and N-[3-methyl-1-(1-methyl-2-phenylethyl)-4-piperidyl]-N-phenylpropanamide. J Med Chem 17:1047–1051.

Vaught JL, Takemori AE (1979): A further characterization of the differential effects of leucine enkephalin, methionine enkephalin and their analogs on morphine-induced analgesia. J Pharmacol Exp Ther 211:280–283.

Vaught JL, Rothman RB, Westfall TC (1982): Mu and delta receptors: Their role in analgesia and in the differential effects of opioid peptides on analgesia. Life Sci 30:1443–1455.

Villiger JW (1984): Binding of buprenorphine to opiate receptors. Neuropharmacology 23:373–375.

Villiger JW, Taylor KM (1981): Buprenorphine: Characteristics of binding sites in the rat CNS. Life Sci 29:2699–2708.

Villiger JW, Taylor KM (1982): Buprenorphine: High affinity binding to the dorsal spinal cord. J Neurochem 38:1771–1773.

Wood PL, Iyengar S (1986): Kappa isoreceptors: Neuroendocrine and neurochemical evidence. NIDA Res Monogr 71:102–108.

Wüster M, Herz A (1976): Significance of physiochemical properties of opiates for *in vitro* testing. In Kosterlitz HW (ed): "Opiates and Endogenous Opioid Peptides." Amsterdam: North-Holland, pp 447–450.

Xu H, Kim C-H, Zhu YC, Weber RJ, Rice KC, Rothman RB (1991): (+)-cis-Methylfentanyl and its analogs bind pseudoirreversibly to the mu opioid binding site: Evidence for pseudoallosteric modulation. Neuropharmacology 30:455–462.

Zhu Y-S, Szeto HH (1989): Morphine-induced tachycardia in fetal lambs: A bell-shaped dose-response curve. J Pharmacol Exp Ther 249:78–82.

Zukin RS, Eghbali M, Olive D, Unterwald EM, Tempel A (1988): Characterization and visualization of rat and guinea pig brain kappa opioid receptors: Evidence for κ_1 and κ_2 opioid receptors. Proc Natl Acad Sci USA 85:4061–4065.

UPDATE ON THE GENERAL PHARMACOLOGY OF BUPRENORPHINE

ALAN COWAN

Department of Pharmacology, Temple University School of Medicine, Philadelphia, PA 19140

INTRODUCTION

Buprenorphine was selected for clinical study (and subsequent commercial use) on the basis of traditional pharmacological approaches and before the advent of modern bioassay, receptor binding, or molecular biology techniques. The agonist and antagonist properties of buprenorphine were described by Martin et al. [1976] after elegant work with the chronic spinal dog preparation and by Cowan et al. [1977a,b], who characterized this oripavine derivative in the following manner:

1. Buprenorphine is 25–40 times more potent than morphine after parenteral administration in mouse stretch and rat tail pressure tests.
2. The dose-response relation for buprenorphine is curvilinear in rodent tail dip tests and in rat catalepsy and gastrointestinal transit assays, but is sigmoidal in mouse stretch and rat tail pressure tests.
3. The physical dependence capacity of buprenorphine is of a low order in mice and monkeys.
4. The morphine antagonist properties of buprenorphine were demonstrated in mouse and rat tail dip tests and in morphine-dependent mice and monkeys.
5. Buprenorphine increased the spontaneous locomotor activity of mice but induced behavioral depression in guinea pigs.
6. Buprenorphine reduced heart rate but had no significant effect on arterial blood pressure in conscious rats and dogs; it caused no major hemodynamic changes in anesthetized, open-chest cats.

Buprenorphine: Combatting Drug Abuse With a Unique Opioid, pages 31–47
© 1995 Wiley-Liss, Inc.

7. Buprenorphine increased arterial P_{CO2} values and P_{O2} values in conscious rats. Submaximal ceiling effects were obtained.
8. Buprenorphine displayed an antitussive action against chemically induced coughing in guinea pigs.
9. Buprenorphine suppressed urine output in water-loaded rats.

These findings were subsequently assessed in relation to a rapidly expanding literature on buprenorphine: For example, Heel et al. [1979] surveyed the animal pharmacology and summarized early clinical results; Lewis et al. [1983] published a similar overview, in which the safety and low abuse potential of buprenorphine were emphasized; and Lewis [1985] reviewed the key animal and human data on buprenorphine that had accumulated by the early 1980s. More recent findings with buprenorphine, particularly those from the laboratories of "rodent pharmacologists," have yet to be assimilated and interpreted against the historical body of knowledge on this unusual analgesic. Such contemporary research is described in the present chapter.

PRECLINICAL RESULTS IN RETROSPECT

Buprenorphine was introduced into medical practice in the United Kingdom (as an intramuscular analgesic) in 1978 and then as a sublingual tablet in 1981. It is now possible to assess the predictive value of the original animal data in the light of 10–15 years of international clinical experience. Buprenorphine has proved to be a safe, effective, and long-lasting analgesic against moderate-to-severe pain in a wide variety of pain states. Analgesic effectiveness has not been limited by a submaximal ceiling, as was the case in several laboratory tests for antinociception. By the same token, the severity of constipation is comparable with that of morphine, and respiratory depression may occur in the therapeutic dose range. Note, however, that high doses of buprenorphine need not necessarily induce severe respiratory depression in patients experiencing postoperative pain [Budd, 1981]. The comment by Jasinski and Preston (this volume) that buprenorphine has "little, if any" physical dependence capacity in human volunteers is in line with predictions from the animal laboratory.

A quick onset of antinociceptive action in mice and a lack of sickness in beagle dogs were two promising features of the pharmacological profile of buprenorphine. These two properties, unfortunately, have not carried over into the clinic.

INTERACTION WITH OPIOID RECEPTORS

Buprenorphine was initially classified as a "mixed agonist–antagonist analgesic" or a "narcotic antagonist analgesic." The work of Martin et al. [1976] with the chronic spinal dog preparation underscored the partial agonist nature of buprenorphine at μ

opioid receptors, and this description was supported in many respects by the compound's general pharmacological profile [Cowan et al., 1977a,b]. Results from receptor binding studies by Villiger [1984], Richards and Sadée [1985a], Su [1985], and Rothman et al. (this volume) indicated that buprenorphine binds with high affinity to μ and $κ_1$ receptors and with lower affinity to δ receptors and other acceptor sites. Recent studies with cloned opioid receptors stress the high affinity of buprenorphine for rat μ and mouse κ and δ receptors. It is of particular interest that buprenorphine shows highest affinity for the cloned δ receptor [Reisine and Bell, 1993; personal communication, T. Reisine].

The high affinity of buprenorphine for different opioid receptors has also been demonstrated through standard bioassays with strips of isolated smooth muscle (despite technical problems sometimes associated with the drug's high lipophilicity and slow washout time) [Lewis, 1985; Kajiwara et al., 1986]. Most recently, Smith et al. [1988] described buprenorphine's comparable affinity for μ and κ receptors in the electrically stimulated mouse vas deferens preparation. Thus, the equilibrium dissociation constant (K_e value) for buprenorphine against normorphine (μ agonist) was 0.04 nM and against ethylketocyclazocine (κ agonist) was 0.08 nM. Buprenorphine also displayed antagonism against [D-Ala^2,D-Leu^5]enkephalin (δ agonist) but, in this case, the K_e value was an order of magnitude higher (0.66 nM).

To what extent does agonism at δ and κ receptors contribute to the overall antinociceptive activity of buprenorphine? The buprenorphine–δ interface in vivo has yet to be examined critically with appropriate pharmacological tools. Sadée et al. [1983] have offered the provocative suggestion that δ receptors mediate, through noncompetitive autoinhibition, the second phase of the buprenorphine biphasic dose-response curve, at least in the rat tail–electroshock test. This hypothesis needs to be tested more fully.

Although Tyers [1980] claimed to identify a κ agonist component in the antinociceptive profile of buprenorphine, it is now generally believed that the drug possesses only limited intrinsic activity at κ receptors. In later work by Tyers and his colleagues [Hayes et al., 1986], the antinociceptive action of buprenorphine was evaluated in the paw pressure test in rats injected s.c. 24 hr previously with 40 mg/kg of β-funaltrexamine (β-FNA), the irreversible antagonist at μ receptors [Ward and Takemori, 1983]. The usual monotonic dose-response relationship for buprenorphine in this procedure now showed a decreased slope and lowered maximum effect. Under the same conditions, the dose-response curve for U-50,488 (the standard κ agonist) was not affected markedly. It seems reasonable to conclude that, in this test, μ receptors rather than κ receptors are mediating partial agonist-induced (buprenorphine-induced) antinociception.

Zimmerman et al. [1987] also examined the influence of β-FNA on the antinociceptive action of buprenorphine; in this instance the mouse acetic acid stretch test was used. Pretreating mice with β-FNA (80 mg/kg s.c. at −24 hr) caused a shift in the linear dose-response curve of buprenorphine to the right in a nonparallel fashion and with a reduced maximum effect. Once again, it was concluded that μ receptors rather than κ receptors are implicated in the antinociceptive action of buprenorphine.

Millan [1989] has proposed the following qualitative guideline when classifying opioid agonists: In the rat tail electroshock test μ, but not κ, agonism is associated with activity. Given the activity (albeit nonmonotonic) of buprenorphine in this pain model [Dum and Herz, 1981], the oripavine should therefore be grouped with μ rather than κ agonists.

Finally, water diuresis in rats (and monkeys) seems to be an inherent agonist property and defining characteristic of standard κ agonists [Cowan and Gmerek, 1986; Brooks et al., 1993]. As noted in the original animal pharmacology, buprenorphine resembles morphine in *suppressing* urine output in water-loaded rats [Cowan et al., 1977a] and this decrease in urination can be attenuated when the animals are pretreated with β-FNA (40 mg/kg s.c. at −24 hr) [Hayes et al., 1987b].

TABLE I. Examples of κ Antagonism Displayed by Buprenorphine

Test/measure	κ agonist	Effect of buprenorphine	Reference[a]
Guinea pig isolated ileum preparation	EKC[b]	Antagonism of EKC-induced inhibition of electrically evoked contractions	[3]
Diuresis (rats)	Bremazocine	Displacement of dose-response curve to right	[9]
Behavior of rats	U-50,488 and tifluadom	Demonstration of neuroadaptation to U-50,488 and tifluadom	[5]
Drug discrimination (rats)	U-50,488	Decrease in U-50,488-appropriate responding	[7]
Multiple fixed-interval/fixed-ratio schedule of food presentation (pigeons)	U-50,488	Antagonism of U-50,488-induced decreases in response rate	[4]
Behavior of patas monkeys	Cyclazocine	Precipitation of κ abstinence syndrome in monkeys receiving cyclazocine chronically	[1]
Behavior of rhesus monkeys	U-50,488	Precipitation of κ abstinence syndrome in monkeys receiving U-50,488 chronically	[2]
Shock titration (squirrel monkeys)	U-50,488	Displacement of dose-response curve to right	[6]
Drug discrimination (squirrel monkeys)	U-50,488	Antagonism of discriminative stimulus effects of U-50,488	[8]
Tail dip test (50° C) (rhesus monkeys)	U-50,488	Displacement of dose-response curve to the right	[10]

[a]Sources: [1] Cowan [1973]; [2] Gmerek et al. [1987]; [3] Kajiwara et al. [1986]; [4] Leander [1988]; [5] Murray and Cowan [1988]; [6] Negus and Dykstra [1988]; [7] Negus et al. [1990]; [8] Negus et al. [1991]; [9] Richards and Sadée [1985b]; [10] Woods et al. [1992].
[b]EKC = ethylketocyclazocine.

There is substantial evidence in support of the κ antagonist actions of buprenorphine. Several of the key references—from in vitro bioassay through antagonism of κ-induced diuresis in rats to applied aspects of monkey behavior—are presented in Table I. It should be mentioned that this κ antagonism occurs with doses of buprenorphine that are often lower than those that are necessary for agonism at μ receptors.

RECENT ANTINOCICEPTIVE DATA

First, some old results. Skingle and Tyers [1980] established that buprenorphine is antinociceptive after sublingual delivery to conscious beagle dogs. This report is important for the 1990s because it is the only one in the preclinical literature that describes the agonist effects of buprenorphine when given by what has clearly become a prime route of administration in the therapeutic application of the oripavine. At sublingual doses of 0.02 and 0.04 mg, buprenorphine markedly increased the nociceptive threshold to electrical stimulation of the tooth pulp of dogs [Skingle and Tyers, 1979, 1980]. The higher dose was associated with sedation, ataxia, and respiratory depression.

Reports in the literature have noted the (usually) higher number of critical κ binding sites in the CNS of guinea pigs relative to rats (e.g., Lahti et al., 1985; Unterwald et al., 1991]. Hayes et al. [1987a] observed that κ agonists are more potent in the guinea pig (noninflamed) paw pressure test than in the corresponding rat paw pressure test. For example, the rat : guinea pig potency ratio for antinociception is 7 for tifluadom and 9.4 for U-50,488 (Table II). Mu agonists such as morphine and fentanyl, though slightly more potent in the test with rats, are essentially equipotent in the two procedures. On the basis of this approach, then, and in keeping with the previous section, buprenorphine (with a potency ratio of 0.33) is appropriately grouped with the μ, rather than the κ, opioid agonists (Table II).

The original antinociceptive profile of buprenorphine was based on traditional tests with rodents that measure *transient* rather than *continuous* pain. The relevance

TABLE II. Comparison of the Antinociceptive Potencies of μ and κ Agonists in Rat and Guinea Pig Paw Pressure Tests[a]

Agonist	Rat paw pressure A_{50} (mg/kg, s.c.)[b]	Guinea pig paw pressure A_{50} (mg/kg, s.c.)[b]	Rat : guinea pig ratio
Buprenorphine	0.001 (0.0005–0.004)	0.003 (0.001–0.01)	0.33
Fentanyl	0.003 (0.002–0.004)	0.004 (0.001–0.01)	0.75
Morphine	0.44 (0.35–0.53)	0.54 (0.17–1.6)	0.81
Tifluadom	0.14 (0.06–0.29)	0.02 (0.01–0.06)	7.0
U-50,488	3.1 (1.2–8.6)	0.33 (0.08–0.99)	9.4

[a]Source: Modified from Hayes et al. [1987a].
[b]The A_{50} value is the dose of agonist that increases the nociceptive threshold 50% above that for the control group.

of these high-intensity, phasic endpoints to clinical pain states and to the proper evaluation of novel partial agonists such as buprenorphine has been increasingly questioned [e.g., Shaw et al., 1988]. How would buprenorphine compare against reference analgesics in animal models of protracted, visceral, or neuropathic types of pain? A start has been made to answer this question. Buprenorphine has been assessed recently in a relatively new model of postinjury pain—the formalin test [Dubuisson and Dennis, 1977; Shibata et al., 1989; Cowan, 1990; Wheeler-Aceto et al, 1990; Wheeler-Aceto and Cowan, 1991b]. In rats, injection of 5% formalin into the dorsum of a hind paw elicits two spontaneous behaviors indicative of pain: flinching/shaking of the paw and/or hindquarters and licking/biting of the injected paw. An acute nociceptive phase over the first 10 min is followed, after a short quiescent period, by a continuous (tonic) background of pain (for about 70 min) that may be neurochemically and neurophysiologically different from the transient (phasic) pain associated with conventional hot plate and tail flick tests [Dennis et al., 1980; Dennis and Melzack, 1983]. The biphasic nature of the response to formalin affords an opportunity to study the influence of pharmacological intervention on acute and tonic pain of common origin.

Buprenorphine, given subcutaneously, was extremely potent in the rat paw formalin test and essentially equipotent against acute and tonic pain (Table III). Morphine, though about 10 times less potent, was also essentially euipotent in both phases of the test. PD 117302, a reference κ agonist [Leighton et al., 1987; Clark et al., 1988], differed from buprenorphine and morphine in being approximately 27 times more potent against flinching in the second phase, emphasizing the importance of stimulus intensity in analgesic evaluation [Millan, 1989; Parsons and Headley, 1989].

Dose-response curves for subcutaneous buprenorphine, morphine, and butorphanol (the reference mixed agonist–antagonist analgesic) against flinching in the second (late) phase of the rat formalin test are shown in Figure 1. Note the steep slope and maximum effect associated with buprenorphine in this chemogenic assay.

Buprenorphine is more potent than morphine in the rat formalin test when given either subcutaneously or intracerebroventricularly (i.c.v.) but is *less* potent after direct intrathecal (i.th.) injection (Table IV). This result is in agreement with the proposition [McQuay et al., 1989; Dickenson et al., 1990], discussed by Rosow in

TABLE III. Antinociceptive Potencies of Test Compounds Against Flinching in the Rat Paw Formalin Test[a]

Agonist	Early phase A_{50} (mg/kg, s.c.)[b]	Late phase A_{50} (mg/kg, s.c.)[b]	Late : early
Buprenorphine	0.019 (0.009–0.034)	0.034 (0.020–0.052)	1.8
Morphine	0.17 (0.09–0.27)	0.23 (0.13–0.36)	1.4
PD 117302	5.2 (3.5–8.9)	0.20 (0.09–0.32)	0.04

[a]Source: Wheeler-Aceto and Cowan (unpublished results).
[b]A_{50} values and 95% confidence limits were determined by linear regression analysis from the percentage antagonism of formalin-induced flinching.

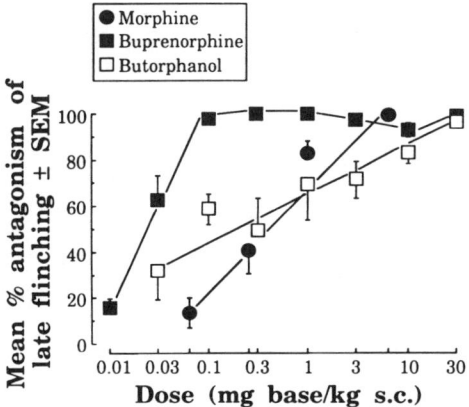

Fig. 1. Morphine, buprenorphine, and butorphanol dose-response curves for antagonism of late-phase flinching induced by formalin in rats. (Wheeler-Aceto and Cowan, unpublished results).

this volume, that an inverse relationship exists between lipophilicity and antinociceptive potency after intrathecal administration of several μ agonists to rats.

INTENSITY OF THE NOXIOUS STIMULUS

The intensity of the noxious stimulus is an experimental variable that has been investigated extensively by analgesic researchers [e.g., Shaw et al., 1988; Millan, 1989; Parsons and Headley, 1989]. The following study illustrates the marked influence of stimulus intensity on the shape of buprenorphine dose-response curves for antinociception. It is well known that the dose-response relation for buprenorphine is submaximal and resembles an inverted U when rats receive the brief, high-intensity noxious stimulus associated with the 50° C tail dip test. This stimulus is of high intensity in terms of the frequency of firing elicited in spinal dorsal horn neurons. By way of contrast, the short early phase and more prolonged late phase of the rat formalin test are associated with intermediate and low firing

TABLE IV. Antinociceptive Potencies of Buprenorphine and Morphine, Given Intrathecally (i.th.) or Intracerebroventricularly (i.c.v.), Against Late-Phase Flinching in the Rat Paw Formalin Test[a]

Agonist	A_{50} (μg i.th.)[b]	A_{50} (μg i.c.v.)[b]
Buprenorphine	0.47 (0.29–0.73)	0.21 (0.12–0.32)
Morphine	0.17 (0.07–0.30)	1.3 (0.47–2.4)

[a]Source: Wheeler-Aceto and Cowan (unpublished results).
[b]A_{50} values and 95% confidence limits were determined by linear regression analysis from the percentage antagonism of formalin-induced flinching.

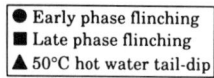
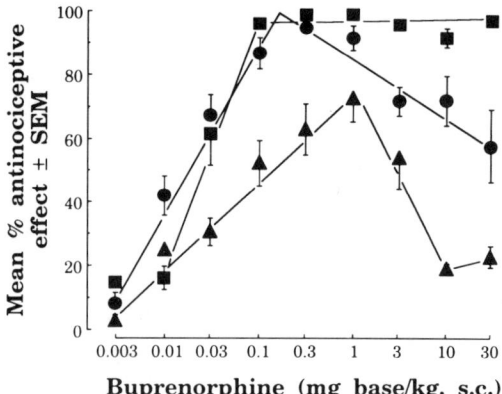

Fig. 2. Effect of pretreating rats with buprenorphine (at −20 min) on early- and late-phase flinching induced by formalin and the tail dip response. Separate groups of rats ($n = 5$–6) were used for each of the three levels of noxious stimulation. (Wheeler-Aceto and Cowan, unpublished results).

rates, respectively [Dickenson and Sullivan, 1987]. To what extent will the bell-shaped dose-response curve of buprenorphine change in response to the intermediate-intensity stimulus or in response to the low-intensity stimulus? The results are brought together in Figure 2. A curvilinear dose-response relation is associated with the intermediate stimulus but the maximum effect is greatly increased. With the low stimulus, a sigmoidal, full-efficacy dose-response curve is obtained.

This preliminary study was repeated so that all three endpoints were recorded *in the same rat*. The same results were obtained and it was possible to show that buprenorphine displays low efficacy against one type of noxious stimulus while *simultaneously* displaying full efficacy against another of differing intensity [Wheeler-Aceto and Cowan, 1991a]. Morphine, unlike buprenorphine, was fully efficacious at all three stimulus levels.

BIPHASIC DOSE-RESPONSE CURVES

Buprenorphine is renowned for the shape of its dose-response curve. In a great number of assays, the dose-response relationship is curvilinear. Examples of these procedures are listed in Table V.

The biphasic dose-response curve for buprenorphine was first observed in rodent

TABLE V. Examples of Procedures in Which the Dose-Response Relation for Buprenorphine Is Curvilinear

Test/measure	Dose range (mg/kg, s.c.)	Peak dose (mg/kg, s.c.)	Peak effect	Reference[a]
Mice				
Rotarod	0.3–10	1	Muscular incoordination	[8]
Respiratory rate	0.3–10	2.5	Decreased respiratory rate	[8]
Hot plate (55° C)	0.5–20	3	Antinociception	[13]
Rats				
Drug discrimination	0.003–0.3 (i.p.)	0.03 (i.p.)	Substitution for morphine	[11]
Conditioned place preference	0.005–0.9 (i.p.)	0.15 (i.p.)	Conditioned place preference	[1]
Dopamine release in nucleus accumbens	0.01–5 (i.p.)	0.5 (i.p.)	Increased dopamine release	[9]
Y-maze	0.03–60	0.3	Decreased rears and arm entries	[3]
Catalepsy	0.03–30	0.3	Immobility	[5]
Electrical stimulation of tail	0.03–10	0.5	Antinociception	[7]
FR 40 responding	0.001–60	1	Decrease in operant responding for food pellets	[3]
Charcoal meal	0.01–30	1	Delayed gastrointestinal transit	[4]
Hotplate (55° C)	0.003–10	1	Antinociception	[2]
Formalin-induced flinching	0.003–30	1	Antinociception in early phase	[14]
TSH release from anterior pituitary	0.1–6.6	1.3	Decrease in serum TSH	[12]
Tail dip (55° C)	0.1–30	3	Antinociception	[5]
Blood gases	0.01–30 (i.a.)	1 (i.a.)	Elevated P_{CO_2} level	[6]
EEG/EMG/gross behavior	0.3–30 (i.v.)	1 (i.v.)	Longest duration of behavioral depression; maximum EEG power output	[10]
Monkeys				
Tail dip (50° C)	0.1–3.2	1	Antinociception	[15]

[a] [1] Brown et al. [1991]; [2] Bryant et al. [1983]; [3] Cowan and Watson [1980]; [4] Cowan et al. [1977a]; [5] Cowan et al. [1977b]; [6] Doxey et al. [1982]; [7] Dum and Herz [1981]; [8] Hayes and Tyers [1983]; [9] Holman et al. [1993]; [10] Kareti et al. [1980]; [11] Negus et al. [1990]; [12] Pechnick et al. [1985]; [13] Tyers [1980]; [14] Wheeler-Aceto and Cowan [1991a]; [15] Woods et al. [1992].

tail dip/flick tests. The antinociception associated with μ agonists in this model is enhanced if the rats are subjected to the stress of restraint (immobility in a Plexiglas cylinder) [Appelbaum and Holtzman, 1984]. Calcagnetti et al. [1990] found that this is also the case with buprenorphine, the subcutaneous ED_{50} value being 0.21 (0.05–0.79) mg/kg in restrained rats and > 1 mg/kg in unrestrained animals. Of immediate interest is the finding that, whatever the pharmacological mechanism(s) behind the potentiation [Appelbaum and Holtzman, 1985; Fleetwood and Holtzman, 1989; Woolfolk and Holtzman, 1993], the dose-response relation for buprenorphine in the stressed rats still maintained the inverted U shape, as consistently reported for the oripavine in this standard assay.

The influence of naloxone or naltrexone on the shape and position of buprenorphine biphasic curves has been investigated using only one endpoint—antinociception—in rat tail flick [Rance et al., 1980] and rat tail electroshock [Dum and Herz, 1981] tests. In both cases, the biphasic curve was displaced to the right in a symmetrical fashion. An account of the naloxone–buprenorphine interaction was recently published [Cowan, 1992] in which a second classical property of opioids—slowing of transit along the gastrointestinal tract [Manara and Bianchetti, 1985]—was studied. The results are illustrated in Figure 3. As expected, buprenorphine

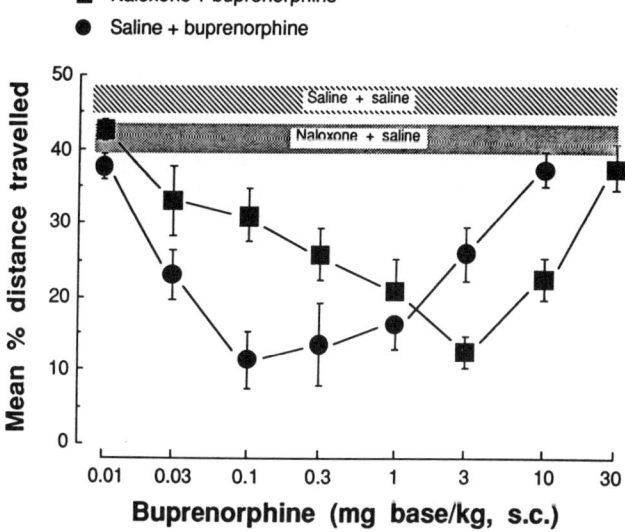

Fig. 3. Effect of buprenorphine on gastrointestinal transit in rats ($n = 8–10$) in the presence and absence of a fixed dose of naloxone (0.30 mg/kg, s.c.). Each point gives the mean distance) ($\pm$ SEM) traveled by a charcoal meal along the intestine, expressed as a percentage of the total length of the small intestine. The shaded bands represent corresponding distances traveled by the meal in (saline + saline) and (naloxone + saline) control rats ($n = 10$). Reproduced from Cowan [1992], with permission of the publisher.

slowed the passage of a charcoal meal along the small intestine of rats. The dose-response curve is U-shaped over the range 0.01–10 mg/kg. When rats were pretreated with 0.30 mg/kg of naloxone, the buprenorphine dose-response curve was shifted to the right without a change in maximal effect. As with antinociception [Rance et al., 1980], naloxone enhanced, rather than attenuated, the antipropulsive action of the higher doses of buprenorphine (3 and 10 mg/kg).

Over the years, the biphasic dose-response curve for buprenorphine-induced antinociception has been explained in terms of a two-receptor model and noncompetitive autoinhibition [Cowan et al., 1977b; Sadée et al., 1982; Richards and Sadée, 1985a]. This hypothesis could also apply to the antitransit assay with naloxone acting as a competitive antagonist at both binding sites. The availability of selective antagonists at κ and δ receptors makes it feasible to probe the nature of the second (low-affinity, inhibitory) binding site.

ABRUPT WITHDRAWAL OF BUPRENORPHINE FROM RATS

Buprenorphine is known to dissociate very slowly from opioid receptors [Hambrook and Rance, 1976; Boas and Villiger, 1985], and this maintenance of homeostasis may help to counter the development of an overt withdrawal syndrome [Dum et al., 1981]. Jasinski et al. [1978] reported that abrupt withdrawal of subcutaneous buprenorphine from three humans who were not dependent on opioids led to a mild syndrome of delayed onset (2–3 days) that took a long time to peak: 14–15 days after termination of buprenorphine. Fudala et al. [1990] administered buprenorphine sublingually to 19 heroin addicts and described a mild to moderate withdrawal syndrome that peaked as soon as 3–5 days and lasted a further 5 days. In view of these clinical studies, two experiments were conducted in rats in which the animals were monitored over a longer than usual period after abrupt termination of buprenorphine injections. In the first experiment, male Sprague Dawley rats (200–220 g; $n = 8$) were injected subcutaneously at 16:00 hr daily with either saline or buprenorphine (0.10 mg/kg) for 28 consecutive days. Weight loss after abrupt termination of buprenorphine was taken as an objective index of withdrawal. Day-to-day mean weight changes were similar over the 15 days following withdrawal. The experiment was repeated, this time by administrating 3 mg/kg of buprenorphine to the rats for 28 days. The results are displayed in Figure 4.

Buprenorphine-withdrawn rats showed slightly greater weight increases than control animals until day +6. Thereafter, the mean weight gain in control rats was impressively greater than that in animals withdrawn from buprenorphine. The rats were observed for overt signs of opiate-like withdrawal but none were noted. In this work, then, neuroadaptation by rats to 28 injections of 3 mg/kg of buprenorphine was demonstrated through the delayed appearance of one, and only one, recognized component of morphine withdrawal: body weight loss. Furthermore, this work provides a standard protocol for in vitro or in vivo study of the "week 1" and "week 2" buprenorphine-withdrawn rat.

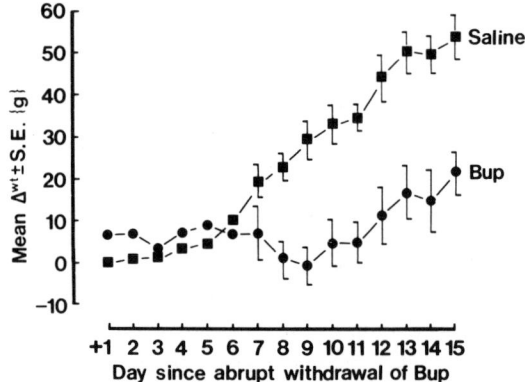

Fig. 4. Delayed appearance of a sign of withdrawal in rats. Two groups of rats ($n = 8$) received either saline or buprenorphine (Bup, 3 mg/kg, s.c.) daily for 28 days and thereafter a daily subcutaneous injection of saline. Each point represents the mean change daily in body weight since the abrupt termination of buprenorphine. (Cowan, unpublished results).

PERSPECTIVE

In proposing a role for buprenorphine in the pharmacotherapy of opioid dependence, Jasinski et al. [1978] concluded their article by describing the pharmacology of buprenorphine as being "unique." Is the pharmacological profile of the compound still "without equal or like" (Collins Dictionary of the English Language, Collins, Glasgow, 1979) in the 1990s? The unusual mix of characteristics—high affinity at opioid receptors, partial agonism at μ receptors, κ antagonism, lipophilicity, slow elimination kinetics, long duration of action, safety, sublingual delivery and omnipresence in basic research—certainly goes a long way to earning the "unparalleled" label.

TABLE VI. Affinities of Test Agents for Cloned Opioid Receptors[a]

Ligand	IC_{50} value (nM)[b]		
	μ	κ	δ
Buprenorphine	0.3	1.1	0.1
Methadone	7.4	>1,000	>1,000
Morphine	52	886	>1,000
Naloxone	1.7	0.7	68
[D-Pen^2,D-Pen^5]enkephalin	>1,000	>1,000	122
U-50,488	1,000	1.1	>1,000

[a]Source: Modifed from Reisine and Bell (1993), and personal communication from T. Reisine.
[b]COS-7 cells, PC12 cells, and CHO cells expressing the cloned rat μ, mouse κ, and mouse δ receptor, respectively, were used. The μ receptors were labeled with [^{3}H]DAMGO, the κ receptors with [^{3}H]U-69,593 and the δ receptors with [^{3}H]naltrindole. Inhibition constant (IC_{50}) values were calculated from inhibition curves for 6–8 different ligand concentrations.

Even the current lack of an effective buprenorphine antagonist can be viewed as an odd situation, especially in the opioid field. This could change with the advent of compounds like clocinnamox [Comer et al., 1992], which may be able to reverse *established* actions of buprenorphine, at least in animals [Aceto et al., 1992].

As to the future, it is clear from comparing affinities of standard opioids at newly unveiled receptor clones (Table VI) that the buprenorphine profile remains perversely different, provocative, and intellectually challenging.

REFERENCES

Aceto MD, Tucker SM, Ji Z (1992): Antagonist activity of 14β-(p-chlorocinnamoyl-amino)-7,8-dihydro-N-cyclopropylmethylnormorphinone mesylate (CCAM) versus buprenorphine. FASEB J 6:A994.

Appelbaum BD, Holtzman SG (1984): Characterization of stress-induced potentiation of opioid effects in the rat. J Pharmacol Exp Ther 231:555–565.

Appelbaum BD, Holtzman SG (1985): Restraint stress enhances morphine-induced analgesia in the rat without changing apparent affinity of receptor. Life Sci 36:1069–1074.

Boas RA, Villiger JW (1985): Clinical actions of fentanyl and buprenorphine. The significance of receptor binding. Br J Anaesth 57:192–196.

Brooks DP, Giardina G, Gellai M, Dondio G, Edwards RM, Petrone G, DePalma PD, Sbacchi M, Jugus M, Misiano P, Wang Y-X, Clarke GD (1993): Opiate receptors within the blood-brain barrier mediate kappa agonist-induced water diuresis. J Pharmacol Exp Ther 266:164–171.

Brown EE, Finlay JM, Wong JTF, Damsma G, Fibiger HC (1991): Behavioral and neurochemical interactions between cocaine and buprenorphine: Implications for the pharmacotherapy of cocaine abuse. J Pharmacol Exp Ther 256:119–126.

Bryant RM, Olley JE, Tyers MB (1983): Antinociceptive actions of morphine and buprenorphine given intrathecally in the conscious rat. Br J Pharmacol 78:659–663.

Budd K (1981): High dose buprenorphine for postoperative analgesia. Anaesthesia 36:900–903.

Calcagnetti DJ, Fleetwood SW, Holtzman SG (1990): Pharmacological profile of the potentiation of opioid analgesia by restraint stress. Pharmacol Biochem Behav 37:193–199.

Clark CR, Birchmore, B, Sharif NA, Hunter JC, Hill RG, Hughes J (1988): PD 117302: A selective agonist for the kappa-opioid receptor. Br J Pharmacol 93:618–626.

Comer SD, Burke TF, Lewis JW, Woods JH (1992): Clocinnamox: A novel, systemically active, irreversible opioid antagonist. J Pharmacol Exp Ther 262:1051–1056.

Cowan A (1973): Evaluation of the physical dependence capacities of oripavine–thebaine partial agonists in patas monkeys. In Braude MC, Harris LS, May EL, Smith JP, Villarreal JE (eds): "Narcotic Antagonists." New York: Raven Press, pp. 427–438.

Cowan A (1990): Recent approaches in the testing of analgesics in animals. In Adler MW, Cowan A (eds): "Testing and Evaluation of Drugs of Abuse." New York: Wiley-Liss, pp 33–42.

Cowan A (1992): Buprenorphine and gastrointestinal transit in rats: Effect of naloxone on the biphasic dose-response curve. Clin Exp Pharmacol Physiol 19:47–49.

Cowan A, Gmerek DE (1986): In vivo studies on kappa opioid receptors. Trends Pharmacol Sci 7:69–72.

Cowan A, Watson T (1980): Buprenorphine gives uncommon, biphasic dose-response curves in Y-maze and operant behavioral procedures. Fed Proc 39:759.

Cowan A, Doxey JC, Harry EJR (1977a): The animal pharmacology of buprenorphine, an oripavine analgesic agent. Br J Pharmacol 60:547–554.

Cowan A, Lewis JW, Macfarlane IR (1977b): Agonist and antagonist properties of buprenorphine, a new antinociceptive agent. Br J Pharmacol 60:537–545.

Dennis SG, Melzack R (1983): Effects of cholinergic and dopaminergic agents on pain and morphine analgesia measured by three pain tests. Exp Neurol 81:167–176.

Dennis SG, Melzack R, Gutman S, Boucher F (1980): Pain modulation by adrenergic agents and morphine as measured by three pain tests. Life Sci 26:1246–1259.

Dickenson AH, Sullivan AF (1987): Subcutaneous formalin-induced activity of dorsal horn neurones in the rat: Differential response to an intrathecal opiate administered pre or post formalin. Pain 30:340–360.

Dickenson AH, Sullivan AF, McQuay HJ (1990): Intrathecal etorphine, fentanyl and buprenorphine on spinal nociceptive neurones in the rat. Pain 42:227–234.

Doxey JC, Everitt JE, Frank LW, MacKenzie JE (1982): A comparison of the effects of buprenorphine and morphine on the blood gases of conscious rats. Br J Pharmacol 75:118P.

Dubuisson D, Dennis SG (1977): The formalin test: A quantitative study of the analgesic effects of morphine, meperidine and brain stem stimulation in rats and cats. Pain 4:161–174.

Dum JE, Herz A (1981): In vivo receptor binding of the opiate partial agonist, buprenorphine, correlated with its agonistic and antagonistic actions. Br J Pharmacol 74:627–633.

Dum J, Bläsig J, Herz A (1981): Buprenorphine: Demonstration of physical dependence liability. Eur J Pharmacol 70:293–300.

Fleetwoood SW, Holtzman SG (1989): Stress-induced potentiation of morphine-induced analgesia in morphine-tolerant rats. Neuropharmacology 28:563–567.

Fudala PJ, Jaffe JH, Dax EM, Johnson RE (1990): Use of buprenorphine in the treatment of opioid addiction. II. Physiologic and behavioral effects of daily and alternate-day administration and abrupt withdrawal. Clin Pharmacol Ther 47:525–534.

Gmerek DE, Dykstra LA, Woods JH (1987): Kappa opioids in rhesus monkeys. III. Dependence associated with chronic administration. J Pharmacol Exp Ther 242:428–436.

Hambrook JM, Rance MJ (1976): The interaction of buprenorphine with the opiate receptor: Lipophilicity as a determining factor in drug-receptor kinetics. In Kosterlitz HW (ed): "Opiates and Endogenous Opioid Peptides." Amsterdam: Elsevier, pp 295–301.

Hayes AG, Tyers MB (1983): Determination of receptors that mediate opiate side effects in the mouse. Br J Pharmacol 79:731–736.

Hayes AG, Skingle M, Tyers MB (1986): Reversal by beta-funaltrexamine of the antinociceptive effect of opioid agonists in the rat. Br J Pharmacol 88:867–872.

Hayes AG, Sheehan MJ, Tyers MB (1987a): Differential sensitivity of models of antinociception in the rat, mouse and guinea pig to mu and kappa opioid receptor agonists. Br J Pharmacol 91:823–832.

Hayes AG, Skingle M, Tyers MB (1987b): Evaluation of the receptor selectivities of opioid drugs by investigating the block of their effect on urine output by beta-funaltrexamine. J Pharmacol Exp Ther 240:984–988.

Heel RC, Brogden RN, Speight TM, Avery GS (1979): Buprenorphine: A review of its pharmacological properties and therapeutic efficacy. Drugs 17:81–110.

Holman RB, Lewis JW, Lalies MD (1993): Acute and chronic buprenorphine treatment: Dopamine release in the nucleus accumbens of conscious rat. NIDA Res Monogr 132:394.

Jasinski DR, Pevnick JS, Griffith JD (1978): Human pharmacology and abuse potential of the analgesic buprenorphine. Arch Gen Psychiatry 35:501–516.

Kajiwara M, Aoki K, Ishii K, Numata H, Matsumiya T, Oka T (1986): Agonist and antagonist actions of buprenorphine on three types of opioid receptor in isolated preparations. Jpn J Pharmacol 40:95–101.

Kareti S, Moreton JE, Khazan N (1980): Effects of buprenorphine, a new narcotic–antagonist analgesic, on the EEG, power spectrum and behavior of the rat. Neuropharmacology 19:195–201.

Lahti RA, Mickelson MM, McCall JM, VonVoigtlander PF (1985): [^{3}H]U-69593 a highly selective ligand for the opioid kappa receptor. Eur J Pharmacol 109:281–284.

Leander JD (1988): Buprenorphine is a potent kappa opioid receptor antagonist in pigeons and mice. Eur J Pharmacol 151:457–461.

Leighton GE, Johnson MA, Meecham KG, Hill RG, Hughes J (1987): Pharmacological profile of PD 117302, a selective kappa-opioid agonist. Br J Pharmacol 92:915–922.

Lewis JW (1985): Buprenorphine. Drug Alcohol Depend 14:363–372.

Lewis JW, Rance MJ, Sanger DJ (1983): The pharmacology and abuse potential of buprenorphine: A new antagonist analgesic. In Mello NK (ed): "Advances in Substance Abuse–Behavioral and Biological Research," Vol. 3. Greenwich, CT: JAI Press, pp 103–154.

Manara L, Bianchetti A (1985): The central and peripheral influences of opioids on gastrointestinal propulsion. Annu Rev Pharmacol Toxicol 25:249–273.

Martin WR, Eades CG, Thompson JA, Huppler RE, Gilbert PE (1976): The effects of morphine- and nalorphine-like drugs in the nondependent and morphine-dependent chronic spinal dog. J Pharmacol Exp Ther 197:517–532.

McQuay HJ, Sullivan AF, Smallman K, Dickenson AH (1989): Intrathecal opioids, potency and lipophilicity. Pain 36:111–115.

Millan MJ (1989): Kappa-opioid receptor-mediated antinociception in the rat. I. Comparative actions of mu- and kappa-opioids against noxious thermal, pressure and electrical stimuli. J Pharmacol Exp Ther 251:334–341.

Murray C, Cowan A (1988): Neuroadaptation of rats to kappa agonists U-50,488 and tifluadom. NIDA Res Monogr 81:136–142.

Negus SS, Dykstra LA (1988): Kappa antagonist properties of buprenorphine in the shock titration procedure. Eur J Pharmacol 156:77–86.

Negus SS, Picker MJ, Dykstra LA (1990): Interactions between mu and kappa opioid agonists in the rat drug discrimination procedure. Psychopharmacology 102:465–473.

Negus SS, Picker MJ, Dykstra LA (1991): Interactions between the discriminative stimulus effects of mu and kappa opioid agonists in the squirrel monkey. J Pharmacol Exp Ther 256:149–158.

Parsons CG, Headley PM (1989): Spinal antinociceptive actions of mu- and kappa-opioids:

The importance of stimulus intensity in determining "selectivity" between reflexes to different modalities of noxious stimulus. Br J Pharmacol 98:523–532.

Pechnick RN, George R, Poland RE (1985): The effects of the acute administration of buprenorphine hydrochloride on the release of anterior pituitary hormones in the rat: Evidence for the involvement of multiple opiate receptors. Life Sci 37:1861–1868.

Rance MJ, Lord JAH, Robinson T (1980): Biphasic dose response to buprenorphine in the rat tail flick assay: Effect of naloxone pretreatment. In Way EL (ed): "Endogenous and Exogenous Opiate Agonists and Antagonists." New York: Pergamon Press, pp. 387–390.

Reisine T, Bell GI (1993): Molecular biology of opioid receptors. Trends Neurosci 16:506–510.

Richards ML, Sadée W (1985a): In vivo opiate receptor binding of oripavines to mu, delta and kappa sites in rat brain as determined by an ex vivo labeling method. Eur J Pharmacol 114:343–353.

Richards ML, Sadée W (1985b): Buprenorphine is an antagonist at the kappa opioid receptor. Pharm Res 2:178–181.

Sadée W, Rosenbaum JS, Herz A (1982): Buprenorphine: Differential interaction with opiate receptor subtypes in vivo. J Pharmacol Exp Ther 223:157–162.

Sadée W, Richards ML, Grevel J, Rosenbaum JS (1983): In vivo characterization of four types of opioid binding sites in rat brain. Life Sci 33(Suppl 1):187–189.

Shaw JS, Rourke JD, Burns KM (1988): Differential sensitivity of antinociceptive tests to opioid agonists and partial agonists. Br J Pharmacol 95:578–584.

Shibata M, Ohkubo T, Takahashi H, Inoki R (1989): Modified formalin test: Characteristic biphasic pain response. Pain 38:347–352.

Skingle M, Tyers MB (1979): Evaluation of antinociceptive activity using electrical stimulation of the tooth pulp in the conscious dog. J Pharmacol Methods 2:71–80.

Skingle M, Tyers MB (1980): Further studies on opiate receptors that mediate antinociception: Tooth pulp stimulation in the dog. Br J Pharmacol 70:323–327.

Smith CFC, Waldron C, Brook NA (1988): Opioid receptors in the mouse ileum. Arch Int Pharmacodyn 291:122–131.

Su T-P (1985): Further demonstration of kappa opioid binding sites in the brain: Evidence for heterogeneity. J Pharmacol Exp Ther 232:144–148.

Tyers MB (1980): A classification of opiate receptors that mediate antinociception in animals. Br J Pharmacol 69:503–512.

Unterwald EM, Knapp C, Zukin RS (1991): Neuroanatomical localization of kappa$_1$ and kappa$_2$ opioid receptors in rat and guinea pig brain. Brain Res 562:57–65.

Villiger JW (1984): Binding of buprenorphine to opiate receptors. Regulation by guanyl nucleotides and metal ions. Neuropharmacology 23:373–375.

Ward SJ, Takemori AE (1983): Relative involvement of mu, kappa and delta receptor mechanisms in opiate-mediated antinociception in mice. J Pharmacol Exp Ther 224:525–530.

Wheeler-Aceto H, Cowan A (1991a): Buprenorphine and morphine cause antinociception by different transduction mechanisms. Eur J Pharmacol 195:411–413.

Wheeler-Aceto H, Cowan A (1991b): Standardization of the rat paw formalin test for the evaluation of analgesics. Psychopharmacology 104:35–44.

Wheeler-Aceto H, Porreca F, Cowan A (1990): The rat paw formalin test: Comparison of noxious agents. Pain 40:229–238.

Woods JH, France CP, Winger GD (1992): Behavioral pharmacology of buprenorphine: Issues relevant to its potential in treating drug abuse. NIDA Res Monogr 121:12–27.

Woolfolk DR, Holtzman SG (1993): Restraint stress potentiates analgesia induced by 5'-N-ethylcarboxamidoadenosine: Comparison with morphine. Eur J Pharmacol 239:177–182.

Zimmerman DM, Leander JD, Reel JK, Hynes MD (1987): Use of beta-funaltrexamine to determine mu opioid receptor involvement in the analgesic activity of various opioid ligands. J Pharmacol Exp Ther 241:374–378.

BEHAVIORAL AND PHARMACOLOGICAL DETERMINANTS OF BUPRENORPHINE'S EFFECTS ON SCHEDULE-CONTROLLED BEHAVIOR

LINDA A. DYKSTRA
Departments of Psychology and Pharmacology and the Curriculum in Neurobiology, University of North Carolina, Chapel Hill, NC 27599–3270

S. STEVENS NEGUS
Alcohol and Drug Abuse Research Center, McLean Hospital, Belmont, MA 02178–2746

INTRODUCTION

Buprenorphine has been characterized as a partial μ opioid agonist. According to this classification, buprenorphine has affinity for μ opioid receptors, but limited intrinsic efficacy at those receptors. Data supporting this classification come from a number of behavioral assays. For example, in the morphine-dependent dog, buprenorphine suppresses signs of withdrawal but is less effective than morphine in this respect [Martin et al., 1976]. In morphine-dependent rhesus monkeys that are withdrawn from morphine, buprenorphine does not suppress signs of morphine withdrawal at all [Woods and Gmerek, 1985]. Buprenorphine also produces less than maximal effects in some analgesia tests in rats [Dum and Herz, 1981; Cowan et al., 1977] and is self-administered at a rate lower than that of morphine [Young et al., 1984; Mello and Mendelson, 1985; Lukas et al., 1986].

This chapter explores buprenorphine's effects in yet another set of behavioral preparations, collectively known as operant behavior of schedule-controlled responding. The particular behaviors examined here are responding maintained by the presentation of food and responding maintained by the termination (or titration) of shock. Buprenorphine's behavioral effects will be examined against the background of evolving concepts of opioid receptor theory so as to expand our understanding of

Buprenorphine: Combatting Drug Abuse With a Unique Opioid, pages 49–69
© 1995 Wiley-Liss, Inc.

both the behavioral and the pharmacological determinants of buprenorphine's action.

BACKGROUND: DRUG RECEPTOR INTERACTIONS

A major area of research in pharmacology is devoted to understanding the relationship between the magnitude of a drug effect and its interaction at the receptor [Ruffolo, 1982]. Two important characteristics of drug–receptor interaction that have been particularly important in determining buprenorphine's classification as a partial agonist are affinity and efficacy.

Affinity

The first of these characteristics, affinity, is the ability of a drug molecule to bind a receptor. It is generally accepted that opioids produce their effects by binding to at least one of three different types of opioid receptors, the μ, κ, and δ opioid receptor types [Martin et al., 1976; Lord et al., 1977]. Moreover, opioid agonists vary in their affinity or selectivity for these three receptor types. For example, although buprenorphine has been shown to bind μ, κ, and δ opioid receptor types [Villiger and Taylor, 1981; Sadée et al., 1982; Su, 1985], the μ opioid receptor is thought to play a prominent role in buprenorphine's behavioral effects.

Efficacy

The second characteristic of importance is intrinsic efficacy. Intrinsic efficacy is the ability of a drug molecule to produce an effect once it is bound to a receptor. A drug that fully activates a receptor and produces maximal effects in an assay that measures effects mediated by that receptor would be classified as a full agonist. For example, morphine produces maximal effects in most analgesic assays and thus has been classified as a full mu agonist in these assays. A drug with little or no intrinsic efficacy, but some affinity for a given receptor, would bind to the receptor but produce no effect, as is the case with opioid antagonists such as naloxone. Importantly, antagonists also interfere with or prevent the binding of full agonists and thus block or antagonize their activity.

A drug that produces submaximal activation of the receptor is classified as an intermediate-efficacy, or partial, agonist. Whether a drug meets this criterion depends on the efficacy requirements of the particular assay employed. In assays with low efficacy requirements, all drugs may produce maximal effects and thus appear as full agonists. In assays with higher efficacy requirements, however, partial agonists may either have submaximal effects or no effects at all. If a partial agonist does not produce any effects alone, it would be expected to antagonize the effects of a full agonist in that assay. Similarly, if a partial agonist produces submaximal effects when administered alone, it would be expected to reverse, at least partially, the effects of a full agonist in that assay. In this context, it is important to emphasize that opioids vary in their intrinsic efficacy at opioid receptors of various types as well as in their affinity for these receptors [Miller et al., 1986]. As a result, a drug such as buprenorphine, which binds more than one type of opioid receptor, may produce agonist effects through one type of receptor, at which it has high intrinsic efficacy, and antagonist effects through a second type of receptor, at which it has low intrinsic efficacy.

Buprenorphine as a Partial Agonist

Given these characteristics of opioid receptor interaction, buprenorphine's classification as a partial (or intermediate-efficacy) agonist becomes clear, since buprenorphine generally produces submaximal effects in high-efficacy assays such as the tail flick analgesia test in rats [Cowan et al., 1977; Calcagnetti et al., 1990] and suppression tests in morphine-dependent animals [Martin et al., 1976; Woods and Gmerek, 1985; France and Woods, 1990], and in most self-administration procedures [Young et al., 1984; Mello and Mendelson, 1985; Lukas et al., 1986]. In contrast, buprenorphine produces maximal effects in some low-efficacy assays such as the phenylquinone-induced stretching test [Cowan et al., 1977], the drug discrimination assay in nondependent animals [France et al., 1984; Shannon et al., 1984; Young et al., 1984; France and Woods, 1990], and tests of its subjective effects in humans [Jasinski et al., 1978].

Investigations of buprenorphine's effects on operant behavior in our own laboratory also indicate that the behavioral assay used for evaluation plays an important role in determining buprenorphine's effects. For example, the effects of buprenorphine on responding maintained by food are very similar to those of a full mu agonist such as morphine. In contrast, under the shock titration procedure, an assay that generally reveals activity only for high-efficacy opioid agonists, buprenorphine's effects are submaximal and more in keeping with its characterization as a partial agonist.

BEHAVIOR MAINTAINED BY SCHEDULES OF FOOD PRESENTATION

Schedule-controlled responding or operant behavior is behavior that is controlled by its consequences. If the consequences that follow a response increase the probability of recurrence of that response, then those consequences are called reinforcers and the resultant behavior is said to have been operantly conditioned. Food, water, intracranial stimulation, and drug administration can all function as positive reinforcers within operant conditioning paradigms. Most investigations of operant conditioning employ situations in which animals are required to perform a simple response such as pressing a lever or pecking a key in order to receive food. When every response is followed by food, the organism is said to be responding under a continuous reinforcement schedule; however, it is more common to deliver a reinforcer according to some intermittent schedule, for example, a schedule in which more than one response is required for food presentation (fixed-ratio schedules or variable-ratio schedules) or a schedule in which the first response after a specified period of time is followed by food presentation (fixed-interval schedule). The schedule with which food is delivered has been shown to be an important determinant of drug–behavior interactions [Seiden and Dykstra, 1977].

Although a variety of schedules of reinforcement have been used to examine drug action, examination of responding under a multiple fixed-ratio, fixed-interval (mult FR FI) schedule of food presentation is one of the most common. Under this procedure, two different schedules of food presentation are employed: a fixed-ratio (FR) schedule, in which a given number of responses are required for food presentation; and a fixed-interval (FI) schedule, in which the first response after a given period of time elapses is followed by food presentation. Each of these so-called schedule components are signaled by a distinct stimulus, such as a red or green light, and in a typical experimental session they are presented in alternation. One of the distinguishing features of this schedule is that it generates two different rates and patterns of responding, the FR component generally producing high rates of responding and the FI component producing lower rates of responding.

Prototypic Mu Opioid Agonists

Morphine Alone. Morphine generally decreases response rates maintained by both FR and FI schedules of food presentation in pigeons, squirrel monkeys, and rhesus monkeys; however, increases have sometimes been reported at very low doses on behavior maintained by fixed-interval schedules. Similar effects have been reported for other µ opioid agonists such as methadone and codeine, as well as for opioids of the kappa type [McMillan and Morse, 1967; McMillan et al., 1970; Downs and Woods, 1976; Goldberg et al., 1976, 1981; Harris, 1980; Milar and Dykstra, 1983; Craft et al., 1989; Doty et al., 1989; Picker et al., 1990]. Representative effects of morphine on responding maintained by a mult FR FI schedule of food presentation in squirrel monkeys are illustrated in Fig. 1. It can be seen that morphine produces dose-dependent decreases in rates of responding under both components of the schedule.

Antagonism of Morphine. It is important to note that opioids are not the only class of drugs that decrease rates of responding maintained by food presentation. Indeed, almost all behaviorally active drugs decrease rates of responding at some dose, although some drugs such as cocaine and amphetamine also produce prominent increases in low rates of responding [Seiden and Dykstra, 1977]. What is unique about morphine-like drugs and allows us to classify these effects as opioid-mediated is the fact that the rate-decreasing effects of morphine can be completely antagonized by doses of opioid antagonists that have no behavioral effects of their own. For example, Goldberg et al. [1981] examined morphine's effects on responding, in both squirrel monkeys and pigeons, under a mult FR FI schedule of food presentation. The morphine dose-effect curve was determined alone and in the presence of increasing doses of the opioid antagonists naloxone and naltrexone. Both antagonists produced dose-dependent, parallel shifts to the right in the mor-

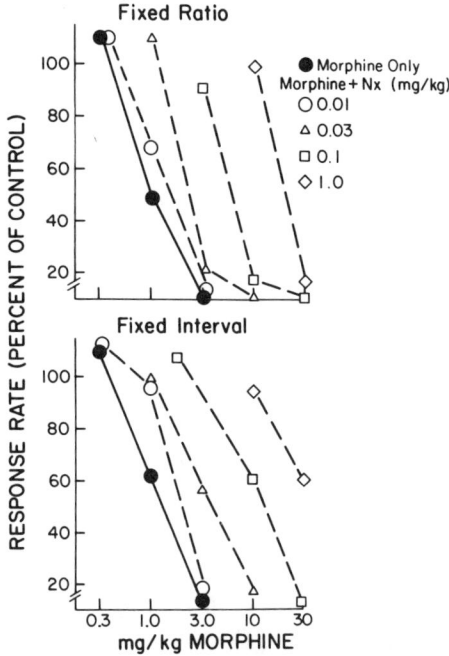

Fig. 1. Effects of morphine alone and in combination with several doses of naloxone in a representative squirrel monkey responding under a multiple fixed-ratio/fixed-interval schedule of food presentation. Abscissa: Dose of morphine (mg/kg). Ordinates: Mean response rates expressed as a percentage of mean response rates during control sessions. For the dose-effect curve for morphine alone, each point is the mean of three determinations made before, during, and after testing of morphine in combination with naloxone. For the dose-effect curves for morphine plus antagonist, each point is a single determination. Adapted from Goldberg et al. [1981].

phine dose-effect curve, indicating that these antagonists competitively antagonized morphine's effects (Fig. 1). Since naloxone and naltrexone are selective for μ opioid receptors at the doses that were effective in antagonizing morphine's effects, these data also indicate that morphine's effects on responding maintained by the presentation of food are mediated by μ opioid receptor types.

Tolerance to Morphine. Repeated administration of opioid agonists typically results in an attenuation in their effects; that is, tolerance develops. For example, tolerance develops to morphine's rate-decreasing effects in responding by squirrel monkeys under a FR schedule of food presentation, as evidenced by a 1–1.5 log unit shift to the right in the morphine dose-effect curve following chronic morphine administration [Doty et al., 1989]. Moreover, morphine-tolerant monkeys are cross-tolerant to the effects of other μ opioid agonists. Tolerance also develops to the effects of other μ opioid agonists such as methadone [Heifetz and McMillan, 1971].

Summary. The majority of the data indicate that μ agonists such as morphine decrease rates of responding maintained by schedules of food presentation. Moreover, these effects are readily reversed by μ-selective opioid antagonists. Finally, tolerance develops not only to morphine's effects on responding maintained by food, but morphine-tolerant animals are generally cross-tolerant to the effects of other μ opioids. Taken together, these results indicate that the effects of morphine-like drugs on responding maintained by food are mediated by μ opioid receptors.

Buprenorphine as a Mu Agonist

Buprenorphine Alone. Given, then, this background information about the effects of morphine as a prototypical mu agonist, it is interesting to compare buprenorphine's effects on schedule-controlled responding to those of morphine. In general, acute administration of buprenorphine decreases rates of responding maintained by food presentation in squirrel monkeys [Dykstra, 1983b; DeRossett and Holtzman, 1984], in rhesus monkeys [Mello et al., 1985], in rats [Moerschbaecher, et al., 1984; Bronson and Moerschbaecher, 1987; Berthold and Moerschbaecher, 1988]; and, in some cases, in pigeons [Cleary et al., 1988; Macenski et al., 1990].

In one particular study, the effects of buprenorphine were examined in squirrel monkeys responding under a mult FR FI schedule of food presentation [Dykstra, 1983b]. In the fixed-interval (FI) component, signaled by a red light, the first response that occurred after 5 min had elapsed produced a food pellet. In the fixed-ratio (FR) component, signaled by a white light, the 30th response produced a food pellet. Dose-effect curves were determined first for buprenorphine alone. Under these conditions, buprenorphine decreased rates of responding under both the FR and FI components of the multiple schedule as shown in Figure 2. In general, decreases in rates of responding were of the same magnitude under both components of the schedule, with complete cessation of responding occurring at a 0.03-mg/kg dose of buprenorphine. Larger doses of buprenorphine (0.1–1.0 mg/kg) not only decreased rates of responding immediately subsequent to administration, but continued to decrease rates of responding for as long as 2 days after drug administration. Under nearly identical conditions, morphine [Goldberg et al., 1976, 1981] also decreased rates of responding in squirrel monkeys (Fig. 1). A comparison of the dose at which morphine decreased rates of responding and its duration of action indicates that buprenorphine is about 100 times more potent than morphine in decreasing rates of responding and has a much longer duration of action. Interestingly, buprenorphine's rate-decreasing effects on responding maintained by food in squirrel monkeys are shared by a number of other partial opioid agonists, including nalorphine, nalbuphine, butorphanol, and pentazocine [Oliveto et al., 1991].

In another study in which buprenorphine's effects on schedule-controlled responding were examined in squirrel monkeys [DeRossett and Holtzman, 1984], a similar dose relationship was observed with decreases in rates of responding occurring over a dose range of 0.01–0.1 mg/kg of buprenorphine and lasting for several days. The same study reported that buprenorphine was equally potent in decreasing rates of responding that had been punished by the presentation of electric shock,

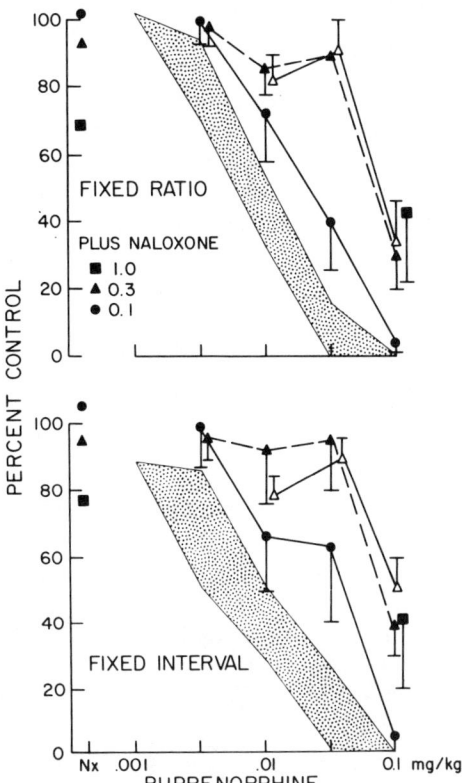

Fig. 2. Effects of buprenorphine alone and in combination with several doses of naloxone in squirrel monkeys responding under a multiple fixed-ratio/fixed-interval schedule of food presentation. Abscissa: Dose of buprenorphine (mg/kg). Ordinates: Mean response rates expressed as a percentage of mean response rates during the immediately preceding control sessions. The shaded area is the range of values for two determinations of the buprenorphine dose-effect curve, one made before and one after buprenorphine was examined in combination with naloxone. Points connected by lines represent dose-effect curves for buprenorphine in the presence of different doses of naloxone. Each point is the mean of one observation in four monkeys and brackets show SEs. The filled triangle represents the effects obtained when 0.3 mg/kg of naloxone was given at the same time as buprenorphine; the open triangle represents effects obtained when 0.3 mg/kg of naloxone was given 15 min before buprenorphine administration. Points above Nx represent the effects of naloxone alone. From Dykstra [1983b], with permission of the publisher.

another effect that is also characteristic of μ opioids such as morphine [Seiden and Dykstra, 1977].

Similar effects have been reported in rhesus monkeys. That is, buprenorphine over a dose range of 0.1–0.3 mg/kg decreased both the number of food pellets earned and the rate of responding under a second-order schedule that required an average of 64 responses for food delivery [Mello et al., 1985]. Although these effects complement those reported in the squirrel monkey, the dose that suppressed food-maintained responding in the rhesus monkeys was higher than in the squirrel monkey and the duration of buprenorphine's effects in the rhesus monkey was shorter. In squirrel monkeys, a dose as low as 0.003 mg/kg decreased response rates, whereas responding in the rhesus monkey was not decreased until a dose of 0.1 mg/kg was administered. A similar dose differential has been reported for baboons [Lukas et al., 1986]; suggesting that the squirrel monkey is particularly sensitive to the effects of buprenorphine. Indeed, the squirrel monkey appears to be exceptionally sensitive to the response-rate-decreasing effects of all the oripavines, including etorphine and diprenorphine as well as buprenorphine [DeRossett and Holtzman, 1984].

One exception to these overall decreases in rates of responding are data from a study in pigeons in which buprenorphine (0.08–5.0 mg/kg) increased rates of responding under a FI schedule without altering rates of responding under a FR schedule of reinforcement, even at doses as high as 40 mg/kg [Leander, 1983]. Moreover, increases under the FI schedule were still apparent 1–2 days following buprenorphine administration. Although morphine also increases FI rates of responding in the pigeon, morphine increases response rate only over a limited dose range, and morphine-induced increases in rates of responding are small compared with those observed with buprenorphine in the pigeon [McMillan and Morse, 1967]. In another study in pigeons, buprenorphine produced only modest decreases in rate of responding under a variable-interval schedule of food presentation [Macenski et al., 1990].

Antagonism of Buprenorphine. In general, the effects of buprenorphine on behavior maintained by food presentation can be antagonized by opioid antagonists such as naloxone. At 0.3 mg/kg, naloxone shifts the buprenorphine dose-effect curve approximately 1 log unit to the right in squirrel monkeys responding under a mult FR FI schedule of food presentation (Fig. 2) [Dykstra, 1983b]. Similarly, doses of naloxone of 0.1–1.0 mg/kg produce rightward shifts in the dose-response curve for buprenorphine under a variable-interval schedule of food presentation [DeRossett and Holtzman, 1984].

Tolerance to Buprenorphine. Tolerance develops to buprenorphine's effects on food-maintained responding. Mello and colleagues [Mello et al., 1985] examined buprenorphine's effects in rhesus monkeys in which responding was maintained both by a schedule of food presentation and by a schedule of buprenorphine self-administration. With this procedure, the dose of buprenorphine available for self-administration was gradually increased, thereby increasing the daily chronic dose of buprenorphine taken by the monkeys. Recall from the discussion above that acute doses of 0.1 and 0.3 mg/kg of buprenorphine decrease food intake as well as rates of responding maintained by food presentation in rhesus monkeys. As the daily dose of buprenorphine increased from 0.3 mg/kg/day to as high as 2.8 mg/kg/day of buprenorphine, food intake was not decreased. In another study in rhesus monkeys [Lukas et al., 1988], buprenorphine also decreased rates of food-maintained responding initially, but recovery occurred within 4 days, whereas decreases in responding induced by heroin self-administration persisted for at least 25 days. Further confirmation for the development of tolerance to buprenorphine's effects on schedule-controlled behavior comes from studies in rats [Berthold and Moerschbaecher, 1988] and pigeons [Cleary et al., 1988]. Although tolerance develops to buprenorphine's rate-decreasing effects in squirrel monkeys [Dykstra, 1983b], tolerance is not as great as that observed in other species, perhaps owing to the marked sensitivity squirrel monkeys display to the oripavines.

Another way in which tolerance to buprenorphine can be examined is through cross-tolerance studies. It has been shown in other situations that tolerance to the effects of one opioid agonist often confers tolerance to other opioids with similar pharmacological properties. For example, animals made tolerant to morphine are cross-tolerant to other μ opioid agonists but not to opioid agonists whose actions are mediated by the κ type of opioid receptor [Dykstra, 1983a; Gmerek et al., 1987; Craft et al., 1989; Doty et al., 1989; Craft and Dykstra, 1990; Picker et al., 1991]. Thus, if morphine and buprenorphine both produce their effects through mu opioid receptors, repeated administration of morphine would be expected to produce cross-tolerance to buprenorphine.

In a study employing cross-tolerance techniques, Negus et al. [1989] determined dose-effect curves for morphine and buprenorphine in rats responding under a FR schedule of food presentation. Rats were then exposed to a chronic regimen of morphine administration in which the daily dose of morphine was gradually increased up to a dose of 40 mg/kg/day. After rates of responding stabilized under this regimen, dose-effect curves were redetermined for morphine and buprenorphine. During the chronic regimen, the morphine dose-effect curve shifted 0.5 log unit to the right. That is, prior to the chronic morphine regimen, 5.6 mg/kg of morphine decreased rates of responding below 50% of control. When the morphine dose-effect curve was redetermined during the chronic regimen, a dose of 17.5 mg/kg of morphine was required to produce a comparable decrease in rates of responding. During the chronic-morphine regimen, the buprenorphine dose-effect curve also shifted to the right. That is, prior to chronic morphine a dose of 0.1 mg/kg buprenorphine decreased rates of responding below 50% of control values, whereas no dose of buprenorphine up to 1.7 mg/kg produced rate-decreasing effects during the chronic regimen. (See top right panel of Fig. 3 below.)

Summary. Taken together, these studies reveal that buprenorphine, like morphine, generally decreases rates of schedule-controlled responding. Under these conditions, buprenorphine is much more potent than morphine and has a longer duration of action. These findings accord well with other studies that indicate that buprenorphine is also much more potent than morphine as an analgesic and that its duration of action exceeds that of morphine on a number of other behavioral measures [Cowan et al., 1977; Jasinski et al., 1978; Mello et al., 1981; Dykstra, 1985]. In addition, buprenorphine's effects are readily inhibited by opioid antagonists such as naloxone and show tolerance upon chronic administration. Moreover, the presence of cross-tolerance between morphine and buprenorphine provides evidence that buprenorphine's rate-decreasing effects in this procedure are mediated by μ opioid receptors.

Buprenorphine as an Antagonist

Mu Antagonist Properties. Another question important in the characterization of buprenorphine as a μ partial agonist pertains to its potential antagonist properties. In this context it is important to note that buprenorphine generally produces maximal effects in assays employing food-maintained responding, and that these effects are similar to those obtained with a full opioid agonist such as morphine. Given that buprenorphine has full efficacy in assays in which responding is maintained by the presentation of food, it would not be expected to function as a mu antagonist under these conditions, although it might very well display antagonist properties in other assays in which it had submaximal effects. In consonance with this prediction, buprenorphine at doses ranging from 0.001 to 0.01 mg/kg was not an effective morphine antagonist in squirrel monkeys responding under a schedule of food presentation. The lack of antagonist activity in this situation was probably due to the fact that higher doses of buprenorphine could not be examined, since buprenorphine displayed full agonist effects at doses as low as 0.03 mg/kg [Dykstra, 1983b].

Although results from the squirrel monkey suggest that buprenorphine's own rate-decreasing effects limit the expression of its antagonist effects, it is important to note that buprenorphine did display μ antagonist properties in pigeons responding under a schedule of food presentation. Since buprenorphine alone did not alter responding under this procedure, it is not surprising that it was able to antagonize morphine's effects in the pigeon [Leander, 1983]. This, then, is in keeping with results from other high-efficacy assays in which buprenorphine has mu antagonist properties [Martin et al., 1976; Cowan et al., 1977; Dum and Herz, 1981, Woods and Gmerek, 1985].

Kappa Antagonist Properties. Although buprenorphine generally does not antagonize the effects of full μ agonists on responding maintained by food, buprenorphine has been shown to have κ antagonist properties under a wide range of conditions. For example, buprenorphine precipitates withdrawal in monkeys rendered physically dependent on the κ agonist U-50,488 [Gmerek et al., 1987] and antagonizes the analgesic and the discriminative stimulus properties of κ agonists in both squirrel monkeys and rodents [Leander, 1988; Negus and Dykstra, 1988; Negus et al., 1990, 1991]. It also antagonizes κ-induced urination in rats, but unlike other opioids with affinity for the κ opioid receptor, buprenorphine itself does not increase urine output [Richards and Sadée, 1985; Leander, 1987]. Thus, buprenorphine is a κ antagonist, rather than a κ agonist in the urine output assay.

Given these demonstrations of kappa antagonist properties for buprenorphine, it is interesting to determine whether buprenorphine might also display κ antagonist properties in situations involving schedule-controlled responding. Because buprenorphine alone produces marked decreases in rates of food-maintained responding, it is difficult to evaluate its potential antagonist effects under these conditions. To

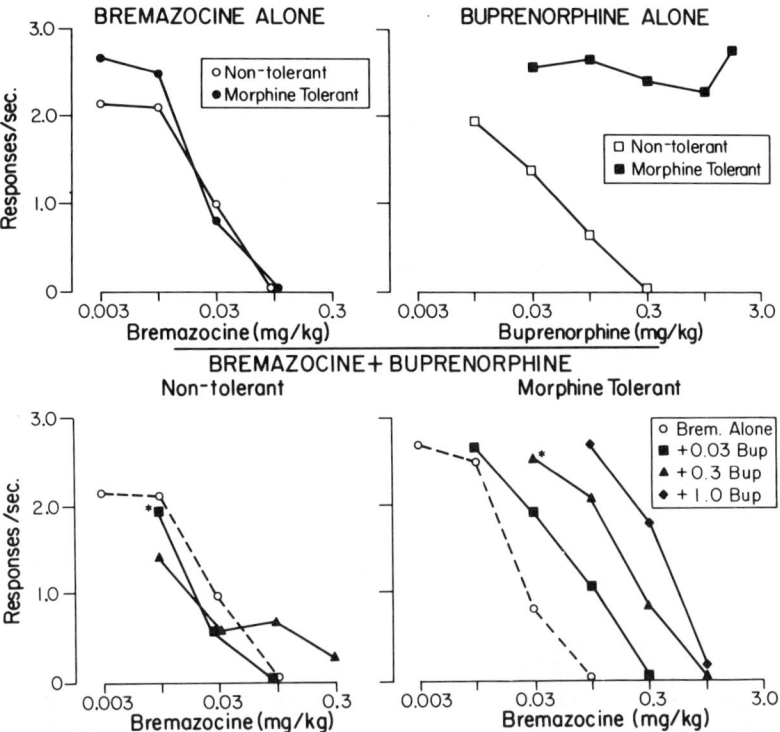

Fig. 3. Effects of bremazocine alone, buprenorphine alone, and bremazocine in combination with buprenorphine both before and after the induction of morphine tolerance in rats responding under a fixed ratio 30 schedule of food presentation. Abscissa: Dose of bremazocine or buprenorphine (mg/kg). Ordinates: Response rate in responses per second. Dose-response curves marked by an asterisk represent the mean of a single determination in each of three rats. All other dose-response curves represent the mean of a single determination in each of four rats. From Negus et al. [1989], with permission of the publisher.

avoid this difficulty, Negus et al. [1989] conducted a study in which rats were made tolerant to buprenorphine's rate-decreasing effects. Briefly, rats were trained to respond under a fixed ratio 30 schedule of food presentation. Initially, dose-effect curves were determined for buprenorphine, morphine, and the κ agonist bremazocine, and then for bremazocine in combination with buprenorphine. Under these conditions, buprenorphine did not antagonize bremazocine's effects. Rats were then made tolerant to morphine by daily administration of 40 mg/kg of morphine, and dose-effect curves for morphine, bremazocine, and buprenorphine were redetermined in the morphine-tolerant animals. As shown in Figure 3, the dose-effect curve for bremazocine was not altered as the result of chronic morphine administration, whereas the buprenorphine curve shifted 1.5 log units to the right, suggesting that morphine-tolerant rats were cross-tolerant to buprenorphine but were not cross-tolerant to bremazocine. Larger doses of buprenorphine could be

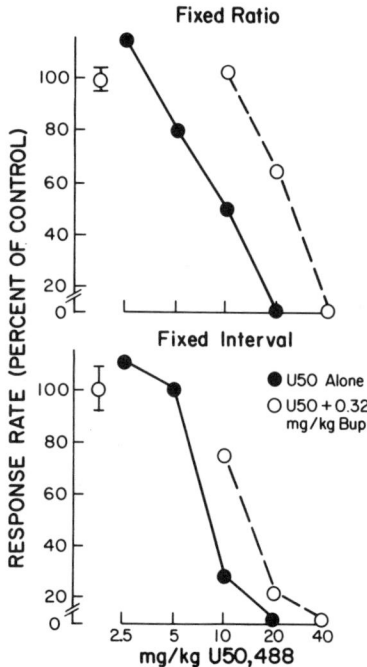

Fig. 4. Effects of U-50,488 in pigeons responding under a multiple fixed-ratio/fixed-interval schedule of food presentation. The effects of U-50,488 are shown alone and in combination with 0.32 mg/kg of buprenorphine. Abscissa: Dose of U-50,488 (mg/kg). Ordinates: Response rate expressed as percentage of control. The points at 0 represent the effects of 0.32 mg/kg of buprenorphine alone. Adapted from Leander [1988].

administered in tolerant rats and these doses of buprenorphine (i.e., 0.03, 0.3, and 1.0 mg/kg) shifted the bremazocine dose-effect curve to the right by approximately 0.5, 1.0, and 1.25 log units, respectively.

Similarly, Leander [1988] found that buprenorphine produced approximately a twofold shift in the dose-effect curve for the kappa agonist U-50,488 in pigeons responding under a mult FR FI schedule of food presentation, as shown in Figure 4. Recall that buprenorphine alone generally does not decrease rates of responding under this procedure; thus, its antagonist effects are not obscured by its own rate-decreasing effects.

Summary. Although buprenorphine generally does not antagonize the effects of mu agonists on responding maintained by food, it does antagonize the effects of kappa agonists in this procedure. These data provide evidence that buprenorphine is an antagonist at the κ opioid receptor and complement data collected in other behavioral assays that suggest that buprenorphine has κ antagonist properties.

BEHAVIOR MAINTAINED BY SCHEDULES OF SHOCK TERMINATION OR SHOCK TITRATION

Shock termination schedules have also been used to examine the behavioral effects of buprenorphine and other opioid agonists. In situations in which responding is maintained by the avoidance or termination of an electric shock, buprenorphine's effects are similar to those of morphine. For example, Shannon [1983] found that both buprenorphine and morphine increased rates of responding maintained under a schedule of shock avoidance in rats; however, buprenorphine was approximately 100 times more potent than morphine in this respect and displayed its effects over a much wider dose range.

The shock titration procedure has also been used to examine the effects of opioid agonists, in particular their analgesic effects. Under this procedure, squirrel monkeys are trained to respond on a lever in order to titrate the intensity of shock delivered to their tails. Briefly, the shock is scheduled to increase from 0 to 2.0 mA or from 0 to 3.7 mA in 30 steps. If the monkey does not respond during the shock period, the shock remains on for the duration of the shock period and increases by one increment during the next 15-sec shock period. If, however, five responses are made during the shock period, the shock terminates for a 15-sec time-out period without shock. After the 15-sec time-out period, the shock resumes at the next lower intensity. Following several months of training, the intensity at which each monkey maintains the shock stabilizes and drug effects can be examined. The data obtained include rates of responding in the presence of shock and the number of times the five-response requirement is completed as a function of shock intensity. Median shock levels (the shock intensity below which the shock was maintained 50% of the time) are derived from these data.

It has been shown that several μ agonists, including morphine, l-methadone and fentanyl as well as κ agonists such as U-50,488, tifluadom, bremazocine, and ethylketocyclazocine increase the level at which shock is maintained under this procedure [Dykstra, 1983a; Genovese and Dykstra, 1986; Dykstra and Massie, 1988]. Moreover, it has been shown that the actions of κ opioids under the titration procedure are primarily mediated by the κ opioid receptor type, whereas morphine and l-methadone's actions are primarily mediated by the μ type of opioid receptor [Dykstra and Massie, 1988; Craft and Dykstra, 1990].

In addition to its sensitivity to both mu and kappa agonist activity, the shock titration procedure is generally more sensitive to the analgesic effects of compounds with a high degree of efficacy than to compounds with intermediate efficacy [Bloss and Hammond, 1985; Dykstra, 1990], especially when the higher range of shock intensities (0–3.7 mA) is employed. Thus, partial agonists such as buprenorphine would be expected to produce submaximal effects under this procedure.

Agonist Action

Morphine and Buprenorphine Alone. Both morphine and buprenorphine increase the the level at which shock is maintained under the shock titration proce-

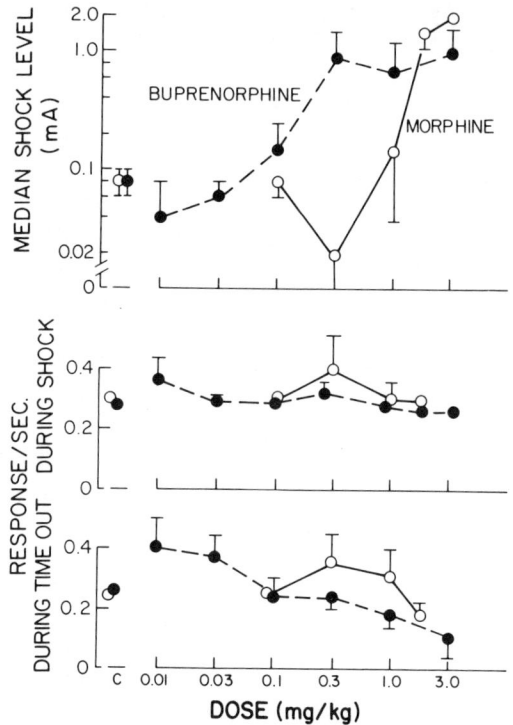

Fig. 5. Effects of morphine and buprenorphine on median shock level and rate of responding in the presence of shock and in the absence of shock in squirrel monkeys responding under a schedule of shock titration. Abscissa: Dose of buprenorphine or morphine (mg/kg). Ordinates: Top graph, median shock level in milliamperes; middle and bottom graphs, rates of resonding in responses per second. The brackets at C represent the range of control values and the circles show the effects of water injections. Each point is the mean of a single injection in at least three monkeys and the brackets represent SEs. Adapted from Dykstra [1985].

dure; however, increases in median shock level following buprenorphine administration are not as great as those seen with morphine [Dykstra, 1985]. Figure 5 shows the effects of morphine and buprenorphine in monkeys responding under a shock titration procedure in which the shock ranged from 0 to 2.0 mA. Under these conditions, 3.0 mg/kg of morphine increased shock intensity to its peak intensity (i.e., 2.0 mA), whereas buprenorphine displayed a shallower dose effect curve with shock intensity increasing to 1.0 mA across a wide dose range (i.e., 0.3–3.0 mg/kg).

Similar results were obtained in another shock titration study in which shock ranged either between 0 and 2.0 mA or between 0 and 3.7 mA [Negus and Dykstra, 1988]. When the shock increased from 0 to 2.0 mA, buprenorphine produced maximal increases in medial shock level in two monkeys, whereas increases in medial shock level were submaximal in two other monkeys. When shock increased from 0 to 3.7 mA, buprenorphine produced submaximal effects in all monkeys. Thus, although buprenorphine increased median shock level in the monkey titration procedure, its efficacy in this assay was less than that of morphine. Taken together, these data indicate that buprenorphine displays partial agonist action under these conditions. It is interesting that other opioids that have been classified as partial agonists also produce submaximal effects under the shock titration procedure. These include nalbuphine, nalorphine, butorphanol, and levallorphan [Dykstra, 1990].

Antagonism of Morphine and Buprenorphine. The increases in shock intensity produced by both morphine and buprenorphine under the shock titration procedure were attenuated by prior administration of diprenorphine, naloxone, and β-FNA (Fig. 6). Since β-FNA is a μ-selective antagonist [Portoghese et al., 1980; Ward et al., 1982], these data indicate that the effects of both morphine and buprenorphine under this procedure are mediated by μ opioid receptors; however, the fact that higher doses of β-FNA were required to antagonize the effects of buprenorphine than the effects of morphine, leaves open the possibility that buprenorphine's analgesic effects are mediated differently than are those of morphine.

Tolerance to Morphine and Buprenorphine. Tolerance has been shown to develop to the effects of several μ opioid agonists under the shock titration procedure, including morphine and *l*-methadone [Dykstra, 1983a, 1985; Craft and Dykstra, 1990]. For example, Craft and Dykstra [1990] showed that daily administration of morphine for approximately 6 weeks shifted the morphine dose-effect curve to the right approximately ³/₄ log unit. Similarly, tolerance has also been shown to develop to the effects of buprenorphine in squirrel monkeys responding under the shock titration procedure. Interestingly, when the morphine dose-effect

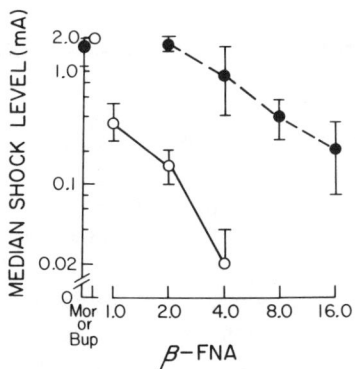

Fig. 6. Effects of morphine (○) or buprenorphine (●) in combination with β-FNA. Abscissa: Dose (mg/kg). Ordinate: Median shock level in milliamperes. The points at Mor and Bup represent the effects of a dose of morphine and a dose of buprenorphine that increased shock to its maximal level. Each point is the mean of a single injection in three monkeys and the brackets represent SEs. From Dykstra [1985], with permission of the publisher.

curve was redetermined in buprenorphine-tolerant monkeys, the morphine curve had shifted to the right at least 10-fold. Thus, not only does tolerance develop to buprenorphine under these conditions, but cross-tolerance between buprenorphine and the μ agonist morphine also occurs.

Summary. Taken together, these data indicate that both morphine and buprenorphine increase the level at which monkeys maintain a shock in the shock titration procedure. Whereas morphine produces full effects under this procedure, buprenorphine's effects are submaximal and its dose-effect curve is shallower than that of morphine. Moreover, the effects of both morphine and buprenorphine are antagonized by the μ-selective antagonist β-FNA and show cross-tolerance with each other, which is suggestive of mu agonist action. Thus, the effects of buprenorphine can be characterized as that of a μ partial agonist under the shock titration procedure.

Antagonist Action

Buprenorphine as a Mu Antagonist. Given that buprenorphine revealed submaximal activity in the shock titration assay, it is interesting to examine its antagonist effects under these conditions. Recall that a drug that produces submaximal effects when administered alone should partially reverse the effects of a full agonist under these conditions. Also recall that buprenorphine produced maximal effects in some monkeys under the shock titration procedure but not in others. Interestingly, buprenorphine was an effective μ antagonist in those monkeys in which its agonist actions were submaximal [Negus and Dykstra, 1988]. Conversely, buprenorphine did not reveal μ antagonist properties in those monkeys in which it produced maximal increases in shock level. It is interesting that other partial agonists such as nalbuphine, levallorphan, nalorphine, and butorphanol, which are less efficacious than buprenorphine as agonists in the shock titration procedure, are very effective μ antagonists in this assay [Dykstra, 1990].

Buprenorphine as a Kappa Antagonist. In addition to the μ antagonist activity displayed by buprenorphine under the shock titration procedure, buprenorphine also has kappa antagonist properties in this procedure. Negus and Dykstra [1988] examined the effects of buprenorphine in combination with doses of the kappa agonist U-50,488, which increased median shock levels under the shock titration procedure. When the U-50,488 dose-effect curve was redetermined in the presence of buprenorphine, doses of buprenorphine as low as 0.01 mg/kg shifted the U-50,488 dose-effect curve to the right, as shown in Figure 7. Moreover, buprenorphine also antagonized the effects of a dose of U-50,488 that increased the shock to its maximal intensity. Thus, although buprenorphine can antagonize the effects of a μ agonist under the shock titration procedure, it is generally more effective as a κ antagonist in this procedure.

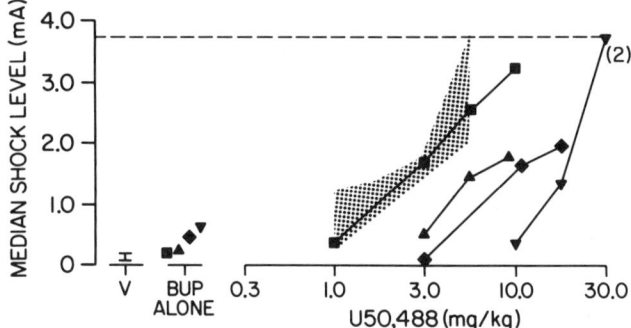

Fig. 7. Effects of U-50,488 alone and in combination with buprenorphine in monkeys responding under a shock titration procedure. Abscissa: Dose of U-50,488 (mg/kg). Ordinate: median shock level in mA. The brackets at V represent the range of median shock levels averaged across three monkeys during vehicle control sessions. The points at Bup alone represent the effects of 0.003 (■), 0.01 (▲), 0.03 (♦) and 0.1 (▼) mg/kg buprenorphine alone. The remaining points represent the effects of U-50,488 in combination with four doses of buprenorphine. The shaded area is the range of values for two determinations of the U-50,488 dose-effect curve, with one dose-effect curve obtained prior to the antagonist interactions and one obtained after the antagonist interactions. Each point is the mean of a single determination in three monkeys. From Negus and Dykstra [1988], with permission of the publisher.

SUMMARY

The data presented in this chapter, and summarized in Table 1, provide evidence that buprenorphine's effects on schedule-controlled responding depend on the behavioral assay used for evaluation. In assays in which responding is maintained by schedules of food presentation, buprenorphine produces effects much like those of a full μ opioid agonist such as morphine. That is, buprenorphine, like morphine, decreases rates of schedule-controlled responding, and these effects are readily reversed by opioid antagonists and show tolerance upon chronic administration. Whereas neither morphine nor buprenorphine displays μ antagonist effects on responding maintained by food, buprenorphine does display κ antagonist properties under these conditions, provided that tolerance has developed to its μ agonist effects.

In the shock titration procedure, both morphine and buprenorphine increase the level at which shock is maintained; however, buprenorphine is less efficacious than morphine in this respect. The effects of both morphine and buprenorphine are antagonized by opioid antagonists and show tolerance upon repeated administration as well as cross-tolerance with each other. Interestingly, buprenorphine, but not morphine, displays prominent κ antagonist activity in this situation. These differential effects across assays are in keeping with buprenorphine's characterization as a μ partial agonist and also with growing evidence that buprenorphine possesses κ

TABLE I. Comparison of the Effects of Morphine, Buprenorphine, and Butorphanol On Schedule-Controlled Responding

Effect	Morphine	Buprenorphine	Butorphanol
Responding maintained by food			
Agonist effects (decreases rate of responding)	Maximal	Maximal	Maximal[a]
Mu antagonist effects	No	No	No
Kappa antagonist effects	No	Yes	Yes[b]
Responding maintained by shock titration			
Agonist effects (increases median shock level)	Maximal	Submaximal	Submaximal[c]
Mu antagonist effects	No	Limited	Yes[c]
Kappa antagonist effects	No	Yes	Yes[c]

[a]Oliveto et al. [1991].
[b]Picker et al. [1990].
[c]Dykstra [1990].

antagonist properties. These findings also point to the importance of behavioral variables in buprenorphine's effects and are in keeping with other situations in which buprenorphine's behavioral profile depends on the assay used for evaluation [e.g., Martin et al., 1976; Cowan et al., 1977; France and Woods, 1990].

REFERENCES

Berthold CW III, Moerschbaecher JM (1988): Tolerance to the effects of buprenorphine on schedule-controlled behavior and analgesia in rats. Pharmacol Biochem Behav 29:393–396.

Bloss JL, Hammond DL (1985): Shock titration in the rhesus monkey: Effects of opiate and nonopiate analgesics. J Pharmacol Exp Ther 235:423–429.

Bronson ME, Moerschbaecher JM (1987): Effects of mu, kappa and sigma opioids on fixed consecutive number responding in rats. Pharmacol Biochem Behav 27:733–743.

Calcagnetti DJ, Fleetwood SW, Holtzman SG (1990): Pharmacological profile of the potentiation of opioid analgesia by restraint stress. Pharmacol Biochem Behav 37:193–199.

Cleary J, Ho B, Nader N, Thompson T (1988): Effects of buprenorphine, methadone and naloxone on acquisition of behavioral chains. J Pharmacol Exp Ther 247:569–575.

Cowan A, Lewis JW, Macfarlane IR (1977): Agonist and antagonist properties of buprenorphine, a new antinociceptive agent. Br J Pharmacol 60:537–545.

Craft RM, Dykstra LA (1990): Differential cross-tolerance to opioids in squirrel monkeys responding under a shock titration schedule. J Pharmacol Exp Ther 252:945–952.

Craft RM, Picker MJ, Dykstra LA (1989): Differential cross-tolerance to opioid agonists in morphine-tolerant pigeons responding under a schedule of food presentation. J Pharmacol Exp Ther 249:386–393.

DeRossett SE, Holtzman SG (1984): Effects of naloxone, diprenorphine, buprenorphine and etorphine on unpunished and punished food-reinforced responding in the squirrel monkey. J Pharmacol Exp Ther 228:669–675.

Doty P, Picker MJ, Dykstra LA (1989): Differential cross-tolerance to opioid agonists in morphine-tolerant squirrel monkeys responding under a schedule of food presentation. Eur J Pharmacol 174:171–180.

Downs DA, Woods JH (1976): Morphine, pentazocine and naloxone effects on responding under a multiple schedule of reinforcement in rhesus monkeys and pigeons. J Pharmacol Exp Ther 196:298–306.

Dum JE, Herz A (1981): *In vivo* receptor binding of the opiate partial agonist, buprenorphine, correlated with its agonistic and antagonistic actions. Br J Pharmacol 74:627–633.

Dykstra LA (1983a): Effects of ketocyclazocine and ethylketocyclazocine on electric shock titration. Eur J Pharmacol 94:19–26.

Dykstra LA (1983b): Behavioral effects of buprenorphine and diprenorphine under a multiple schedule of food presentation in squirrel monkeys. J Pharmacol Exp Ther 226:317–323.

Dykstra LA (1985): Effects of buprenorphine on shock titration in squirrel monkeys. J Pharmacol Exp Ther 235:20–25.

Dykstra LA (1990): Butorphanol, levallorphan, nalbuphine and nalorphine as antagonists in the squirrel monkey. J Pharmacol Exp Ther 254:245–252.

Dykstra LA, Massie CA (1988): Antagonism of the analgesic effects of mu and kappa opioid agonists in the squirrel monkey. J Pharmacol Exp Ther 246:813–821.

France CP, Woods JH (1990): Discriminative stimulus effects of opioid agonists in morphine-dependent pigeons. J Pharmacol Exp Ther 254:626–632.

France CP, Jacobson AE, Woods JH (1984): Discriminative stimulus effects of reversible and irreversible opiate agonists: Morphine, oxymorphazone and buprenorphine. J Pharmacol Exp Ther 230:652–657.

Genovese RF, Dykstra LA (1986): Tifluadom's effects under electric shock titration and tail-immersion procedures in squirrel monkeys. Life Sci 39:1713–1719.

Gmerek DE, Dykstra LA, Woods JH (1987): Kappa opioids in rhesus monkeys. III. Dependence associated with chronic administration. J Pharmacol Exp Ther 242:428–436.

Goldberg SR, Morse WH, Goldberg DM (1976): Some behavioral effects of morphine, naloxone and nalorphine in the squirrel monkey and the pigeon. J Pharmacol Exp Ther 196:625–636.

Goldberg SR, Morse WH, Goldberg DM (1981): Acute and chronic effects of naltrexone and naloxone on schedule-controlled behavior of squirrel monkeys and pigeons. J Pharmacol Exp Ther 216:500–509.

Harris RA (1980): Interactions between narcotic agonists, partial agonists and antagonists evaluated by schedule-controlled behavior. J Pharmacol Exp Ther 213:497–503.

Heifetz SA, McMillan DE (1971): Development of behavioral tolerance to morphine and methadone using the schedule-controlled behavior of the pigeon. Psychopharmacolgia 19:40–52.

Jasinski DR, Pevnick JS, Griffith JD (1978): Human pharmacology and abuse potential of the analgesic buprenorphine. Arch Gen Psychiatry 35:501–516.

Leander JD (1983): Opioid agonist and antagonist behavioural effects of buprenorphine. Br J Pharmacol 78:607–615.

Leander JD (1987): Buprenorphine has potent kappa opioid receptor antagonist activity. Neuropharmacology 26:1445–1447.

Leander JD (1988): Buprenorphine is a potent κ-opioid receptor antagonist in pigeons and mice. Eur J Pharmacol 151:457–461.

Lord JAH, Waterfield AA, Hughes J, Kosterlitz HW (1977): Endogenous opioid peptides: Multiple agonists and receptors. Nature 267:495–499.

Lukas SE, Brady JV, Griffiths RR (1986): Comparison of opioid self-administration and disruption of schedule-controlled performance in the baboon. J Pharmacol Exp Ther 238:924–931.

Lukas SE, Mello NK, Bree MP, Mendelson JH (1988): Differential tolerance development to buprenorphine-, diprenorphine-, and heroin-induced disruption of food maintained responding in macaque monkeys. Pharmacol Biochem Behav 30:977–982.

Macenski MJ, Cleary J, Thompson T (1990): Effects on opioid-induced rate reductions by doxepin and buproprion. Pharmacol Biochem Behav 37:247–252.

McMillan DE, Morse WH (1967): Some effects of morphine and morphine antagonists on schedule-controlled behavior. J Pharmacol Exp Ther 157:175–184.

McMillan DE, Wolf PS, Carchman RA (1970): Antagonism of the behavioral effects of morphine and methadone by narcotic antagonists in the pigeon. J Pharmacol Exp Ther 175:443–458.

Martin WR, Eades CG, Thompson JA, Huppler RE, Gilbert PE (1976): The effects of morphine- and nalorphine-like drugs in the non-dependent and morphine-dependent chronic spinal dog. J Pharmacol Exp Ther 197:517–732.

Mello NK, Mendelson JH (1985): Behavioral pharmacology of buprenorphine. Drug Alcohol Depend 14:283–303.

Mello NK, Bree MP, Mendelson JH (1981): Buprenorphine self-administration by rhesus monkey. Pharmacol Biochem Behav 15:215–225.

Mello NK, Bree MP, Lukas SE, Mendelson JH (1985): Buprenorphine effects on food-maintained responding in macaque monkeys. Pharmacol Biochem Behav 23:1037–1044.

Milar KS, Dykstra LA (1983): Effects of ketocyclazocine alone and in combination with naloxone on schedule-controlled responding in squirrel monkeys. Pharmacol Biochem Behav 18:395–400.

Miller L, Shaw JS, Whiting EM (1986): The contribution of intrinsic activity to the action of opioids in vitro. Br J Pharmacol 87:595–601.

Moerschbaecher JM, Mastropaolo J, Winsauer PJ, Thompson DM (1984): Effects of opioids on accuracy of a fixed-ratio discrimination in monkeys and rats. J Pharmacol Exp Ther 230:541–549.

Negus SS, Dykstra LA (1988): Kappa antagonist properties of buprenorphine in the shock titration procedure. Eur J Pharmacol 156:77–86.

Negus SS, Picker MJ, Dykstra LA (1989): Kappa antagonist properties of buprenorphine in non-tolerant and morphine-tolerant rats. Psychopharmacology 98:141–143.

Negus SS, Picker MJ, Dykstra LA (1990): Interactions between mu and kappa opioid agonists in the rat drug discrimination procedure. Psychopharmacology 102:465–473.

Negus SS, Picker MJ, Dykstra LA (1991): Interactions between the discriminative stimulus effects of mu and kappa opioid agonists in the squirrel monkey. J Pharmacol Exp Ther 256:149–158.

Oliveto AH, Picker MJ, Dykstra LA (1991): Acute and chronic morphine administration: Effects of mixed-action opioids in rats and squirrel monkeys responding under a schedule of food presentation. J Pharmacol Exp Ther 257:8–18.

Picker MJ, Negus SS, Craft RM (1990): Butorphanol's efficacy at mu and kappa opioid receptors: Inferences based on the schedule-controlled behavior of nontolerant and morphine-tolerant rats and on the responding of rats under a drug discrimination procedure. Pharmacol Biochem Behav 36:563–568.

Picker MJ, Negus SS, Powell KR (1991): Differential cross-tolerance to mu and kappa opioid agonists in morphine-tolerant rats responding under a schedule of food presentation. Psychopharmacology 103:129–135.

Portoghese PS, Larson DL, Sayre LM, Fried DS, Takemori AE (1980): A novel opioid receptor site directed alkylating agent with irreversible narcotic antagonistic and reversible agonistic activities. J Med Chem 23:233–234.

Richards ML, Sadée W (1985): Buprenorphine is an antagonist at the kappa opioid receptor. Pharm Res 2:178–181.

Ruffolo RR (1982): Important concepts of receptor theory. J Auton Pharmacol 2:277–295.

Sadée W, Rosenbaum JS, Herz A (1982): Buprenorphine: Differential interaction with opiate receptor sybtypes in vivo. J Pharmacol Exp Ther 223:157–162.

Seiden LS, Dykstra LA (1977): "Psychopharmacology: A Biochemical and Behavioral Approach". New York: Van Nostrand Reinhold.

Shannon HE (1983): Stimulation of avoidance behavior by buprenorphine in rats. Psychopharmacology 80:19–23.

Shannon HE, Cone DJ, Gorodetzky CW (1984): Morphine-like discriminative stimulus effects of buprenorphine and demethoxybuprenorphine in rats: Quantitative antagonism by naloxone. J Pharmacol Exp Ther 229:768–774.

Su TP (1985): Further demonstration of kappa opioid binding sites in the brain: Evidence for heterogeneity. J Pharmacol Exp Ther 232:144–148.

Villiger JW, Taylor KM (1981): Buprenorphine: Characteristics of binding sites in the rat central nervous system. Life Sci 29:2699–2708.

Ward SJ, Portoghese PS, Takemori AE (1982): Pharmacological characterization *in vivo* of the novel opiate, β-funaltrexamine. J Pharmacol Exp Ther 220:494–498.

Woods JH, Gmerek DE (1985): Substitution and primary dependence studies in animals. Drug Alcohol Dep 14:233–247.

Young AM, Stephens KR, Hein DW, Woods JH (1984): Reinforcing and discriminative stimulus properties of mixed agonist–antagonist opioids. J Pharmacol Exp Ther 229:118–126.

REINFORCING EFFECTS, DISCRIMINATIVE STIMULUS EFFECTS, AND PHYSICAL DEPENDENCE LIABILITY OF BUPRENORPHINE

S. STEVENS NEGUS and JAMES H. WOODS
Department of Pharmacology, University of Michigan Medical School, Ann Arbor, MI 48109

INTRODUCTION

Opiates such as morphine produce clinically useful effects, principally analgesia and an inhibition of gastrointestinal transit, but their use is limited by side effects. The liability of many opiates to abuse and dependence is one of these side effects. A goal of opiate research has been to identify compounds that retain the clinically useful effects of classical opiates such as morphine, but that reduce or eliminate the side effects such as liability to abuse. The more promising compounds have fallen into three general categories. Compounds in the first category produce clinical effects qualitatively similar to those produced by morphine but act at different types of receptors and therefore produce different profiles of side effects. Opiates are thought to produce their effects by binding to at least three types of opioid receptors, the μ, δ, and κ opioid receptors [Martin et al., 1976; Lord et al., 1977]. Morphine is thought to produce both its clinical effects and principal side effects by acting at μ receptors [Martin et al., 1976; Jaffe and Martin, 1990]. However, agonists at κ receptors share morphine's ability to produce analgesia [Wood and Iyengar, 1988; Jaffe and Martin, 1990] but have lower abuse liability than morphine [Woods and Winger, 1987]. Unfortunately, some κ agonists produce psychotomimetic effects that may limit their clinical utility [Kumor et al., 1986; Pfeiffer et al., 1986].

Compounds in the second category differ from morphine primarily in their pharmacokinetics. Following systemic administration, morphine distributes throughout the body and crosses the blood–brain barrier to reach the central nervous system

[Jaffe and Martin, 1990]. Thus, morphine acts at both peripheral and central receptors. More recently developed drugs such as loperamide [Niemegeers et al., 1974; Wüster and Herz, 1978] and quaternary opiate derivatives [Smith et al. 1982; Ferreira et al., 1984] do not distribute as readily across the blood–brain barrier and are confined to some extent to the periphery following systemic administration. Interestingly, the activation of peripheral opioid receptors with these compounds appears to be sufficient to produce an inhibition of gastrointestinal transit, and possibly analgesia [Niemegeers et al., 1974; Wüster and Herz, 1978; Smith et al., 1982; Ferreira et al., 1984], with less abuse liability than more centrally active drugs [Yanagita et al., 1980].

Buprenorphine has emerged as one of the most promising compounds of the third category. Buprenorphine and other drugs in this category act principally at μ opioid receptors, as does morphine, but they are characterized by having a relatively low efficacy at these receptors [Martin et al., 1976; Cowan et al., 1977b]. A drug's ability to interact with a receptor and produce an effect is governed by two drug-dependent factors: affinity and efficacy [Kenakin, 1987]. Affinity is a measure of a drug's ability to bind with a receptor, and both buprenorphine and morphine have a high affinity for μ receptors. Efficacy is a measure of a drug's ability to activate transduction mechanisms associated with the receptor, and current evidence reviewed in this volume and elsewhere indicates that buprenorphine has a lower efficacy than morphine at μ receptors. This difference in efficacy can be exploited, since different effects appear to require different levels of receptor activation. Thus, a drug could have sufficient efficacy to produce effects requiring low levels of receptor activation, but not enough efficacy to produce other effects requiring higher levels of receptor activation. Among opiate effects, the level of receptor activation required for analgesia depends in part on the intensity and type of the noxious stimulus presented to the subject, but results with drugs such as buprenorphine suggest that relatively low levels of μ receptor activation can produce a clinically useful level of analgesia in humans [Heel et al., 1979; Lewis, this volume]. Life-threatening levels of opiate-induced respiratory depression, in contrast, appear to require higher levels of μ receptor activation, and low-efficacy agonists such as buprenorphine do not produce a profound respiratory depression [Cowan et al., 1977a; Heel et al., 1979].

In addition to its low efficacy at μ receptors, buprenorphine features three other salient pharmacological characteristics. First, buprenorphine dissociates unusually slowly from its receptors [Hambrook and Rance, 1976; Boas and Villiger, 1985]. This slow rate of dissociation is thought to contribute to buprenorphine's long duration of action. In addition, buprenorphine's slow rate of dissociation may account for the finding that buprenorphine's effects can be prevented by prior or simultaneous administration of narcotic antagonists but are frequently difficult to reverse with narcotic antagonists administered after buprenorphine. Second, buprenorphine has virtually the same high affinity for κ receptors as for μ receptors and an affinity for δ receptors that is about 10-fold lower than its affinity for μ and κ receptors [Villiger and Taylor, 1981; Wood et al., 1981; Sadée et al., 1982]. Buprenorphine has very low efficacy at κ receptors, such that it generally functions as a κ antagonist in roughly the same dose ranges in which it produces its μ-mediated effects [Kajiwara et al., 1986; Leander, 1987; Negus et al., 1989]. Buprenorphine probably has low efficacy at δ receptors as well [Kajiwara et al., 1986]. Finally, in several assays the buprenorphine dose-effect curve has an inverted-U shape such that intermediate doses produce bigger effects than higher doses [Cowan et al., 1977b]. Thus, at high doses buprenorphine seems to antagonize its own effects. The mechanism of this autoantagonist effect is not clear; however, it may have clinical relevance, since buprenorphine's autoantagonism, together with its low efficacy at μ receptors, serves to limit the toxic effects of its administration in high doses.

Some of the earliest studies describing buprenorphine's effects suggested that its low efficacy at μ receptors and other unusual pharmacological properties might limit its abuse liability and qualify it as a potential maintenance drug for the treatment of opiate dependence [Jasinski et al., 1978; see Woods et al., 1992, for review]. Since then, this hypothesis has been largely confirmed by studies evaluating buprenorphine's effects in a number of preclinical behavioral assays designed to predict the abuse liability of drugs. In addition, it has become apparent that buprenorphine might have clinical value as a maintenance drug not only for the treatment of opiate dependence, but for the treatment of dependence on other drugs as well [Woods et al., 1992].

THE REINFORCING EFFECTS OF BUPRENORPHINE

Self-Administration of Buprenorphine

Drug-taking behavior in humans can be considered an example of operant behavior in which the delivery of a stimulus (the drug effect) is contingent on the performance of certain behaviors (obtaining, preparing, and administering the drug). By definition, drugs that increase the probability of the behaviors that lead to their delivery are considered to act as reinforcers of behavior and to produce reinforcing effects. These reinforcing effects can be most directly evaluated by using drug self-administration [Brady et al., 1987]. In the drug self-administration assay, as in the real world outside the laboratory, drug delivery is made contingent on behavior; however, both the behaviors and the consequences of those behaviors are simplified to maximize the likelihood of identifying a drug's reinforcing effects. Specifically, subjects are typically placed alone in an operant conditioning chamber and allowed to effect a simple and quantifiable response such as pressing a lever to receive unit doses of drug by intravenous administration. Drugs that maintain higher rates of responding than their vehicles are considered to be reinforcing. There is a high correlation between drugs that produce reinforcing effects in self-administration assays and drugs that are abused by humans, indicating that this procedure serves as a good predictor of a drug's abuse potential [Brady et al., 1987; but see Woods, 1983].

Studies examining the reinforcing effects of a test drug in the self-administration procedure employ one of two general strategies. The first and most direct of these strategies evaluates the ability of a test drug to induce and maintain responding in a drug-naive subject. In a less direct but quicker strategy, subjects are initially trained to self-administer a drug, such as morphine or cocaine, with known reinforcing effects. The test drug is then substituted for the training drug and evaluated for its ability to maintain responding in these drug-experienced subjects.

Published reports have described the self-administration of buprenorphine in only three drug-naive subjects. Yanagita et al. [1982] allowed two drug-naive rhesus monkeys to respond for i.v. doses of buprenorphine (4.0–60.0 µg/kg per injection) under a fixed-ratio (FR) 1 schedule during daily 4-hour sessions. Both monkeys eventually responded at much higher rates for buprenorphine than for saline, although one of the two monkeys had to be primed with forced administration of buprenorphine over a period of 2 weeks before voluntary self-administration began. Mello et al. [1981] similarly reported that a drug-naive rhesus monkey rapidly acquired self-administration of 5.0 µg/kg/inj buprenorphine under a second-order schedule that required an average of 48 responses for each buprenorphine injection. These results suggest that buprenorphine produces sufficient reinforcing effects to induce and maintain self-administration behavior in drug-naive subjects.

Buprenorphine can also maintain responding in drug-experienced subjects initially trained to self-administer another drug. Figure 1 shows that buprenorphine was self-administered by rhesus monkeys initially trained to respond under a multiple FR 30/time-out 600-sec schedule of codeine administration [Woods, 1977; Young et al., 1984]. Under the baseline conditions of codeine self-administration, monkeys responded at rates of 2.0–2.5 responses per second, whereas substitution of saline for codeine yielded responses rates of about 0.2 responses per second. When buprenorphine was substituted for codeine, it produced a dose-dependent increase in response rates; monkeys responded for the lowest unit dose of buprenorphine (0.1 µg/kg/inj) at rates no different than the rates maintained by saline, but the highest unit dose of buprenorphine (100.0 µg/kg/inj) maintained a rate of approximately 1.5 responses per second. Hoffmeister [1988] reported a similar dose-response function for buprenorphine in rhesus monkeys trained to self-administer codeine under much different experimental conditions. Buprenorphine has also been shown to produce reinforcing effects in rhesus monkeys initially trained to self-administer lefetamine [Yanagita et al., 1982], morphine [Mello et al. 1981] and alfentanil [Winger et al., 1992].

These studies concur in suggesting the general rule that buprenorphine produces reinforcing effects, but as with any rule, there is an exception. Lukas et al. [1986] found that buprenorphine at doses up to 3.2 mg/kg maintained self-administration in only one of four baboons initially trained to self-administer cocaine under an FR 160/time-out 3-hr schedule. A number of other opiates, including codeine, maintained self-administration in all four baboons. It was suggested by the authors that the unusual failure of buprenorphine to produce reinforcing effects in this study resulted from buprenorphine's long duration of action, and in support of this argument they pointed out that long-acting benzodiazepines maintained lower rates of

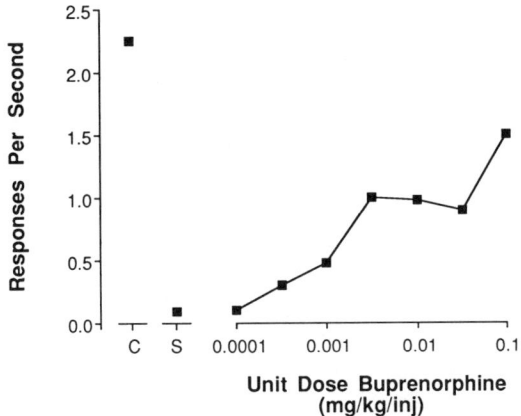

Fig. 1. Self-administration of buprenorphine in rhesus monkeys under a fixed-ratio 30 time-out 600-sec schedule of intravenous drug administration. Ordinate: Response rate (responses per second). Abscissa: Unit dose per injection of buprenorphine (mg/kg/inj), log scale. Monkeys were initially trained to self-administer codeine (0.32 mg/kg/inj). Point over C shows the mean response rate during two representative codeine self-administration sessions. Point over S shows the mean response rate during two representative saline substitution sessions. The remaining points show the response rates during sessions in which various unit doses of buprenorphine were substituted for codeine. Each point represents the mean data from four monkeys. Adapted from Young et al. [1984].

self-administration in baboons than short-acting benzodiazephines [Griffiths et al., 1981; 1985].

However, several other factors might also have contributed to this lapse in buprenorphine reinforcement. First, the study by Lukas et al. [1986] has been the only one to examine the reinforcing effects of buprenorphine in baboons, and buprenorphine's reinforcing effects may be species-dependent. Second, whereas most studies examining the reinforcing effects of buprenorphine in substitution procedures have used an opiate as the baseline drug, this is the only study to report on the reinforcing effects of buprenorphine in subjects for whom cocaine was the baseline drug. These results suggest that buprenorphine's reinforcing effects in a substitution procedure may depend on the baseline drug. In consonance with this hypothesis is the finding by Hoffmeister and Schlichting [1972] that monkeys self-administering codeine as their baseline drug were more sensitive to the reinforcing effects of other opiates than monkeys self-administering cocaine as their baseline drug. Finally, the Lukas et al. [1986] study employed a schedule of drug self-administration with a higher FR and longer time-out than have been used in any other study examining the reinforcing effects of buprenorphine, which suggests that buprenorphine's reinforcing effects may be schedule-dependent.

Regardless of the reason for buprenorphine's failure to produce reinforcing effects in baboons, the finding that it produces reinforcing effects in some but not all self-administration experiments is consistent with the characterization of buprenorphine as an agonist of intermediate efficacy. Other features of buprenorphine's reinforcing effects are also consistent with this characterization. For example, the dose-effect relationship for buprenorphine's reinforcing effects is often shallow compared with the dose-effect curves for higher-efficacy compounds [Woods, 1977; Hoffmeister, 1988; but see Winger et al., 1992].

Other evidence suggesting that buprenorphine has relatively low efficacy as a reinforcer has come from studies that used a progressive-ratio schedule of reinforcement. Under the progressive-ratio schedule, the ratio requirement (e.g., the number of discrete responses required to produce a single administration of drug) increases over time. For example, a single press of a lever may be sufficient to produce the first drug injection in a session, but subsequent injections require progressively higher numbers of lever presses. The ratio at which the subject stops self-administering drug is referred to as the *breaking point,* and drugs that maintain high breaking points are considered to have higher reinforcing efficacy than drugs that maintain low breaking points. Mello et al. [1988] used a progressive-ratio schedule to compare the reinforcing efficacy of different unit doses of buprenorphine, heroin, and methadone. In their study, monkeys were initially trained to self-administer some dose of the test drug under a second-order schedule requiring an average of 64 responses to earn one drug delivery. Once responding stabilized, the number of responses required to earn a reinforcer was increased every 2 days until the monkey failed to earn a reinforcer for two consecutive days. The response requirement for the final reinforcer defined the breaking point. Responding was then restabilized at the original response requirement before a new dose of the test drug was evaluated. Under this procedure, the average breaking points maintained by different unit doses of buprenorphine were 403 (10.0 μg/kg/inj); 331 (30.0 μg/kg/inj), 384 (50.0 μg/kg/inj), and 552 (100.0 μg/kg/inj). In contrast, the highest breaking points maintained by other treatments were 1,067 for heroin (100.0 μg/kg/inj), 480 for methadone (100.0 μg/kg/inj), and 282 for saline. These results suggest that the reinforcing efficacy of buprenorphine is less than that of heroin, more than that of saline, and roughly equivalent to that of methadone. Yanagita et al. [1982] compared the reinforcing efficacy of buprenorphine and pentazocine in a somewhat different progressive-ratio procedure. As in the Mello et al. [1988] study, Yanagita et al. [1982] found that a low unit dose of buprenorphine (15.0 μg/kg/inj) generally maintained higher breaking points than an intermediate dose (60.0 μg/kg/inj). Furthermore, pentazocine maintained a slightly higher average breaking point than buprenorphine, suggesting that pentazocine is more reinforcing that buprenorphine.

Other Effects of Buprenorphine Related to Reinforcement

Buprenorphine has also been shown to produce effects related to reinforcement in drug-naive subjects by means of two other assays, the place-conditioning procedure [Brown et al. 1991] and the facilitation of electrical brain stimulation [Hubner and Kornetsky, 1988]. In the place-conditioning procedure, subjects are typically confined to one place following administration of drug and to a second, distinct place following administration of vehicle. Subsequently, each subject is allowed to move freely between the two places. If the subjects spend more time in the drug-paired place than in the vehicle-paired place, then the drug is considered to produce a place preference, and this place preference is thought to reflect the drug's reinforcing effects [van der Kooy, 1987]. Brown et al. [1991] showed that buprenorphine at doses of 30.0–300.0 μg/kg (i.p.) produced significant place preferences in rats. These investigators also used microdialysis techniques to demonstrate that buprenorphine caused a release of dopamine in the nucleus accumbens. Although the presence of dopamine in the nucleus accumbens may not be necessary for opiate reinforcement [Koob and Bloom, 1988; Wise, 1989], it is interesting that buprenorphine's ability to release dopamine in this site is shared by many other drugs of abuse [Di Chiara and Imperato, 1988].

Hubner and Kornetsky [1988] evaluated the effects of buprenorphine in a model of electrical brain stimulation. In this procedure, rats were first trained to press a lever for the administration of electrical stimulation to the medial forebrain bundle at the level of the lateral hypothalamus. The intensity of the stimulation was then varied to determine the threshold level that would maintain self-stimulation. Drugs that lower the threshold of stimulation required to maintain behavior are considered to facilitate electrical brain stimulation, and this effect may be related to a drug's reinforcing effects [Reid, 1987]. Buprenorphine in doses of 4.0–60.0 μg/kg (s.c.) facilitated electrical brain stimulation in each of the four rats tested.

Effects of Buprenorphine on the Reinforcing Effects of Other Drugs

In addition to producing reinforcing effects of its own, buprenorphine has also been shown to alter behavior reinforced by other drugs. Most of this work has been conducted using self-administration as the measure of drug reinforcement, and buprenorphine has been demonstrated to suppress the self-administration of heroin, dilaudid, alfentanil, cocaine, ethanol, and acetaldehyde [Mello and Mendelson, 1980; Martin et al., 1983; Mello et al., 1982, 1983, 1989, 1990; Myers et al., 1984; Domanque et al., 1991; Kamien et al., 1991; Winger et al., 1992]. For example, Winger et al. (1992) evaluated the effects of buprenorphine on responding maintained by various doses of either alfentanil or cocaine under a FR 30 time-out 45-sec schedule. The dose-effect curves for alfentanil, cocaine, and buprenorphine in this procedure are shown in Fig. 2. The dose-effect relationship relating unit dose to response rate for both alfentanil and cocaine had an inverted-U shape, with maximum rates maintained by 1.0 μg/kg/inj alfentanil and 10.0 μg/kg/inj cocaine. Buprenorphine also maintained high rates of responding when substituted for alfentanil in the alfentanil-trained monkeys, with 3.2 μg/kg/inj maintaining maximum rates. Interestingly, in this procedure using a short time-out period and alfentanil as the baseline drug, the buprenorphine dose-effect curve had a steeper slope and a higher peak than in codeine-maintained monkeys responding under a FR 30 time-out 600-sec schedule [Young et al., 1984] (Fig. 1).

Figure 3 shows that pretreatment with buprenorphine produced a dose-dependent flattening of both the alfentanil and cocaine dose-effect curves; however, buprenorphine was approximately 1,000-fold more potent in suppressing alfentanil-

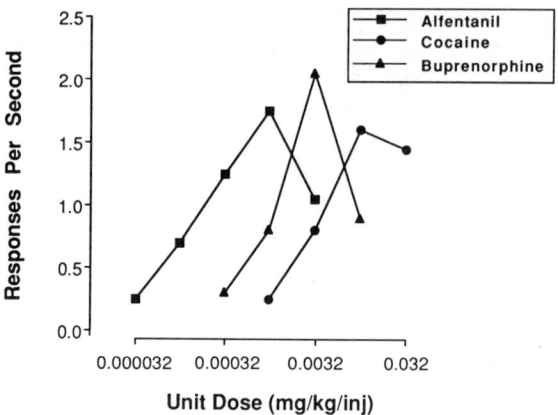

Fig. 2. Self-administration of alfentanil, cocaine, and buprenorphine in rhesus monkeys under a fixed-ratio 30 time-out 45-sec schedule of intravenous drug administration. Ordinate: Response rate (responses per second). Abscissa: Unit dose per injection (mg/kg/inj), log scale. The alfentanil and buprenorphine dose-effect curves were determined in monkeys initially trained to self-administer alfentanil, whereas the cocaine dose-effect curve was determined in monkeys initially trained to self-administer cocaine. Response rates in both groups of monkeys during saline substitution sessions were approximately 0.1 responses per second. Each point represents the mean data from three monkeys. Adapted from Winger et al. [1992].

maintained responding than in suppressing cocaine-maintained responding. A dose of 3.2 μg/kg buprenorphine (the same dose that maintained maximum response rates when buprenorphine was self-administered) suppressed the maximum response rates maintained by alfentanil by more than 50%, whereas a dose of 3,200.0 μg/kg buprenorphine was required to produce an equivalent suppression of cocaine-maintained responding. Mello and colleagues have reported that buprenorphine produces a qualitatively similar suppression of responding maintained by opiates and cocaine in both monkeys and humans [Mello and Mendelson, 1980; Mello et al., 1982, 1983, 1989, 1990]; however, in these studies buprenorphine was roughly equipotent in suppressing both opiate and cocaine self-administration. Furthermore, this group has also shown that daily buprenorphine administration continued to suppress cocaine self-administration for up to 4 months in monkeys, indicating that tolerance did not develop to buprenorphine-induced suppression of cocaine self-administration [Kamien et al., 1991]. Finally, administration of 30.0–3,000.0 μg/kg buprenorphine reduced the self-administration of both ethanol and the ethanol metabolite acetaldehyde in rats [Martin et al., 1983; Myers et al., 1984; Domanque et al., 1991], although the effect of buprenorphine on ethanol self-administration may depend on a complex interaction between dose and pretreatment time [Domanque et al., 1991].

The finding that buprenorphine suppresses drug-maintained responding, especially when only one dose of the self-administered drug is examined, can be inter-

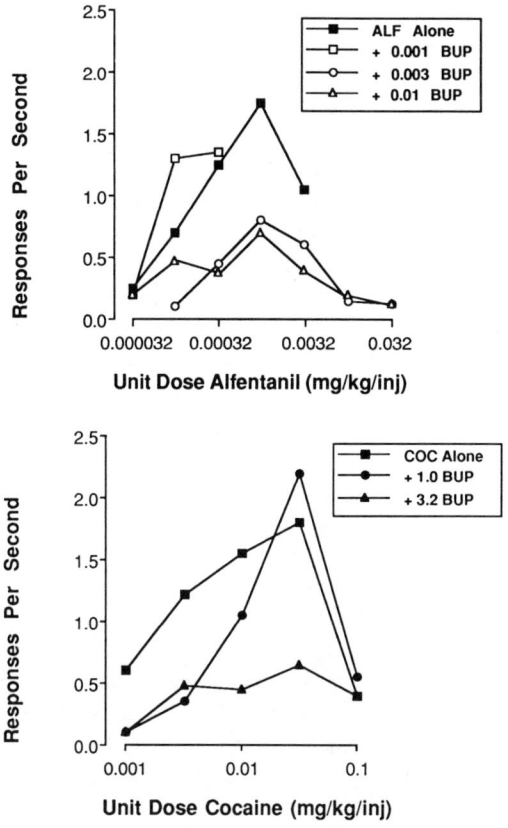

Fig. 3. Suppression of both alfentanil- and cocaine-maintained responding by pretreatment with buprenorphine in rhesus monkeys. Ordinate: Response rate (responses per second). Abscissa: Unit dose per injection (mg/kg/inj), log scale. Dose-effect curves for alfentanil in alfentanil-trained monkeys and for cocaine in cocaine-trained monkeys were determined either after no pretreatment (control) or after pretreatment with various doses of buprenorphine. Buprenorphine was administered i.v. 30 min before the beginning of each self-administration session. Response rates during saline substitution sessions in both groups of monkeys were about 0.1 responses per second. Each point represents the mean data from three monkeys. Adapted from Winger et al. [1992].

preted in a number of different ways. As noted above, the dose-effect curves describing the reinforcing effects of self-administered drugs often have an inverted-U shape in which intermediate unit doses maintain the highest rates of responding, whereas both lower and higher unit doses maintain lower rates of responding. Thus, the consistent finding that buprenorphine suppresses self-administration of opiates, cocaine, and possibly other drugs could be interpreted as suggesting that buprenorphine (1) antagonizes the reinforcing effects of the self-administered drug to produce a rightward shift in the dose-effect curve, or (2) potentiates the reinforcing effects of the self-administered drug to produce a leftward shift in the dose-effect curve. However, the finding that buprenorphine shifted both the alfentanil and cocaine dose-effect curves downward, and not laterally, in monkeys trained to self-administer either alfentanil or cocaine [Winger et al., 1992] (Fig. 3) argues against either of these interpretations. Additionally, the effects of buprenorphine do not appear to depend on its potential for producing opiate antagonist effects, since other opiate antagonists typically produce effects on opiate-maintained responding consistent with a rightward shift in the opiate dose-effect curve [Koob et al., 1984; Winger et al., 1992] but have little or no effect on cocaine-maintained responding [Mello et al., 1990; Winger et al., 1992]. A third possible interpretation—that the buprenorphine-induced suppression of self-administration reflects a general suppression of all behavior—has been ruled out by most studies (but see Myers et al., 1984) by demonstration that buprenorphine suppressed drug-maintained responding at doses that had lesser effects on responding maintained by other reinforcers such as food.

This leads to a fourth possibility: that buprenorphine obscures the reinforcing effects of other drugs by producing agonist effects of its own. In drug self-administration, a contingency is established such that a response by the subject results in the delivery of a reinforcing stimulus—the drug. The ability of the subject to detect that stimulus depends not only on the quality and intensity of the stimulus, but also on the "background noise" against which that stimulus is superimposed [Green and Swets, 1966]. Thus, a subject could fail to detect a stimulus either because the stimulus intensity is too low or because the background noise is too high. Under control conditions in self-administration, the stimulus effects of a self-administered dose of drug are superimposed against a low level of background noise insofar as no other drug effects are present. However, pretreatment with buprenorphine may produce agonist effects that effectively raise the level of background noise to the point at which the stimulus effects of the self-administered drug are obscured and no longer capable of reinforcing behavior. Notably, such an effect would not shift the dose-effect curve for the self-administered drug laterally, since the effect of buprenorphine would be to alter the context in which the self-administered drug stimulus is detected rather than the stimulus itself. Instead, the predicted effect of buprenorphine would be to lower the reinforcing effects of all doses of the self-administered drug and produce a downward shift of the dose-effect curve. These hypothetical agonist effects of buprenorphine appear to be mediated by μ opioid receptors, since other μ agonists have also been shown to suppress the self-administration of both opiates and cocaine [Hoffmeister and Schlichting, 1972; Stretch, 1977; Negus et al., 1992; Winger et al., 1992].

Brown et al. [1991] have recently investigated the interaction between buprenorphine and cocaine in the place-conditioning assay of drug reinforcement. In this study, cocaine and buprenorphine were shown to produce place preferences when administered alone and to produce additive effects when administered in combination. The authors concluded from these data that buprenorphine accentuated the reinforcing effects of cocaine. These results agree with the self-administration studies in demonstrating that buprenorphine produces reinforcing effects on its own and does not produce a traditional antagonism of the reinforcing effects of cocaine. Brown et al. [1991] go on, however, to speculate that buprenorphine's suppression of cocaine self-administration in published self-administration studies resulted from an additive effect between the reinforcing effects of buprenorphine and cocaine that essentially shifted the cocaine dose-effect curve to the left. This conclusion is unwarranted. As demonstrated by Winger et al. [1992] (Fig. 3), buprenorphine shifted the cocaine dose-effect curve in a self-administration assay downward rather than to the left.

Furthermore, direct comparisons between this place-conditioning study and the published self-administration studies should be made cautiously, since different treatment protocols have been employed in the two types of experiments. In the self-administration studies cited above, buprenorphine was administered as a pretreatment before the subsequent evaluation of the reinforcing effects of cocaine; thus, the administration of buprenorphine *was not* contingent on the subjects' behavior. The study by Brown et al. [1991], in contrast, used a different treatment protocol, in which one place was paired with buprenorphine + cocaine, whereas a second place was paired only with saline; thus, the delivery of buprenorphine in this study *was* contingent on place. A more appropriate place-conditioning study for comparison with published self-administration studies would pair one place with buprenorphine + cocaine and the second place with buprenorphine + saline. (This experiment has not been done, but under these circumstances buprenorphine might suppress cocaine-induced place preferences.) Alternatively, the additivity of the reinforcing effects of buprenorphine and cocaine could be evaluated in a self-administration experiment by establishing a contingency in which responding by the subject produced a combined injection of both buprenorphine and cocaine.

THE DISCRIMINATIVE STIMULUS EFFECTS OF BUPRENORPHINE

Generalization of Buprenorphine to Other Drugs

In addition to producing reinforcing effects, drugs also produce discriminative stimulus effects. In the terms of operant conditioning, a discriminative stimulus signals the presence or absence of a behavior–reinforcer contingency. In procedures developed to evaluate the discriminative stimulus effects of drugs [Holtzman, 1985; Overton, 1987], subjects are typically given access to two levers and trained to press these levers for food. Following this initial training, subjects are injected daily with either a set dose of a training drug or with vehicle, and the contingencies are altered such that pressing one lever (the drug-appropriate lever) produces food following drug injection, whereas pressing the other lever (the vehicle-appropriate lever) produces food following vehicle injection. If subjects learn to press the drug-appropriate lever following drug injection and to press the vehicle-appropriate lever following vehicle injection, then the subject is considered to discriminate between the presence and absence of the drug, and the drug is considered to produce discriminative stimulus effects. Test drugs can then be administered to the subject, and if the subject presses the drug-appropriate lever, then the test drug is said to generalize to the training drug.

In a typical drug discrimination experiment, human or laboratory animal subjects "describe" the drug effects they experience using a limited "vocabulary" of two or three levers. However, humans can describe drug effects using the much richer vocabulary of words such as good, bad, happy, sick and excited. These verbal descriptions of drug effects constitute the "subjective" effects of drugs, and these subjective effects have been categorized with questionnaires such as the Addiction Research Center Inventory (ARCI) [Haertzen and Hickey, 1987]. There is a high correlation between drugs that share discriminative stimulus effects as measured in two-lever and related discrimination assays and drugs that share subjective effects as measured with various questionnaires [Schuster et al., 1981]; in fact, a drug's verbally described subjective effects can be thought of as one manifestation of a drug's discriminative stimulus effects.

The role of a drug's discriminative stimulus effects in generating drug-taking behavior is not clear [Schuster et al., 1981; Holtzman, 1985; Overton, 1987; Lamb et al., 1991]. Whereas nearly all drugs that produce reinforcing effects also have abuse liability, many drugs that produce discriminative stimulus effects are not abused. Thus, the demonstration that laboratory animals or humans can be trained to discriminate a drug does not provide a useful prediction of the drug's abuse liability. However, a drug's discriminative stimulus effects do display a marked degree of pharmacological selectivity that can be exploited. Test drugs that generalize to a training drug often share some pharmacological mechanism of action with the training drug and produce many other effects in common with the training drug [Holtzman, 1985; Overton, 1987]. Thus, if a test drug generalizes to a known drug of abuse such as morphine, then it is likely that the test drug acts as an agonist at μ opioid receptors and produces a profile of morphine-like effects, including high abuse potential. The value of the drug-discrimination assay in predicting abuse liability, then, lies in its utility in identifying compounds pharmacologically similar to known drugs of abuse.

There have been no published reports in which buprenorphine has served as the training drug in a drug-discrimination experiment. However, several investigators have trained subjects to discriminate other, usually opiate, compounds and evaluated the ability of buprenorphine to generalize to the training stimulus. By this procedure, it has consistently been found that buprenorphine generalizes completely to μ agonists such as morphine, codeine, etorphine, and fentanyl in rats, pigeons, and nonhuman primates [Colpaert, 1978; Leander, 1983; France et al., 1984; Shannon et al., 1984; Young et al., 1984, 1991; Holtzman, 1985; Hoffmeister, 1988; Picker and Dykstra, 1989; France and Woods, 1990; Negus et al., 1990, 1991; Versage et al., 1990]. For example, Figure 4 shows that both morphine (0.3–3.0 mg/kg) and buprenorphine (3.0–30.0 μg/kg) produce a dose-dependent increase in drug-appropriate responding in rats trained to discriminate morphine (3.0 mg/kg, s.c.) from saline [Shannon et al., 1984]. Furthermore, Figure 4 shows that naloxone antagonizes the discriminative stimulus effects of both morphine and buprenorphine in a quantitatively similar manner; Schild analysis yielded apparent pA_2 values of

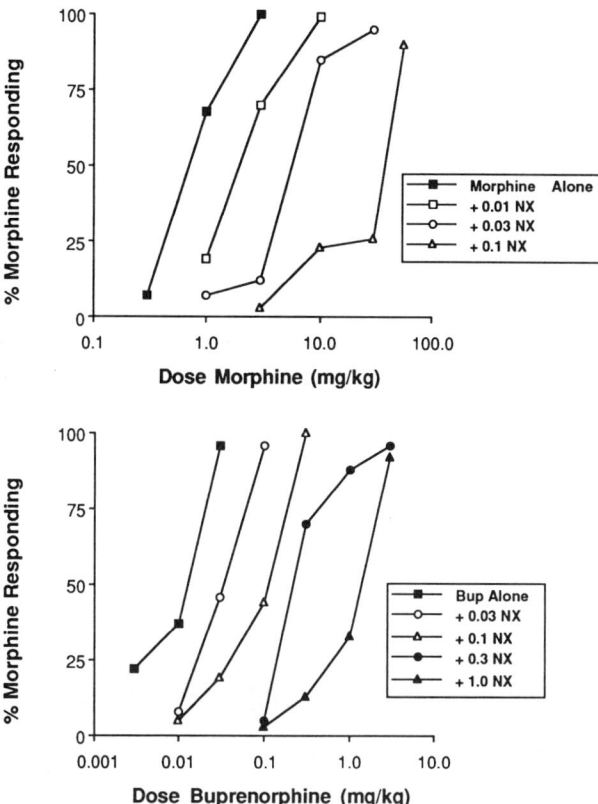

Fig. 4. Discriminative stimulus effects of morphine (upper panel) and buprenorphine (lower panel) administered either alone or in combination with naloxone in rats trained to discriminate between 3.0 mg/kg morphine and saline. Ordinate: Percentage morphine-appropriate responding measured as percentage trials completed on the morphine-appropriate lever. Abscissa: Dose of test drug (mg/kg), log scale. Each point represents the mean data from five rats. Adapted from Shannon et al. [1984].

7.85 ± 0.36 for naloxone in combination with morphine and 7.45 ± 0.16 for naloxone in combination with buprenorphine [Shannon et al., 1984]. These results indicate that morphine and buprenorphine produce similar discriminative stimulus effects, probably by acting at μ opioid receptors. Buprenorphine also produces morphine-like subjective effects in humans with a history of opiate dependence [Jasinski et al., 1978].

In most assays of buprenorphine's discriminative stimulus effects in subjects trained to discriminate a μ agonist from vehicle, buprenorphine has produced full agonist effects in so far as it has completely generalized to the training drug. However, there is evidence from drug discrimination studies that buprenorphine has less efficacy at μ opioid receptors than some other μ agonists. The best evidence from a drug discrimination study for buprenorphine's low efficacy at μ opioid receptors was reported by France and Woods [1990]. These investigators maintained pigeons on chronic morphine (10.0 mg/kg/day) and trained a three-key discrimination between morphine (10.0 mg/kg, i.m.), saline, and naltrexone (0.032 mg/kg, i.m.). The pigeons were tested under two states: a morphine-dependent state, in which the daily dose of morphine was administered 6 hr prior to the discrimination testing, and a morphine-abstinent state, in which saline was administered instead of the daily morphine dose such that the pigeons entered spontaneous morphine abstinence. In the morphine-dependent state, treatment with saline produced responding primarily on the saline lever, whereas morphine and naltrexone produced a dose-dependent increase in responding on their respective levers. Figure 5 compares the effects of buprenorphine and the higher-efficacy μ agonist etonitazene in this procedure. Buprenorphine (0.1–32.0 mg/kg) administered to pigeons in the morphine-dependent state produced primarily saline-appropriate responding, whereas etonitazene produced a dose-dependent increase in morphine-appropriate responding. Thus, in the morphine-dependent pigeons, buprenorphine produced discriminative stimulus effects unlike (perhaps between) those of morphine or naltrexone, whereas etonitazene generalized completely to morphine. These results contrast with the finding that buprenorphine (0.01–0.32 mg/kg) produced complete generalization to morphine (5.6 mg/kg) in pigeons not maintained on chronic morphine [France et al., 1984], but agree with the finding that buprenorphine produces no subjective effects in methadone-maintained human volunteers [Strain et al., 1991].

In the morphine-abstinent state, treatment with saline produced responding on the naltrexone lever, whereas low doses of morphine produced a dose-dependent switch to saline-appropriate responding, and higher doses of morphine produced a switch from saline- to morphine-appropriate responding. Figure 6 compares the effects of buprenorphine and etonitazene administered to pigeons in the morphine-abstinent state. As with morphine, etonitazene produced a dose-dependent switch from the naltrexone to the saline to the morphine lever. Buprenorphine (0.1–0.32 mg/kg) produced a dose-dependent switch from the naltrexone lever to the saline lever, but higher doses of buprenorphine up to 32.0 mg/kg failed to produce the subsequent switch to the morphine lever. In morphine-abstinent pigeons, then, the results indicated that buprenorphine had sufficient efficacy to produce a morphine-like switch from the naltrexone lever to the saline lever, but not enough efficacy to produce a subsequent switch to the morphine lever.

Other evidence of buprenorphine's relatively low efficacy as a μ agonist comes from studies using different training doses of morphine. In the drug discrimination paradigm, the intensity of the training stimulus can be increased by increasing the training dose of the training drug. As the intensity of the training stimulus is raised by increasing the training dose (assuming that the training drug is a high-efficacy agonist), test drugs must produce progressively more agonist activity to produce generalization. The effect of increasing the training dose on high-efficacy test drugs is to shift the dose effect curve to the right. For lower-efficacy test drugs, however, the effect of increasing the training dose may be to decrease the maximum effect, since a low-efficacy drug may have sufficient efficacy to generalize completely to a

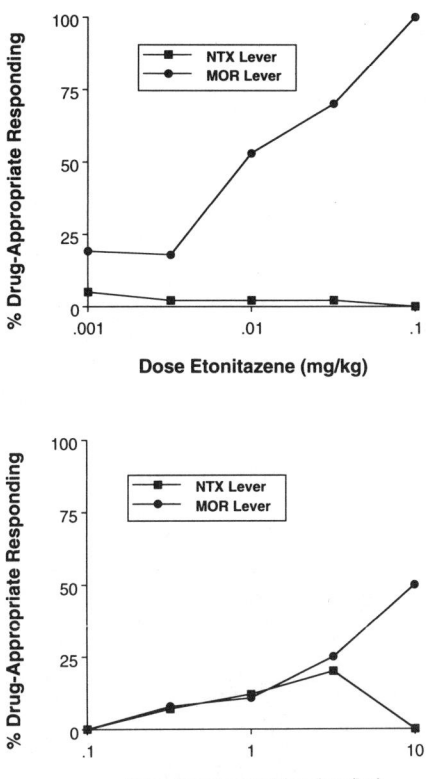

Fig. 5. Discriminative stimulus effects of etonitazene (upper panel) and buprenorphine (lower panel) in morphine-dependent pigeons trained to discriminate morphine, naltrexone, and saline. Ordinate: Percentage drug-appropriate responding on the morphine key (circles) and naltrexone key (squares). Percentage responding on the saline key is not shown but equals 100 − (% morphine responding + % naltrexone responding). Abscissa: Dose of test drug (mg/kg), log scale. Each point represents the mean data from five pigeons. Adapted from France and Woods [1990].

low dose of the training drug but insufficient efficacy to generalize completely to a higher dose of the training drug. In rats trained to discriminate 3.0 mg/kg [Shannon et al., 1984] or 3.2 mg/kg [Young et al., 1991] morphine from saline, buprenorphine produced complete generalization at doses of 0.03 mg/kg and above. In rats trained to discriminate 10.0 mg/kg morphine, buprenorphine also produced its peak effect at 0.03 mg/kg; however, this dose of buprenorphine produced only about 80% morphine-appropriate responding [Negus et al., 1989]. Furthermore, higher doses of buprenorphine produced primarily saline-appropriate responding so that the buprenorphine dose-effect curve had the inverted-U shape characteristic of many other assays in which buprenorphine acts as a lower-efficacy agonist [Cowan et al., 1977b].

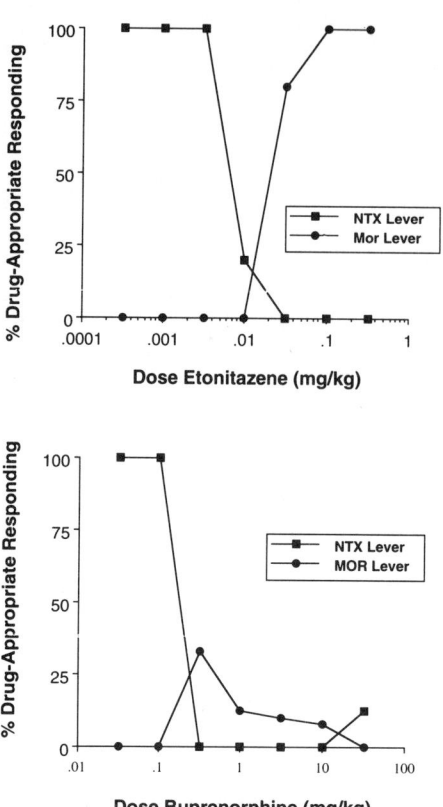

Fig. 6. Discriminative stimulus effects of etonitazene (upper panel) and buprenorphine (lower panel) in morphine-abstinent pigeons trained to discriminate morphine, naltrexone, and saline. Ordinate: Percentage drug-appropriate responding on the morphine key (circles) and naltrexone key (squares). Percentage responding on the saline key is not shown but equals 100 − (% morphine responding + % naltrexone responding). Abscissa: Dose of test drug (mg/kg), log scale. Each point represents the mean data from five pigeons. Adapted from France and Woods [1990].

Another approach to characterizing the efficacy of buprenorphine has been to evaluate its discriminative stimulus effects in subjects trained to discriminate other low-efficacy agonists. For example, Walker and Young [1990; and personal communication] trained rats to discriminate nalbuphine (3.2 mg/kg) from saline and tested five opiate agonists with varying efficacies and selectivities at opioid receptors. The high-efficacy μ agonist etorphine produced a dose-dependent and complete generalization to nalbuphine, as did morphine. Buprenorphine, in contrast, elicited drug-appropriate responding in only half of the rats tested. U-50,488, a selective κ agonist, did not generalize to nalbuphine in any rats. These results suggest that the nalbuphine stimulus in rats is selective for μ agonists but not selective for low-efficacy μ agonists.

A similar pattern of partial generalization for buprenorphine was found in human volunteers with a history of opiate dependence trained to discriminate between the higher efficacy μ agonist hydromorphone (0.043 mg/kg), the less selective and lower-efficacy agonist pentazocine (0.64 mg/kg) and saline [Preston et al., 1989]. In this assay, buprenorphine (0.0016–0.013 mg/kg) was identified as hydromorphone by 50% of the subjects and as pentazocine by the other 50%. These discriminative stimulus effects paralleled the subjective effects produced by buprenorphine, hydromorphone, and pentazocine as measured by the Addiction Research Center Inventory. The ARCI is a true–false questionnaire that allows drugs to be classified under five subscales that include an MBG scale of drug-induced euphoria and an LSD subscale of dysphoria. Hydromorphone produced a dose-dependent increase in the MBG subscale score, whereas pentazocine produced a dose-dependent increase in the LSD subscale score. Buprenorphine increased the score on both scales.

Buprenorphine's effects in subjects trained to discriminate opiate antagonists has also been evaluated. As noted above, buprenorphine did not generalize to the opiate antagonist naltrexone in morphine-dependent pigeons [France and Woods, 1990; Versage et al., 1990]. Buprenorphine (0.001–0.01 mg/kg) did generalize completely to the nonselective opiate antagonist diprenorphine in squirrel monkeys [DeRosset and Holtzman, 1986]; however, a series of μ and κ agonists including etorphine and morphine also generalized completely to the diprenorphine stimulus, whereas opiate antagonists such as naloxone and naltrexone elicited primarily vehicle-appropriate responding. Interestingly, naloxone antagonized the discriminative stimulus effects of all the μ and κ agonists, including buprenorphine, in this study, but naloxone did not antagonize the discriminative stimulus effects of diprenorphine. Thus, the nature of the diprenorphine stimulus was unusual both in its low selectivity for μ versus κ agonists and in its resistance to naloxone antagonism. At any rate, diprenorphine was not functioning as a simple opiate antagonist in this assay, and buprenorphine's generalization to diprenorphine cannot be taken as an indication of buprenorphine's antagonist effects.

The slow dissociation of buprenorphine from its receptor(s) appears to account for the finding that buprenorphine's μ-like discriminative stimulus effects can be both long-lasting and difficult to reverse with opiate antagonists. The long duration of buprenorphine's discriminative stimulus effects have been especially apparent in pigeons, since large doses can be administered to this species with minimal decrements in response rates or sign of toxic effects [Leander, 1983; France et al., 1984; Picker and Dykstra, 1989]. For example, Leander [1983] found that buprenorphine at doses of 0.08 mg/kg and above generalized completely to fentanyl (0.05 mg/kg) in pigeons, whereas doses of buprenorphine up to 10.0 mg/kg failed to reduce response rates. Furthermore, 10.0 mg/kg buprenorphine elicited 100% drug-appropriate responding 24 hr after its administration, and 75% drug-appropriate responding after 48 hr. France et al. [1984] reported a similarly long duration of action for the discriminative stimulus effects of high-dose buprenorphine in pigeons, with a dose of 56.0 mg/kg producing morphine-like discriminative stimulus effects for up to 5 days. France et al. [1984] also demonstrated that the discriminative stimulus effects of a maximally effective dose of buprenorphine could be prevented by administration of naltrexone 10 min before buprenorphine but could not be reversed by administration of naltrexone 25 min or more after buprenorphine. Similarly, in rats, 1.0 mg/kg naloxone completely antagonized the discriminative stimulus effects of a simultaneously administered dose of buprenorphine, whereas doses of naloxone up to 100.0 mg/kg produced only a partial antagonism when administered 30 min after buprenorphine [Shannon et al., 1984].

Buprenorphine has not generalized to a variety of κ agonists in pigeons [Picker and Dykstra, 1989], rats [Shearman and Herz, 1982; Negus et al., 1990], or nonhuman primates [Young et al., 1984; Negus et al., 1991]. Similarly, buprenorphine failed to generalize to PCP in rats [Shannon, 1982] or to the psychomotor stimulant cocaine in rats [Doty et al., 1991; Pettit et al., 1991] or squirrel monkeys [Kamien et al., 1990; Spealman and Bergman, 1991]. These results suggest that buprenorphine's agonist effects in these discriminative stimulus studies arise exclusively from its action at μ opioid receptors.

Effects of Buprenorphine on the Discriminative Stimulus Effects of Other Drugs

Buprenorphine usually generalizes completely to μ agonists in subjects trained to discriminate a μ agonist from saline, and under these circumstances, buprenorphine does not antagonize the effects of other μ agonists. For example, buprenorphine (0.01–0.1 mg/kg) generalized completely to fentanyl (0.1 mg/kg) in rats, and pretreatment with buprenorphine (0.01–0.1 mg/kg) produced only leftward or upward shifts in the fentanyl dose-effect curve (Negus, unpublished results). However, when buprenorphine fails to generalize completely, it may antagonize the effects of higher-efficacy μ agonists. For example, buprenorphine produced primarily saline-appropriate responding in morphine-dependent pigeons trained to discriminate morphine, naltrexone, and saline, and pretreatment with 1.0 mg/kg buprenorphine prevented the discriminative stimulus effects of both morphine and naltrexone [France and Woods, 1990]. These findings agree with the characterization of buprenorphine as an intermediate-efficacy μ agonist.

Buprenorphine has more consistently been shown to antagonize the discriminative stimulus effects of κ agonists [Negus et al., 1990, 1991]. For example, buprenorphine (0.003–0.1 mg/kg) produced vehicle-appropriate responding in squirrel monkeys trained to discriminate the κ agonist U-50,488 (0.75 or 1.7 mg/kg) from water; however, pretreatment with buprenorphine (0.003–0.1 mg/kg) produced a dose-dependent reduction in drug-appropriate responding induced by the training dose of U-50,488. Furthermore, pretreatment with 0.03 mg/kg buprenorphine produced a 5- to 10-fold rightward shift in the U-50,488 dose-effect curve. The selective μ agonist fentanyl did not affect the discriminative stimulus effects of U-50,488. These findings agree with the characterization of buprenorphine as an antagonist at κ opioid receptors.

Finally, given the recent demonstration that buprenorphine suppresses the self-administration of cocaine, there has been intense interest in evaluating the effects of buprenorphine on the discriminative stimulus effects of cocaine. In squirrel monkeys, buprenorphine (3.0–5.6 μg/kg), as well as the other μ agonists morphine, methadone, and levorphanol, failed to generalize to cocaine but produced leftward shifts in the cocaine dose-effect curve [Kamien et al., 1990; Spealman and Bergman, 1991]. Naltrexone had no effect on the cocaine dose-effect curve, but blocked the potentiating effects of both morphine and buprenorphine [Spealman and Bergman, 1991]. In rats, buprenorphine, morphine, and naltrexone also failed to generalize to cocaine, but in contrast with the findings in squirrel monkeys, none of these compounds significantly affected the discriminative stimulus effects of cocaine [Doty et al., 1991; Pettit et al., 1991]. These contrasting results may reflect differences in the species or procedures used.

As is apparent, the effects of buprenorphine on the discriminative stimulus effects of μ agonists and cocaine have been variable. This contrasts with the more consistent finding that buprenorphine suppresses the self-administration of both μ agonists and cocaine. However, comparisons between the effects of buprenorphine on the reinforcing versus the discriminative stimulus effects of other drugs must again consider an important difference in treatment protocols that has distinguished the two types of studies. As noted above, in studies evaluating the effects of buprenorphine on the reinforcing effects of other drugs, buprenorphine is generally administered as a pretreatment before the beginning of a self-administration session during which responding produces injections of the test drug. Thus, the reinforcing effects of the test drug are evaluated against a background of effects induced by buprenorphine. Importantly, buprenorphine does not serve as a reinforcing stimulus in this situation, and responding cannot be considered to reflect the combined reinforcing effects of buprenorphine and the test drug. In discrimination studies, in contrast, buprenorphine is generally administered together with the training drug at some time before a test session. Under these circumstances, buprenorphine does produce discriminative stimulus effects, and the pattern of responding expressed by the subject may reflect the combined discriminative stimulus effects of buprenorphine plus the test drug.

As described above, the only study that has evaluated the combined reinforcing effects of buprenorphine plus a test drug used the place-conditioning procedure [Brown et al., 1991]. These investigators determined that buprenorphine "accentuated" the reinforcing effects of cocaine in rats. This contrasts with the finding that buprenorphine had no effect on the discriminative stimulus effects of cocaine in the rat [Doty et al., 1991; Pettit et al., 1991], although buprenorphine did potentiate the discriminative stimulus effects of buprenorphine in squirrel monkeys [Kamien et al., 1990; Spealman and Bergman, 1991].

In order to evaluate the effect of a "buprenorphine background" on the discriminative stimulus effects of other drugs, subjects would have to be evaluated for their ability to discriminate buprenorphine plus test drug from buprenorphine alone. This experiment has not been done by means of a discriminative stimulus assay; however, the subjective effects of morphine and cocaine both before and during chronic buprenorphine (4.0 or 8.0 mg/day) have been determined in humans with a history of drug dependence [Jasinski et al., 1978; Teoh et al., 1992]. Chronic buprenorphine attenuated the subjective effects of morphine. This attenuation could reflect tolerance to the subjective effect of μ agonists, antagonism by the low-efficacy buprenorphine of the higher-efficacy morphine, or the ability of chronic buprenorphine's own agonist effects to obscure the agonist effects of acute morphine. Regardless of the mechanism, this study suggests that a buprenorphine background appears to attenuate the subjective (discriminative) stimulus properties as well as the reinforcing stimulus properties of μ agonists in humans. The effects of chronic buprenorphine on the subjective effects of cocaine were more complex, with buprenorphine suppressing the subjective effects of cocaine in some subjects and enhancing the subjective effects of cocaine in others [Teoh et al., 1992].

EFFECTS OF BUPRENORPHINE IN ASSAYS OF PHYSICAL DEPENDENCE

Physical Dependence Produced by Chronic Treatment With Buprenorphine

The emergence of an abstinence syndrome following withdrawal from the chronic administration of a drug defines the presence of physical dependence. Withdrawal from the chronic administration of high doses of classical opiates such as morphine, produced by either termination of agonist delivery (spontaneous abstinence) or administration of an antagonist such as naltrexone (precipitated abstinence), results in an abstinence syndrome characterized by severe flu-like signs and symptoms as well as such emotional symptoms as anxiety and, in individuals with histories of self-administration, drug craving [Jaffe, 1990]. Physical dependence appears to be neither necessary nor sufficient to maintain drug-taking behavior [e.g., Yanagita, 1980; Jaffe, 1990]. Thus, the demonstration that a drug does or does not produce physical dependence does not alone predict its abuse liability in man. However, different types of drugs are characterized by distinctly different types of abstinence syndromes, and drugs that produce similar abstinence syndromes may share similar mechanisms of action and produce other effects in common [Woods and Gmerek, 1985]. Thus, the demonstration that withdrawal from the chronic administration of a test drug produces a morphine-like abstinence syndrome suggests that the test drug may produce other morphine-like effects, including a liability to abuse.

The withdrawal from chronic buprenorphine has been shown to produce either no signs of abstinence or a protracted but relatively mild morphine-like abstinence syndrome in rats, dogs, nonhuman primates, and humans [Martin et al., 1976; Cowan et al., 1977b; Jasinski et al., 1978; Dum et al., 1981; Woods and Gmerek, 1985]. For example, Jasinski et al. [1978] maintained human volunteers on 8.0 mg/day (approximately 110.0 μg/kg/day) buprenorphine, a dose of buprenorphine comparable to about 240 mg/day morphine. Figure 7 shows that administration of up to 4.0 mg naloxone, a dose of more than 20 times higher than the dose necessary to precipitate abstinence in morphine-dependent subjects, failed to precipitate absti-

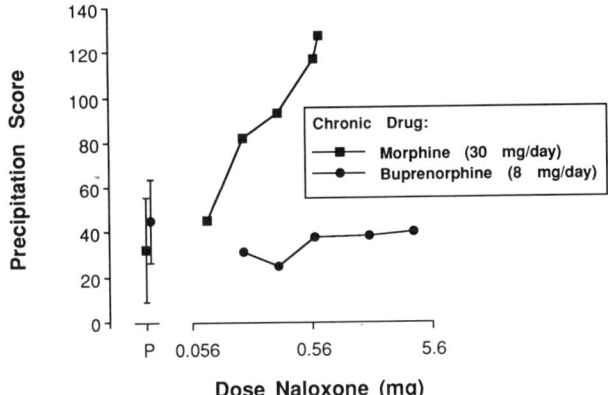

Fig. 7. Abstinence scores precipitated by naloxone in humans dependent on either morphine (squares) or buprenorphine (circles). Ordinate: Intensity of precipitated abstinence as measured by precipitation score. Abscissa: Dose of naloxone (mg, s.c.). Seven morphine-dependent subjects were maintained on 30.0 mg/day morphine, whereas four buprenorphine-dependent subjects were maintained on 8.0 mg/day buprenorphine. The points above P show the precipitation score obtained after placebo administration; brackets show the 95% confidence limits. Adapted from Jasinski et al. [1978].

nence in the buprenorphine-treated subjects. Furthermore, Figure 8 shows that the spontaneous withdrawal from buprenorphine produced a mild, morphine-like abstinence syndrome with a delayed onset of 3–14 days and a duration of about 1 week, whereas spontaneous withdrawal from morphine produced a more profound abstinence syndrome with a shorter time course. This abstinence syndrome produced by withdrawal from chronic buprenorphine could be reversed by the administration of morphine.

The emergence of a morphine-like and morphine-reversible abstinence syndrome in these subjects suggests that buprenorphine can produce physical dependence by acting on μ opioid receptors. The resistance of buprenorphine-treated subjects to naloxone-precipitated abstinence and the delayed onset of spontaneous abstinence are consistent with the slow dissociation of buprenorphine from its receptors. Buprenorphine's low efficacy at μ opioid receptors may also contribute to its relatively mild physical dependence-producing properties, but two pieces of evidence suggest that buprenorphine's efficacy is less important than its time course in accounting for the apparently mild abstinence syndrome observed during withdrawal. First, the withdrawal from chronic administration of other low-efficacy μ agonists that dissociate more rapidly from μ receptors (e.g., nalbuphine) has been shown to produce more severe abstinence syndromes [Woods and Gmerek, 1985]. This suggests that even low-efficacy μ agonists can produce a high degree of physical dependence.

Second, Dum et al. [1981] were able to elicit a more dramatic precipitated abstinence syndrome in rats with a morphine substitution strategy designed to

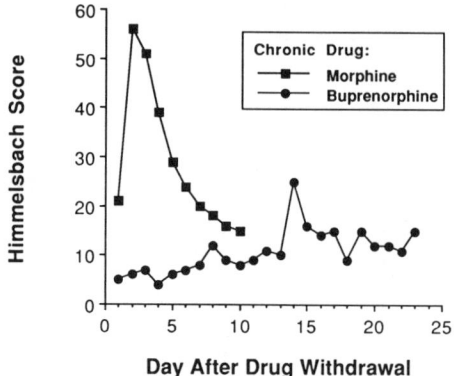

Fig. 8. Intensity of spontaneous abstinence over time in human subjects following withdrawal from chronic treatment with either morphine (various doses and durations) or buprenorphine (8.0 mg/day). Ordinate: Intensity of spontaneous abstinence as measured by the Himmelsbach scoring technique of opiate abstinence [Kolb and Himmelsbach, 1938]. Abscissa: Consecutive days following drug withdrawal. Each point for the morphine group represents the mean Himmelsbach score for 65 subjects who were admitted to treatment for opiate dependence and were maintained for 1 week on four daily injections of morphine prior to the evaluation of withdrawal. Each point for the buprenorphine group represents the mean data from three to five subjects. Adapted from Kolb and Himmelsbach [1938] and Jasinski et al. [1978].

compensate for buprenorphine's unusual kinetics. In their study, rats were treated with buprenorphine (500.0 μg/kg) twice daily for 4 days. This treatment regimen produced tolerance to the antinociceptive effects of buprenorphine and "cross-tolerance" to morphine. On the fifth day, the administration of 10.0 mg/kg naloxone, a dose shown to precipitate a profound abstinence syndrome in morphine-treated rats, precipitated only a mild abstinence syndrome in the buprenorphine-treated rats. However, in a separate group of buprenorphine-treated rats administered 10.0 mg/kg morphine 1 hr before the naloxone challenge, the same dose of naloxone precipitated a more intense abstinence syndrome. Since rats treated only with 10.0 mg/kg morphine and challenged an hour later with naloxone displayed no signs of abstinence, it was concluded that "morphine can prime animals pretreated with buprenorphine so that naloxone can induce withdrawal." The mechanism for this priming is not clear, but these results do suggest that chronic buprenorphine treatment can at least contribute to the development of physical dependence.

Effects of Buprenorphine on the Physical Dependence Produced by the Chronic Administration of Other Drugs

The effects of buprenorphine on the physical dependence induced by the chronic administration of other drugs has been evaluated in suppression and precipitation studies [Woods et al., 1979]. In suppression studies, subjects are withdrawn from the chronic administration of a reference drug such as morphine so that the signs of spontaneous abstinence emerge. The test drug is then administered and evaluated for its ability to suppress these signs of spontaneous abstinence. Test drugs that suppress abstinence are considered to be agonists similar to the reference drug (since the reference drug can also suppress abstinence) and may produce other effects in common with the reference drug. Thus, as with drug discrimination, the value of the suppression test lies in its ability to identify compounds pharmacologically similar to known drugs of abuse; however, it should be noted that the suppression test is not as pharmacologically selective as the drug discrimination test, since drugs quite different from the reference drug can also suppress abstinence signs [Dwoskin et al., 1983; Neal and Sparber, 1986]. In precipitation studies, subjects are maintained on the chronic administration of a reference drug. The test drug is then administered and evaluated for its ability to precipitate an abstinence syndrome. Test drugs that precipitate abstinence are considered to be antagonists of the reference drug.

As is consistent with the hypothesis that buprenorphine has low efficacy at μ receptors, buprenorphine has been shown to both suppress and precipitate abstinence in subjects dependent on μ agonists [Martin et al., 1976; Cowan et al., 1977b; Gmerek, 1984; Woods and Gmerek, 1985; France and Woods, 1990]. The effect in any particular study appears to depend on such variables as the species of the subject and the degree of physical dependence. For example, Martin et al. [1976] studied buprenorphine in chronic spinal dogs made dependent on 125 mg/day of morphine. In suppression studies, buprenorphine (1.0–16.0 μg/kg) produced a dose-dependent suppression of the spontaneous abstinence signs observed in dogs that had not received morphine for 40 hr. In precipitation studies, higher doses of buprenorphine (24.0–384.0 μg/kg) precipitated a mild abstinence syndrome in these same dogs. However, the buprenorphine dose-effect curve was shallow in both experiments, and buprenorphine was neither as effective as morphine in suppressing abstinence nor as effective as naloxone in precipitating abstinence.

The tenacity of buprenorphine's binding to opioid receptors has been revealed in a precipitation study in rhesus monkeys [Gmerek, 1984]. Both buprenorphine (300.0 μg/kg) and naltrexone (10.0 μg/kg) precipitated abstinence syndromes of similar intensity in monkeys maintained on 12.0 mg/kg/day morphine. Furthermore, these precipitated abstinence syndromes could be reversed by morphine. However, naltrexone-precipitated abstinence could be reversed by 17.8 mg/kg morphine, whereas 178.0 mg/kg morphine was required to reverse abstinence precipitated by buprenorphine. It was suggested that the 10-fold lower potency of morphine in reversing buprenorphine- versus naltrexone-precipitated abstinence resulted from buprenorphine's slow dissociation from its receptors.

Buprenorphine has also been shown to precipitate withdrawal in rhesus monkeys maintained on escalating doses of the κ agonist U-50,488 (up to 105.0 mg/kg/day) [Gmerek et al., 1987]. This is consistent with buprenorphine's high affinity but low efficacy at κ receptors. Buprenorphine has not been evaluated for its ability to suppress or precipitate abstinence in subjects chronically maintained on other drugs.

SUMMARY

A goal of opiate pharmacology has been to identify new compounds that retain useful clinical effects such as analgesia without producing side effects such as a high potential for abuse. Buprenorphine is one of the most promising drugs to emerge from this effort. Studies on buprenorphine at both in vitro and in vivo levels have identified three pharmacological characteristics of particular interest. First, buprenorphine binds with high affinity to μ opioid receptors and acts as an intermediate-efficacy μ agonist. Second, buprenorphine dissociates slowly from these receptors. Third, buprenorphine also binds with high affinity to κ and δ opioid receptors and acts primarily as a κ and δ antagonist.

The preclinical evaluation of buprenorphine's potential for abuse has largely confirmed this profile of buprenorphine's pharmacological characteristics. First, buprenorphine produces μ agonist-like reinforcing, discriminative stimulus, and physical dependence-producing effects, indicating that buprenorphine acts at μ opioid receptors. However, buprenorphine appears to be less efficacious than other μ agonists such as morphine in producing these effects. For example, buprenorphine maintains lower breaking points than higher-efficacy μ agonists in progressive-ratio self-administration experiments, fails to generalize to morphine in morphine-dependent subjects, and may precipitate abstinence in subjects maintained on the chronic administration of a μ agonist. These findings agree with the classification of buprenorphine as an intermediate-efficacy agonist at μ opioid receptors.

Buprenorphine has also been shown in several assays to produce effects that are of long duration. For example, buprenorphine generalized to morphine for up to 5 days in pigeons and suppressed both opiate and cocaine self-administration for up to 24 hr in monkeys. Similarly, in human volunteers withdrawn from chronic buprenorphine, abstinence signs did not emerge in most subjects for two weeks. Furthermore, the agonist and antagonist effects of buprenorphine are difficult to reverse with other drugs. The discriminative stimulus effects of buprenorphine could not be reversed by even high doses of naloxone in rats or pigeons, and naloxone does not precipitate withdrawal in subjects maintained on chronic buprenorphine. Similarly, the abstinence precipitated by buprenorphine in rhesus monkeys could be reversed only by high doses of morphine. The long duration and persistence of buprenorphine's effects are consistent with the slow dissociation of buprenorphine from its binding sites.

Finally, buprenorphine has been shown to alter the reinforcing, discriminative stimulus, and physical dependence effects of drugs that do not act at μ opioid receptors. As is consistent with buprenorphine's reported antagonist effects at κ opioid receptors, buprenorphine antagonized the discriminative stimulus effects of κ agonists in squirrel monkeys and rats and precipitated abstinence in κ-dependent rhesus monkeys. Buprenorphine has also been reported to suppress the self-administration of cocaine and ethanol and to potentiate cocaine-induced place preferences and discriminative stimulus effects. These effects on cocaine, and possibly ethanol, may be due to buprenorphine's μ agonist effects since other μ agonists produce similar effects.

The finding that buprenorphine produces reinforcing effects, μ agonist-like discriminative stimulus effects, and μ agonist-like physical dependence predicts that buprenorphine does have potential for abuse in humans. However, buprenorphine's low efficacy in producing these effects indicates that buprenorphine may have less abuse potential than other higher-efficacy μ agonists such as morphine. In findings consistent with this prediction, instances of intravenous buprenorphine abuse have been reported, primarily in persons already dependent on another opiate, but buprenorphine does not appear to be a drug of choice [Lewis, 1985; Chowdhury and Chowdhury, 1990].

The ability of buprenorphine to modify the reinforcing effects, discriminative stimulus effects, and physical dependence states produced by other drugs suggests that buprenorphine may influence the abuse potential of these other drugs. Specifically, buprenorphine shows promise as a potential pharmacotherapy for the treatment of dependence on μ opiates and possibly for dependence on other drugs such as cocaine and ethanol. Importantly, buprenorphine's clinical promise as a pharmacotherapy for drug dependence derives not only from its ability to modify the abuse potential of other drugs, but also from its long duration of action and low toxicity even at high doses.

FUTURE DIRECTIONS

Buprenorphine's own abuse potential and its effect on the abuse potential of other drugs is presumed to derive from its activity at μ opioid receptors. However, the pharmacology of buprenorphine's effects, and especially its reinforcing effects, has not been well examined. For example, it has yet to be shown that buprenorphine's own reinforcing effects or its ability to suppress the reinforcing effects of other drugs (especially cocaine) can be reversed by either general opiate antagonists such as naloxone or selective μ antagonists such as β-funaltrexamine or CTAP (D-Phe-Cys-Tyr-D-Trp-Arg-Thr-Pen-Thr-NH_2). Furthermore, subtypes of the μ receptor have been proposed [Pasternak and Wood, 1986; Rothman et al., 1989], and the role of these μ receptor subtypes in the actions of buprenorphine has not been examined. That such μ receptor subtypes might be important has been suggested by a recent study reporting that morphine and buprenorphine may cause their antinociceptive effects by activating different transduction mechanisms and possibly different receptors [Wheeler-Aceto and Cowan, 1991]. Finally, the role of buprenorphine's antagonist activity at kappa and delta receptors in producing its own reinforcing effects or in modifying the effects of other drugs has not been well evaluated.

Several other issues could also be clarified by future studies. First, evidence described above suggests that the reinforcing effects of buprenorphine in substitution procedures may depend on the baseline drug on which subjects are maintained. The role of baseline drug in determining the reinforcing effects of buprenorphine should be more thoroughly investigated, since this variable could have implications both for the likelihood of buprenorphine abuse among different populations of drug users and for the utility of buprenorphine maintenance in the treatment of drug dependence. Second, the effect of a "buprenorphine background" on the discriminative stimulus effects of other drugs should be evaluated, since in a therapeutic context drug users would probably receive periodic doses of buprenorphine in a clinical environment and would experience the effects of other subsequently used drugs against this buprenorphine background. Third, buprenorphine has been demonstrated to suppress the self-administration of μ agonists, cocaine, and possibly ethanol. It would be useful to examine the generality of buprenorphine's reinforcement-suppressing effects to other drugs, especially other stimulants such as amphetamine and other depressants such as barbiturates or benzodiazepines. Finally, it would be useful to determine whether other drugs that share aspects of buprenorphine's unusual pharmacological profile produce a similar suppression of drug self-administration. For example, codeinones such as NIH 10420 act as μ partial agonists with slightly higher efficacy than buprenorphine, but possibly as a result of metabolism to their respective morphinones, these compounds display long-lasting μ, δ, and κ antagonist effects [Aceto et al., 1989].

REFERENCES

Aceto MD, Bowman ER, May EL, Harris LS, Woods JH, Smith CB, Medzihradsky F, Jacobson AE (1989): Very long-acting narcotic antagonists: The 14β-p-substituted cinnamoylaminomorphinones and their partial mu agonist codeinone relatives. Arzneim Forsch 39:570–575.

Boas RA, Villiger JW (1985): Clinical actions of fentanyl and buprenorphine. Br J Anaesth 57:192–196.

Brady JV, Griffiths RR, Hienz RD, Ator NA, Lukas SE, Lamb RJ (1987): Assessing drugs for abuse liability and dependence potential in laboratory primates. In Bozarth MA (ed): "Methods of Assessing the Reinforcing Effects of Abused Drugs." New York: Springer-Verlag, pp 48–86.

Brown EE, Finlay JM, Wong JTF, Damsma G, Fibiger HC (1991): Behavioral and neurochemical interactions between cocaine and buprenorphine: Implications for the pharmacotherapy of cocaine abuse. J Pharmacol Exp Ther 256:119–126.

Chowdhury AN, Chowdhury S (1990): Buprenorphine abuse: Report from India. Br J Addiction 85:1349–1350.

Colpaert FC (1978): Discriminative stimulus properties of narcotic analgesic drugs. Pharmacol Biochem Behav 9:863–887.

Cowan A, Doxey JC, Harry EJR (1977a): The animal pharmacology of buprenorphine, an oripavine analgesic agent. Br J Pharmacol 60:547–554.

Cowan A, Lewis JW, Macfarlane IR (1977b): Agonist and antagonist properties of buprenorphine, a new antinociceptive agent. Br J Pharmacol 60:537–545.

DeRosset SE, Holtzman SG (1986): Discriminative stimulus effects of the opioid antagonist diprenorphine in the squirrel monkey. J Pharmacol Exp Ther 237:437–444.

Di Chiara G, Imperato A (1988): Drugs abused by humans preferentially increase synaptic dopamine concentrations in the mesolimbic system of freely moving rats. Proc Natl Acad Sci (USA) 85:5274–5278.

Domanque KR, June HL, Smith TD, Dixon JT, Norona A, Lewis MJ (1991): Buprenorphine produces dose- and time-dependent effects on ethanol self-administration. Soc Neurosci Abstr 17:1422.

Doty P, Johnson AB, Picker MJ, Dykstra LA (1991): Discriminative stimulus properties of cocaine: Lack of antagonism by buprenorphine, morphine or naltrexone. Soc Neurosci Abstr 17:1425.

Dum J, Bläsig J, Herz A (1981): Buprenorphine: Demonstration of physical dependence liability. Eur J Pharmacol 70:293–300.

Dwoskin LP, Neal BS, Sparber SB (1983): Yohimbine exacerbates and clonidine attenuates acute morphine withdrawal in rats. Eur J Pharmacol 90:269–273.

Ferreira SH, Lorenzetti BB, Rae GA (1984): Is methylnalorphinium the prototype of an ideal peripheral analgesic? Eur J Pharmacol 99:23–29.

France CF, Woods JH (1990): Discriminative stimulus effects of opioid agonists in morphine-dependent pigeons. J Pharmacol Exp Ther 254:626–632.

France CP, Jacobson AE, Woods JH (1984): Discriminative stimulus effects of reversible and irreversible opiate agonists: Morphine, oxymorphazone and buprenorphine. J Pharmacol Exp Ther 230:652–657.

Gmerek DE (1984): The suppression of deprivation and antagonist-induced withdrawal in morphine-dependent monkeys. Neuropeptides 5:19–22.

Gmerek DE, Dykstra LA, Woods JH (1987): Kappa opioids in rhesus monkeys. III. Dependence associated with chronic administration. J Pharmacol Exp Ther 242:428–436.

Green DM, Swets JA (1966): "Signal Detection Theory and Psychophysics." New York: Wiley.

Griffiths RR, Roache JD, Ator NA, Lamb RJ, Lukas SE (1981): Self-injection of barbiturates and benzodiazepines in baboons. Psychopharmacology 75:101–109.

Griffiths RR, Roache JD, Ator NA, Lamb RJ, Lukas SE (1985): Similarities in reinforcing and discriminative stimulus effects of diazepam, triazolam, and pentobarbital in animals and humans. In Seiden LS, Balster RL (eds): "Behavioral Pharmacology of Psychotropic Agents." New York: Alan R. Liss, pp 419–432.

Haertzen CA, Hickey JE (1987): Addiction research center inventory (ARCI): Measurement of euphoria and other drug effects. In Bozarth MA (ed): "Methods of Assessing the Reinforcing Properties of Abused Drugs." New York: Springer-Verlag, pp 489–524.

Hambrook JM, Rance MJ (1976): The interaction of buprenorphine with the opiate receptor: Lipophilicity as a determining factor in drug–receptor kinetics. In Kosterlitz H (ed): "Opiates and Endogenous Opiate Peptides." Amsterdam: Elsevier/North-Holland, pp 295–302.

Heel RC, Brogden RN, Speight TM, Avery GS (1979): Buprenorphine: A review of its pharmacological properties and therapeutic efficacy. Drugs 17:81–110.

Hoffmeister F (1988): A comparison of the stimulus effects of codeine in rhesus monkeys under the contingencies of a two lever discrimination task and a cross self-administration paradigm: Tests of generalization of pentazocine, buprenorphine, tilidine and different doses of codeine. Psychopharmacology 94:315–320.

Hoffmeister F, Schlichting UU (1972): Reinforcing properties of some opiates and opioids in rhesus monkeys with histories of cocaine and codeine self-administration. Psychopharmacologia 23:55–74.

Holtzman SG (1985): Drug discrimination studies. Drug Alcohol Depend 14:263–282.

Hubner CB, Kornetsky C (1988): The reinforcing properties of the mixed agonist–antagonist buprenorphine as assessed by brain-stimulation reward. Pharmacol Biochem Behav 30:195–197.

Jaffe JH (1990): Drug addiction and drug abuse. In Gilman AG, Rall TW, Nies AS, Taylor P (eds): "The Pharmacological Basis of Therapeutics." New York: Pergamon, pp 522–573.

Jaffe JH, Martin WR (1990): Opioid analgesics and antagonists. In Gilman AG, Rall TW, Nies AS, Taylor P (eds): "The Pharmacological Basis of Therapeutics." New York: Pergamon, pp 485–521.

Jasinski DR, Pevnick JS, Griffith JD (1978): Human pharmacology and abuse potential of the analgesic buprenorphine. Arch Gen Psychiatry 35:501–516.

Kajiwara M, Aoki K, Ishii K, Numata H, Matsumiya T, Oka T (1986): Agonist and antagonist actions of buprenorphine on three types of opioid receptor in isolated preparations. Jpn J Pharmacol 40:95–101.

Kamien J, Bergman J, Madras BK, Spealman RD (1990): Buprenorphine potentiates behavioral effects of cocaine in squirrel monkeys. FASEB J 4:A593.

Kamien JB, Mello NK, Mendelson JH, Lukas SE (1991): Buprenorphine suppresses cocaine self-administration by rhesus monkeys over 1 to 4 months of daily treatment. NIDA Res Monogr 105: 619–620.

Kenakin TP (1987): "Pharmacologic Analysis of Drug–Receptor Interactions." New York: Raven.

Kolb L, Himmelsbach CK (1938): Clinical studies of drug addiction. III. A critical review of the withdrawal treatments with method of evaluating abstinence syndromes. Am J Psychiatry 94:759–797.

Koob GF and Bloom FE (1988): Cellular and molecular mechanisms of drug dependence. Science 242:715–723.

Koob GF, Pettit HO, Ettenberg A, Bloom FE (1984): Effects of opiate antagonists and their quaternary derivatives on heroin self-administration in the rat. J Pharmacol Exp Ther 229:481–486.

Kumor KM, Haertzen CA, Johnson RE, Kocher T, Jasinski D (1986): Human psychopharmacology of ketocyclazocine as compared with cyclazocine, morphine and placebo. J Pharmacol Exp Ther 238:960–968.

Lamb RJ, Preston KL, Schindler CW, Meisch RA, Davis F, Katz JL, Henningfield JE, Goldberg SR (1991): The reinforcing and subjective effects of morphine in post-addicts: A dose-response study. J Pharmacol Exp Ther 259:1165–1173.

Leander JD (1983): Opioid agonist and antagonist behavioural effects of buprenorphine. Br J Pharmacol 78:607–615.

Leander DJ (1987): Buprenorphine has potent kappa opioid receptor antagonist activity. Neuropharmacology 26:1445–1447.

Lewis JW (1985): Buprenorphine. Drug Alcohol Depend 14:363–372.

Lord JAH, Waterfield AA, Hughes J, Kosterlitz HW (1977): Endogenous opioid peptides: Multiple agonists and receptors. Nature 267:495–499.

Lukas SE, Brady JV, Griffiths RR (1986): Comparison of opioid self-injection and disruption of schedule-controlled performance in the baboon. J Pharmacol Exp Ther 238:924–931.

Magnan J, Paterson SJ, Tavani A, Kosterlitz HW (1982): The binding spectrum of narcotic analgesic drugs with different agonist and antagonist properties. Naunyn-Schmiedeberg's Arch Pharmacol 319:197–205.

Martin A, Pilotto R, Singer G, Oei TPS (1983): The suppression of ethanol self-injection by buprenorphine. Pharmacol Biochem Behav 19:985–986.

Martin WR, Eades CG, Thompson JA, Huppler RE, Gilbert PE (1976): The effects of morphine- and nalorphine-like drugs in the nondependent and morphine-dependent dog. J Pharmacol Exp Ther 197:517–532.

Mello NK, Mendelson JH (1980): Buprenorphine suppresses heroin use by heroin addicts. Science 207:657–659.

Mello NK, Bree MP, Mendelson JH (1981): Buprenorphine self-administration by rhesus monkey. Pharmacol Biochem Behav 15:215–225.

Mello NK, Mendelson JH, Kuehnle JC (1982): Buprenorphine effects on human heroin self-administration: an operant analysis. J Pharmacol Exp Ther 223:30–39.

Mello NK, Bree MP, Mendelson JH (1983): Comparison of buprenorphine and methadone effects on opiate self-administration in primates. J Pharmacol Exp Ther 225:378–386.

Mello NK, Lukas SE, Bree MP, Mendelson JH (1988): Progressive ratio performance maintained by buprenorphine, heroin and methadone in macaque monkeys. Drug Alcohol Depend 21:81–97.

Mello NK, Mendelson JH, Bree MP, Lukas SE (1989): Buprenorphine suppresses cocaine self-administration by rhesus monkeys. Science 245:859–862.

Mello NK, Mendelson JH, Bree MP, Lukas SE (1990): Buprenorphine and naltrexone effects on cocaine self-administration by rhesus monkeys. J Pharmacol Exp Ther 254:926–939.

Myers WD, Ng KT, Singer G (1984): Effects of naloxone and buprenorphine on intravenous acetaldehyde self-injection in rats. Physiology Behav 33:449–455.

Neal BS, Sparber SB (1986): Mianserin attenuates naloxone-precipitated withdrawal in rats acutely or chronically dependent upon morphine. J Pharmacol Exp Ther 236:1–9.

Negus SS, Picker MJ, Dykstra LA (1989): Kappa antagonist properties of buprenorphine in non-tolerant and morphine-tolerant rats. Psychopharmacology 98:141–143.

Negus SS, Picker MJ, Dykstra LA (1990): Interactions between mu and kappa opioid agonists in the rat drug discrimination procedure. Psychopharmacology 102:465–473.

Negus SS, Picker MJ, Dykstra LA (1991): Interactions between the discriminative stimulus effects of mu and kappa opioid agonists in the squirrel monkey. J Pharmacol Exp Ther 256:149–158.

Negus SS, Weinger MB, Henriksen SJ, Koob GF (1992): Effect of mu, delta and kappa opioid agonists on heroin-maintained responding in the rat. NIDA Res Monogr 119:410.

Niemegeers CJE, Lenaerts FM, Janssen PAJ (1974): Loperamide (R 18553), a novel type of antidiarrheal agent. II. Arzneim Forsch 24:1636–1641.

Overton DA (1987): Applications and limitations of the drug discrimination method for the study of drug abuse. In Bozarth MA (ed): "Methods of Assessing the Reinforcing Properties of Abused Drugs." New York: Springer-Verlag, pp 291–340.

Pasternak GW, Wood PL (1986): Minireview: Multiple mu opiate receptors. Life Sci 38:1889–1898.

Pettit HO, Smith S, Eckstrom AJ, Pryor GT (1991): Evaluation of the effects of buprenorphine in both a cocaine self-administration and a drug-discrimination paradigm. Soc Neurosci Abstr 17:1426.

Pfeiffer A, Brantl V, Herz A, Emrich HM (1986): Psychotomimesis mediated by κ opiate receptors. Science 233:774–776.

Picker MJ, Dykstra LA (1989): Discriminative stimulus effects of mu and kappa opioids in the pigeon: Analysis of the effects of full and partial mu and kappa agonists. J Pharmacol Exp Ther 249:557–566.

Preston KL, Bigelow GE, Bickel WK, Liebson IA (1989): Drug discrimination in human postaddicts: Agonist antagonist opioids. J Pharmacol Exp Ther 250:184–196.

Reid L (1987): Tests involving pressing for intracranial stimulation as an early procedure for screening likelihood of addiction of opioids and other drugs. In Bozarth MA (ed): "Methods of Assessing the Reinforcing Properties of Abused Drugs." New York: Springer-Verlag, pp 391–420.

Rothman RB, Bykov V, Long JB, Brady LS, Jacobson AE, Rice KC, Holaday JW (1989): Chronic administration of morphine and naltrexone up-regulate μ-opioid binding sites labeled by [^{3}H][D-Ala^2,MePhe^4,Gly-ol^5]enkephalin: Further evidence for two μ-binding sites. Eur J Pharmacol 160:71–82.

Sadée W, Rosenbaum JS, Herz A (1982): Buprenorphine: Differential interaction with opiate receptor subtypes in vivo. J Pharmacol Exp Ther 223:157–162.

Schuster CR, Fischman MW, Johanson CE (1981): Internal stimulus control and subjective effects of drugs. NIDA Res Monogr 37:116–129.

Shannon HE (1982): Pharmacological analysis of the phencyclidine-like discriminative stimulus properties of narcotic derivatives in rats. J Pharmacol Exp Ther 222:146–151.

Shannon HE, Cone EJ, Gorodetzky CW (1984): Morphine-like discriminative stimulus effects of buprenorphine and demethoxybuprenorphine in rats: Quantitative antagonism by naloxone. J Pharmacol Exp Ther 229:768–774.

Shearman GT, Herz A (1982): Discriminative stimulus properties of narcotic and non-narcotic drugs in rats trained to discriminate opiate κ-receptor agonists. Psychopharmacology 78:63–66.

Smith TW, Buchan P, Parsons DN, Wilkinson S (1982): Peripheral antinociceptive effects of N-methyl morphine. Life Sci 31:1205–1208.

Spealman RD, Bergman J (1991): Modulation of the discriminative-stimulus effects of cocaine by mu and kappa opioids. Soc Neurosci Abstr 17:1425.

Strain EC, Bigelow GE, Preston KL, Liebson I (1991): The effects of buprenorphine in methadone-maintained volunteers. NIDA Research Monogr 105:584.

Stretch R (1977): Discrete-trial control of cocaine self-injection behavior in squirrel monkeys: Effects of morphine, naloxone, and chlorpromazine. Can J Physiol Pharmacol 55:778–790.

Teoh SK, Sintavanarong P, Kuehnle J, Mendelson JH, Hallgring E, Rhoades E, Mello NK (1992): Buprenorphine's effects on morphine and cocaine challenges in heroin and cocaine dependent men. NIDA Research Monographs 119:460.

van der Kooy D (1987): Place conditioning: A simple and effective method for assessing the motivational properties of drugs. In Bozarth MA (ed): "Methods of Assessing the Reinforcing Properties of Abused Drugs." New York: Springer–Verlag, pp 229–240.

Versage EM, Goushaw PJ, Young AM (1990): Discriminative stimulus profile of buprenorphine in morphine-dependent pigeons. Pharmacol Biochem Behav 36:436.

Villiger JW, Taylor KM (1981): Buprenorphine: Characteristics of binding sites in the rat central nervous system. Life Sci 29:2699–2708.

Walker EA, Young AM (1990): Discriminative-stimulus profile of the low-efficacy opioid agonist nalbuphine. Psychopharmacology 101:S76.

Wheeler-Aceto H, Cowan A (1991): Buprenorphine and morphine cause antinociception by different transduction mechanisms. Eur J Pharmacol 195:411–413.

Winger G, Skjoldager P, Woods JH (1992): Effects of buprenorphine and other opioid agonists and antagonists on alfentanil- and cocaine-reinforced responding in rhesus monkeys. J Pharmacol Exp Ther 261:311–317.

Wise RA (1989): Opiate reward: Sites and substrates. Neurosci Biobehav Rev 13:129–133.

Wood PL, Iyengar S (1988): Central actions of opiates and opioid peptides: In vivo evidence for opioid receptor multiplicity. In Pasternak GW (ed): "The Opiate Receptors." Clifton, NJ: Humana, pp 307–356.

Wood PL, Charleson SE, Lane E, Hudgin RL (1981): Multiple opiate receptors: Differential binding of μ, κ and δ agonists. Neuropharmacology 20:1215–1220.

Woods JH (1977): Narcotic-reinforced responding: A rapid screening procedure. In "Proceedings of the 39th Meeting of the Committee on Problems of Drug Dependence." Cambridge, MA: National Institute on Drug Abuse, pp 420–437.

Woods JH (1983): Some thoughts on the relations between animal and human drug-taking. Prog Neuro-Psychopharmacol Biol Psychiatry 7:577–584.

Woods JH, Gmerek DE (1985): Substitution and primary dependence studies in animals. Drug Alcohol Depend 14:233–247.

Woods JH, Winger G (1987): Opioids, receptors and abuse liability. In Meltzer HY (ed): "Psychopharmacology: The Third Generation of Progress." New York: Raven, pp 1555–1564.

Woods JH, Smith CB, Medzihradsky, Swain HH (1979): Preclinical testing of new analgesic drugs. In Beers RF, Bassett EG (eds): "Mechanisms of Pain and Analgesic Compounds." New York: Raven, pp 429–445.

Woods JH, France CP, Winger GD (1992): Behavioral pharmacology of buprenorphine: Issues relevant to its potential in treating drug abuse. NIDA Res Monogr 121:12–27.

Wüster M, Herz A (1978): Opiate agonist action of antidiarrheal agents in vitro and in vivo— Findings in support of selective action. Naunyn Schmiedebergs Arch Pharmacol 301:187–194.

Yanagita T (1980): Self-administration studies on psychological dependence. Trends Pharmacol Sci 1:161–164.

Yanagita T, Miyasato K, Sato J (1980): Dependence potential of loperamide studied in rhesus monkeys. NIDA Res Monogr 27:106–113.

Yanagita T, Katoh S, Wakasa Y, Oinuma N (1982): Dependence potential of buprenorphine studied in rhesus monkeys. NIDA Res Monogr 41: 208–214.

Young AM, Stephens KR, Hein DW, Woods JH (1984): Reinforcing and discriminative stimulus properties of mixed agonist–antagonist opioids. J Pharmacol Exp Ther 229:118–126.

Young AM, Kapitsopoulos G, Makhay MM (1991): Tolerance to morphine-like stimulus effects of mu opioid agonists. J Pharmacol Exp Ther 257:795–805.

ASSAY, METABOLISM, AND PHARMACOKINETICS

ANALYSIS

R. ANDREW MOORE
Euro/DPC Ltd., Glyn Rhonwy, Llanberis, Caernarfon, Wales, UK

INTRODUCTION

Analysis of buprenorphine is difficult. This is not because of any inherent concerns with stability or analytical problems with the molecule per se, but rather the circumstances in which analyses of interest are sought.

For a better understanding of the pharmacology, kinetics, and dynamics of buprenorphine, what is sought is the concentration of buprenorphine in biological fluids; most importantly this means in plasma, but also in urine and feces. Buprenorphine abuse has also been reported by a growing number of drug treatment clinics [Hand et al., 1989; San et al., 1989; San, 1990].

Biological samples are complex mixtures of substances from which to recover drugs, especially those like buprenorphine that are lipophilic and tend to bind nonspecifically to a wide range of macromolecules.

To compound the problem further, buprenorphine is potent, being effective at doses of a few hundred micrograms. This initially small amount is then, in effect, diluted in a very large volume, so that concentrations at which it is to be measured are of the order of 1 ng/ml or less.

Finally, the drug is extensively metabolized. The free phenolic hydroxyl group is available for glucuronidation, while N-dealkylation can give rise to conjugated and unconjugated N-dealkylbuprenorphine (Fig. 1).

The challenge for the analyst is to overcome these difficulties to produce accurate values for concentrations of the drug (and metabolites). The aim of this chapter is to review the methods available.

GAS CHROMATOGRAPHY–MASS SPECTROMETRY

Gas chromatography–mass spectrometry (GC-MS) involves the derivatization of a compound into a form volatile at elevated temperatures, separation from other

Buprenorphine: Combatting Drug Abuse With a Unique Opioid, pages 105–112
© 1995 Wiley-Liss, Inc.

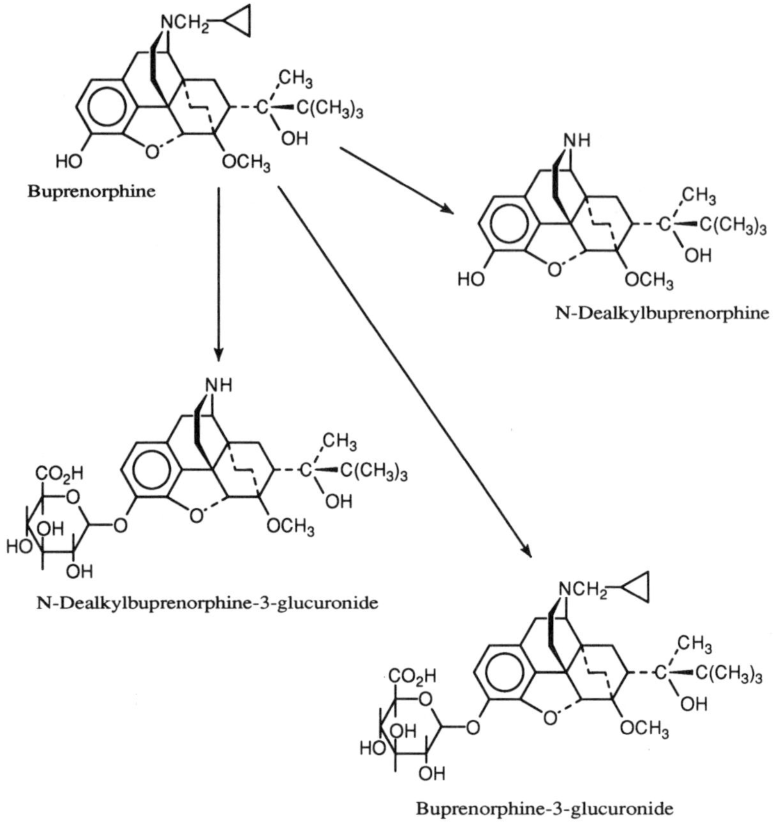

Fig. 1. Metabolism of buprenorphine in humans.

molecules by gas chromatography, and a final determination in a mass spectrometer. The method can be extremely sensitive and very specific, though it is slow and complicated.

The method described by Blom et al. [1985] is probably the best. After a preliminary extraction of buprenorphine into organic solvents, heating the drug in dilute acid at 110°C for 1 hr produced a molecular rearrangement involving loss of methanol followed by ring formation between the side-chain and methoxy group (Fig. 2) [Cone et al., 1984a]. Further organic extraction and treatment with pentafluoropropionic anhydride produced a pentafluoryl derivative suitable for analysis. This complicated procedure was deemed desirable to reduce problems with adsorption of buprenorphine onto the chromatography column, with subsequent low sensitivity.

The extraction efficiency (determined by use of [^{3}H]-buprenorphine) was only about 30%. However, it was shown that both the internal standard, N-propylnorbuprenorphine, and N-dealkylbuprenorphine showed the same acid-catalyzed

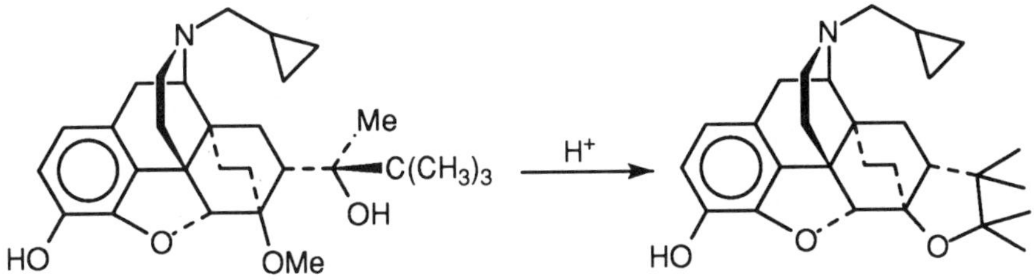

Fig. 2. Acid-catalyzed rearrangement of buprenorphine.

rearrangement. The method could therefore be used with confidence to measure both buprenorphine and the N-dealkyl metabolite.

The analysis was conducted on a Finnegan 4000 GC-MS with an OV-1701 column operated at 280° C and in electron-impact mode using single-ion monitoring. Ions monitored were m/z 581 for buprenorphine, m/z 658 for N-dealkylbuprenorphine, and m/z 569 for N-propyl-norbuprenorphine.

The precision of the method was good, with coefficients of variation between 5 and 10% in plasma and urine, though the precision was only monitored at 5 ng/ml and 50 ng/ml—rather high concentrations for most clinical purposes. The lower-limit of sensitivity was claimed to be 0.15 ng/ml.

The authors describe the concentration time curve for 6 hr in plasma for one healthy volunteer given 0.6 mg buprenorphine intravenously; plasma buprenorphine levels were reported to be about 1 ng/ml at 100 min after injection and about 0.2 ng/ml about 350 min after injection. Cumulative urinary excretions of conjugated and unconjugated buprenorphine over 4 days were measured, most conjugated and unconjugated buprenorphine being excreted during Day 1, and most conjugated and unconjugated N-dealkylbuprenorphine over Days 2–4.

GC-MS determination of N-dealkylbuprenorphine as a metabolite in urine and feces following subcutaneous, sublingual, and oral dosing was performed by Cone et al. [1984b], though in this case GC-MS was used for identification rather than quantitation. Another GC-MS method used for quantitation of buprenorphine in baboons had a detection limit of 20 ng/ml, and was useful only in animals dosed with 5 mg/kg buprenorphine [Lloyd-Jones et al., 1980].

GAS CHROMATOGRAPHY

Gas chromatography is not a method easily applied to analysis of buprenorphine. The method depends upon the partition of a molecule between gaseous and liquid phases, and the high molecular weight of buprenorphine makes it difficult to maintain in the gaseous phase, even at elevated temperatures. Derivatization can help, but losses of molecules in the chromatographic columns are not unusual and have been referred to previously. Extra sensitivity can be obtained by the use of selective detectors for nitrogen atoms (electron capture detectors).

Cone et al. [1984b, 1985] have described an electron-capture gas chromatography method for buprenorphine and N-dealkylbuprenorphine in urine and feces. It involves complicated organic solvent extractions with inevitable losses; use of an internal standard (they used etorphine) is therefore essential. The biggest problem is that of sensitivity; the procedure could not measure less than 10 ng/ml of buprenorphine and 5 ng/ml of N-dealkylbuprenorphine.

Such a degree of sensitivity was sufficient to measure the cumulative excretion of conjugated buprenorphine and N-dealkylbuprenorphine in urine after a 40-mg oral dose, but the method is not suitable for pharmacokinetic purposes.

LIQUID CHROMATOGRAPHY

Liquid chromatography overcomes several problems encountered with gas chromatography; derivatization is not necessary because volatility is not an issue. Simple extraction into organic solvents is needed, but problems with on-column losses of drugs are rarely encountered. However, detectors used in liquid chromatography generally do not have the sensitivity to be useful for the analysis of buprenorphine.

Methods for analysis of buprenorphine in plasma have been described. Tebbett [1985] used an ethereal alkaline extraction of plasma and chromatographed the concentrated extract on a reverse-phase column. The sensitivity was unstated, but probably not much better than 20 ng/ml. This is similar to the sensitivity found for buprenorphine in urine [Hackett et al., 1986].

A similar method was used by Noda and Kojima [1985] to measure buprenorphine in plasma. Their article reports a sensitivity of 0.2 ng/ml with a carbon electrode. While this is theoretically possible under the best conditions, it is about 10 times more sensitive than experienced chromatographers would expect to obtain routinely. In addition, plasma concentrations reported in five patients given an intravenous dose of 6 µg/kg buprenorphine are some 5–10 times higher than generally accepted; ascribing this to the vagaries of general anesthesia is not likely.

IMMUNOASSAY

Immunoassays can be extremely sensitive and usually have the advantage that no extraction of the sample is necessary; but the important issue is always that of cross-reaction with related compounds. For drugs, it is important that the extent of cross-reaction with metabolites be known, as well as their concentration relative to the parent drug. If the metabolite is present in very high relative concentration, low cross-reactivity is essential for sensible analysis; moderate or even high cross-reaction may be tolerated in situations in which the metabolite is present in low relative concentrations.

An antiserum can be made by haptenizing a drug to a large protein (usually bovine serum albumin or keyhole limpet hemocyanin) and injecting the conjugate into a rabbit. For buprenorphine the chemical coupling was achieved by conjugation through either the oxygen at position 3 or the nitrogen atom. As a general rule, specificity of an antiserum is low (and cross-reaction high) to compounds with metabolic changes at or close to the site of conjugation. Antisera made through conjugation at the 3-O position are likely to cross-react with 3-O-glucuronides, and antisera conjugated through the nitrogen with N-dealkylbuprenorphine.

Radioimmunoassay

The first radioimmunoassay (RIA) for buprenorphine was described by Bartlett et al. [1980]. This assay used [^{3}H]buprenorphine as the radioactive label and plasma samples without prior extraction; it was reasonably rapid, being completed within 3 hr for up to 100 samples (though counting times were much longer). It explored cross-reaction of antisera conjugated as described above. Antisera raised by a 3-carboxymethylbuprenorphine antigen had high (100%) cross-reaction with buprenorphine-3-O-glucuronide, while those raised by a N-hemisuccinyl norbuprenorphine had high (80%) cross-reaction with N-dealkylbuprenorphine.

The latter system was used for analysis of human and animal sera, as radiotracer studies had shown little or no N-dealkylbuprenorphine in the serum of dogs given single doses of buprenorphine. The method was sensitive (0.025–0.050 ng/ml) and demonstrated excellent recovery of added buprenorphine and parallelism of diluted samples with high concentrations; precision in the clinical range was good, with coefficients of variation generally below 10%.

Plasma buprenorphine concentrations were measured in a large number of patients given buprenorphine by intravenous or intramuscular injection [Bullingham et al., 1980; Watson et al., 1982] or by sublingual tablets [Bullingham et al., 1982]. After intravenous injection, plasma buprenorphine concentrations at 100 min after injection were measured at about 0.5 ng/ml following a 0.3-mg dose and about 1.0 ng/ml after a 0.6-mg dose; this compared favorably with the single patient reported by Blom et al. [1985] using GC-MS. Comparisons were similar at other time points. After sublingual administration of 0.4 mg and 0.8 mg, peak concentrations at about 200 min were 0.5 ng/ml and 1.0 ng/ml, respectively.

A further refinement of the RIA technique has been reported [Hand et al., 1986; Hand, 1987] that allows the simultaneous measurement of buprenorphine and some of its metabolites and makes use of iodine label with much shorter counting times. This "differential radioimmunoassay" used the differential extraction of buprenorphine (but not the glucuronide) together with 3-CMO-antiserum to measure buprenorphine without interference with buprenorphine glucuronide. Samples measured without extraction gave values for buprenorphine plus the glucuronide, whence the value for the glucuronide was obtained by difference. Similar measurement of unextracted samples with an N-hemisuccinyl-derived antiserum gave values for buprenorphine plus N-dealkylbuprenorphine, whence the value for the N-dealkylbuprenorphine was obtained by difference.

This method indicated undetectable amounts of N-dealkylbuprenorphine in plasma of patients given single doses of buprenorphine [Hand, 1987], thus lending further support to previous RIA stratagems. However, in patients taking daily sublingual doses of buprenorphine, clearly measurable concentrations of buprenorphine, buprenorphine-3-O-glucuronide, and N-dealkylbuprenorphine were shown. This method is not simple, but it could be of value in circumstances of organ dysfunction, or when buprenorphine is given in high doses for long periods, as in treatment for opiate addiction.

A commercial RIA has also been used for measuring buprenorphine and its metabolites in urine for the detection of buprenorphine abuse [Hand et al., 1989]. Cross-reaction to major metabolites was not given, but evidence pointed to major cross-reaction to most likely metabolites. At a cutoff of 1 ng/ml, 62% of 97 patients attending a drug clinic were positive by this method.

Enzyme Immunoassay

Tiong and Olley [1988] have used an alternative strategy. Rather than employ a radioactive tracer, they have coupled N-dealkylbuprenorphine to β-galactosidase and used N-hemisuccinyl norbuprenorphine antiserum to generate a sensitive assay. After the steps of a conventional immunoassay have been carried out, the final signal is generated by releasing 4-methylumbelliferone from a galactoside conjugate and measuring it in a spectrofluorimeter.

The assay has characteristics similar to the RIA described by Bartlett et al. [1980], with a sensitivity of about 0.1 ng/ml. Plasma concentrations after intravenous and sublingual administration of single doses in humans give values similar to the RIA and, where comparable, to the GC-MS procedure.

TABLE I. Comparison of Analytical Methods for Buprenorphine

Method[a]	Sensitivity	Comments
GD-MS	0.15	Difficult to set up, but well characterized. Complicated extraction and derivatization. Plasma curve on one patient and six baboons given high doses.
GC	10.0	Complicated extraction and derivatization. Insensitive. No plasma kinetics.
LC	10.0	Simple extraction, no derivatization. Insensitive. No plasma kinetics.
RIA	0.05	Plasma analyzed directly without extraction. Method well characterized, and capable of high throughput. Apparently specific for single doses and can be used for metabolite measurement. Many kinetic publications on single-dose administration. Also used for screening for abuse.
EIA	0.10	Plasma analyzed directly without extraction. Method well characterized, but used for fewer than 10 patients.

[a]GC, gas chromatography; MS, mass spectrometry; LC, liquid chromatography; RIA, radioimmunoassay; EIA, enzyme immunoassay.

SUMMARY

The salient features of methods of analysis for buprenorphine are summarized in Table I. Whether the purpose of the analysis is furthering the knowledge of the pharmacokinetics of the drug, or detecting it in circumstances of abuse, sensitivity of below 1 ng/ml is required. Only two methods, GC-MS and immunoassays, currently are able to fulfill this requirement. Of these, immunoassays have proved, so far, to be the most easily used in laboratories. There are insufficient data available for thorough comparison of the two techniques in terms of their specificity: Good immunoassays appear to give plasma concentrations very similar to those of GC-MS.

REFERENCES

Bartlett AJ, Lloyd-Jones JG, Rance MJ, Flockhart IR, Dockray GJ, Bennett MRD, Moore RA (1980): The radioimmunoassay of buprenorphine. Eur J Clin Pharmacol 18:339–345.

Blom Y, Bondesson U, Änggård E (1985): Analysis of buprenorphine and its N-dealkylated metabolite in plasma and urine by selected-ion monitoring. J Chromatogr 338:89–98.

Bullingham RES, McQuay HJ, Moore A, Bennett MRD (1980): Buprenorphine kinetics. Clin Pharm Ther 28:667–672.

Bullingham RES, McQuay HJ, Porter EJB, Allen MC, Moore RA (1982): Sublingual buprenorphine used postoperatively: Ten hour plasma drug concentration analysis. Br J Clin Pharmacol 13:665–673.

Cone EJ, Gorodetzky CW, Darwin WD, Buchwald WF (1984a): Stability of the 6,14-endo-ethanotetrahydrooripavine analgesics: Acid-catalyzed rearrangement of buprenorphine. J Pharm Sci 73:243–246.

Cone EJ, Gorodetzky CW, Yousefnejad D, Buchwald WF, Johnson RE (1984b): The metabolism and excretion of buprenorphine in humans. Drug Metab Disp 12:577–581.

Cone EJ, Goredetzky CW, Yousefnejad D, Darwin WD (1985): ^{63}Ni electron-capture gas chromatographic assay for buprenorphine and metabolites in human urine and feces. J Chromatogr 337:291–300.

Hackett LP, Dusci LJ, Ilett KF, Seow SSW, Quigley AJ (1986): Sensitive screening method for buprenorphine in urine. J Chromatogr 374:400–404.

Hand CW (1987): Studies on the Human Pharmacology of the Opiates. Doctoral dissertation, University of Oxford.

Hand CW, Baldwin D, Moore RA, Allen MC, McQuay HJ (1986): Radioimmunoasay of buprenorphine with iodine label: Analysis of buprenorphine and metabolites in human plasma. Ann Clin Biochem 23:47–53.

Hand CW, Ryan KE, Dutt SK, Moore RA, O'Connor J, Talbot D, McQuay HJ (1989): Radioimmunoassay of buprenorphine in urine: Studies in patients and in a drug clinic. J Anal Toxicol 13:100–104.

Lloyd-Jones JG, Robinson P, Henson R, Biggs SR, Taylor T (1980): Plasma concentrations and disposition of buprenorphine after intravenous and intramuscular doses to baboons. Eur J Drug Metab Pharmacokinet 5:233–239.

Noda J, Kojima T (1985): Quantitative determination of buprenorphine in human plasma by high-performance liquid chromatography. Hiroshima J Med Sci 34:261–265.

Tebbett IR (1985): Analysis of buprenorphine by high-performance liquid chromatography. J Chromatogr 347:411–413.

San L, Tremoleda J, Olle JM, Porta Serra M, de la Torre R (1989): Prevalencia del consumo de buprenorphina en heroinomanos en tratemento. Med Clin (Barcelona) 93:645–648.

San L (1990): Interpretation of results of urine drug testing in clinical practice. In Segura J, de la Torre R (eds): "First International Symposium on Current Issues of Drug Abuse Testing." Boca Raton, FL: CRC Press.

Tiong GKL, Olley JE (1988): Enzyme immunoassay of buprenorphine. Naunyn-Schmiedebergs Arch Pharmacol 338:202–206.

Watson PJQ, McQuay HJ, Bullingham RES, Allen MC, Moore RA (1982): Single-dose comparison of buprenorphine 0.3 and 0.6 mg IV given after operation: Clinical effects and plasma concentrations. Br J Anaesth 54:37–43.

ABSORPTION, DISTRIBUTION, METABOLISM, AND EXCRETION OF BUPRENORPHINE IN ANIMALS AND HUMANS

DONALD S. WALTER
Reckitt & Colman Products, Hull HU8 7DS, UK

CHARLES E. INTURRISI
Department of Pharmacology, Cornell University Medical College, New York, NY 10021

INTRODUCTION

This chapter reviews ADME (absorption, distribution, metabolism, and excretion) studies carried out with buprenorphine in animals and humans. The drug was developed in the early 1970s by the pharmaceutical research and development departments of Reckitt & Colman Products, Ltd, UK, leading to its registration in the UK as an analgesic for moderate to severe pain in 1977 (Temgesic Injection® and Temgesic Sublingual® tablets). Since then the products have been registered in over 40 countries. The main findings of ADME studies were summarized in an early review by Heel et al. [1979].

Drug metabolism studies were made difficult because of the high potency of buprenorphine such that at normal therapeutic doses chromatographic techniques were pushed to the limits of sensitivity for measuring plasma and tissue levels of the drug. Much work was carried out to provide a specific and sensitive chemical assay for the drug, with progressive use of gas chromatography (GC), high-performance liquid chromatography (HPLC), and GC/mass spectroscopy. Good, reproducible linear assays have been produced by all three methods, but sensitivity has always been a problem with GC and HPLC and sample throughput is very slow with GC/MS. Parallel development of a radioimmunoassay for buprenorphine gave two antibodies that bind with buprenorphine. Unfortunately, one of the antibodies cross-

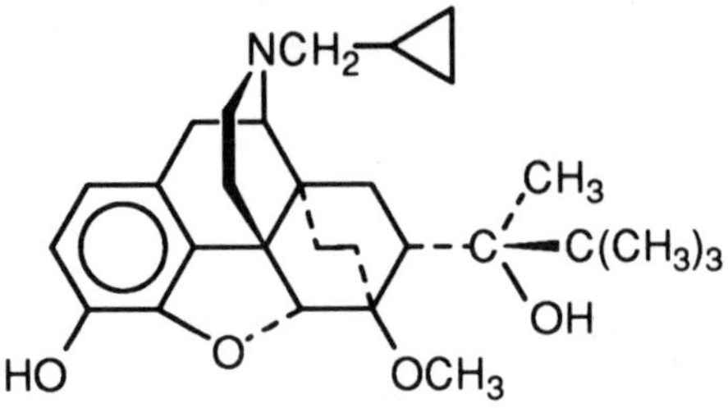

Buprenorphine

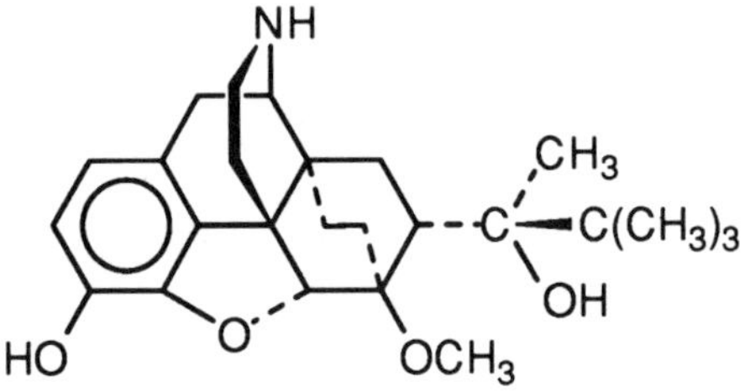

N-Dealkylbuprenorphine

Fig. 1. Structures of buprenorphine and N-dealkyl buprenorphine.

reacted with a major metabolite, N-dealkyl buprenorphine (Fig. 1) and the other antibody cross-reacted with the other major metabolite, buprenorphine glucoronide conjugate. However, radioimmunoassay with the first antibody has been used extensively in clinical research and human bioavailability studies because this was the only feasible assay for providing information about the absorption and pharmacokinetics of the drug in humans. Its use in single-dose studies is justified because the contribution of the N-dealkyl metabolite to total immunoreactivity after a single dose is very low; this is discussed later.

Drug metabolism studies are facilitated if a high-specific-activity, stable radiolabeled form of the drug is available. Various options for radiolabeling buprenorphine were considered; the most satisfactory option was a ^{14}C label but the nonavailability of high-specific-activity carbon-labeled precursors and the poor yield in synthetic steps precluded the production of carbon-labeled buprenorphine. Labeling with ^{125}I has been successfully used for the radioimmunoassay [Hand et al., 1986] but this molecule was considered to be too dissimilar to the parent drug for ADME studies.

A method for tritium labeling of buprenorphine to high specific activity was developed using tritium exchange at the 15,16-positions of buprenorphine [Rance et al., 1976]. The suitability of tritium-labeled buprenorphine was assessed by examining lability of the label in rats over a period of 48 hr after a single intramuscular dose of drug [Brewster et al., 1981a]. Recovery of radioactivity in urine, feces, carcass, and expired air was determined. The lability in each sample was quantified by freeze-drying and distillation to constant specific activity. The total level of lability in the rats ranged from 0.3% to 5.9% of administered dose and the majority of this labile material remained in the carcass after 48 hr, suggesting that it was ^3H_2O. It was concluded from these experiments that the low lability would not be expected to greatly affect the general metabolic picture but could make a significant contribution to total radioactivity in plasma and other tissues when the levels of drug-related material are low, especially at later times after dosing. Therefore, plasma and tissue samples were freeze-dried routinely prior to analysis by combustion.

A comprehensive profile of the metabolism of buprenorphine has been obtained using the tritium-labeled drug in conjunction with chromatographic and radioimmunoassay techniques. This chapter reviews the findings of studies looking first at the results obtained in animals and then examining their similarity to results obtained in humans.

ADME STUDIES IN ANIMALS

Absorption

Most of the studies reported here were carried out with [^3H]buprenorphine as part of the drug development program. Another group [Pontani et al., 1985] has also reported ADME studies with the same radiolabel in the rat.

The absorption of buprenorphine has been studied in rat, dog, rhesus monkey [Brewster et al., 1981a; Numata et al., 1981], rabbit, cynomolgus monkey, and baboon [Lloyd-Jones et al., 1980]. Following intramuscular administration of [^3H]buprenorphine, blood levels of radioactivity peaked at 10–15 min after dosing in all species (Table I), whereas the absorption peak was delayed following oral (except the rat), sublingual, and buccal administration of the drug. In general, peak blood levels of buprenorphine were higher after intramuscular doses than after larger oral doses owing to extensive first-pass metabolism.

In the rat, *in vivo* studies using *in situ* isolated intestinal loops and portal vein cannulation [Castle et al., 1985] showed that buprenorphine administered into the loop was extensively metabolized to a conjugate by rat intestine, and all the absorbed drug material following a 10-µg bolus, and 90% following a 100-µg bolus, appeared as a glucuronide conjugate [Rance and Shillingford, 1977]. The extensive first-pass metabolism was accompanied by marked enterohepatic cycling of buprenorphine following biliary excretion of conjugated buprenorphine and its probable hydrolysis in the lower gut [Brewster et al., 1981a]. Another study in the rat using the same techniques [Brewster et al., 1981b] presented the absorption profiles (Fig. 2) and the bioavailability (Table II) of buprenorphine following intravenous, intrarectal, intrahepatoportal, sublingual, and intraduodenal administration of 200 µg/kg of [^3H]buprenorphine in comparison with the results of intraarterial

TABLE I. Peak Plasma Concentrations of Buprenorphine After Administration of [³H]Buprenorphine to Various Species

Species	Route	Dose (μg/kg)	T_{max} (min)	C_{max} (ng/g)
Rat	IM	20	10–15	4
	IM	5,000	15	370
	PO	100	10	5[a]
	PO	20,000	10	66
	SL	20	30	0.4
	SL	200	60	1.1
Rabbit	IM	5,000	15	132
Dog	IM	20	~15	3[a]
	PO	100	60–120	2–3
	PO	800	30–60	10–14
Baboon	IM	2	~15	0.8
	IM	5,000	30	805
	PO	40	120	0.4
	IV	5,000	4	2,290
Rhesus monkey	IM	2	~15	0.7
	PO	15	60	0.8–3[a]
Cynomolgus monkey	IM	38	15	20[a]
	SL	38	120	4[a]
	BU	38	120	4[a]

[a]Value based on total radioactivity.

administration. The results show that if the contribution of the intestine to the metabolism of buprenorphine is bypassed by intrahepatoportal administration, there is a marked increase in the bioavailability of the drug. Sublingual administration, in the anaesthetized rat, gave a slower absorption profile than other routes, and the bioavailability shown in Table II is an underestimate of the true value. Results presented later show that the sublingual route is a satisfactory noninvasive route of administration of buprenorphine in humans when the bioavailability is of the order of 55% at normal analgesic doses.

The oral bioavailability of buprenorphine was also low in the dog (mean 7.4%; range 1.2%–19.7%) after an oral dose of 1 mg/kg. The low oral bioavailability in animals contrasts with the good bioavailability following intramuscular and subcutaneous injection of the drug. In the baboon the mean intramuscular bioavailability of buprenorphine relative to intravenous administration was 70% [Lloyd-Jones et al., 1980]. A similar level of intramuscular bioavailability was also obtained in humans (see below).

For a centrally acting drug like buprenorphine, systemic availability is a necessary prelude to availability to brain tissue. The latter has been studied in rat and monkey by measuring brain levels of buprenorphine following intramuscular, oral,

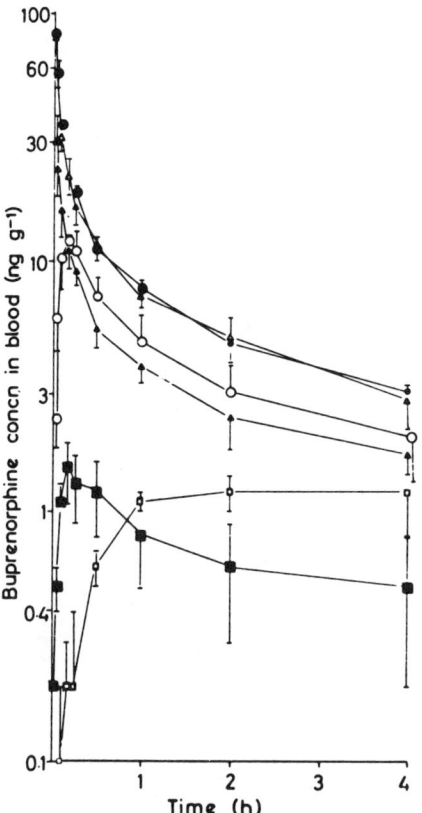

Fig. 2. Blood concentrations of buprenorphine in female rats following administration (200 μg/kg) by various routes. (●) Intraarterial; (△) intravenous; (○) rectal; (▲) intrahepatoportal; (□) sublingual; (■) intraduodenal. Points represent mean values ± SEM of four animals. Reproduced from Brewster et al. [1981b], with permission of the publisher.

and sublingual administration. Higher buprenorphine levels were found in brain following administration by the intramuscular and sublingual routes than by the oral route (Table III). Pontani et al. [1985] carried out a detailed study of the disposition of buprenorphine in the rat and also showed that buprenorphine was readily available to brain tissue following an intravenous dose of 0.2 mg/kg. After 15 min the mean brain level of buprenorphine was 117 ± 12 ng/g compared with a level of 46 ± 15 ng/ml in plasma (Fig. 3). The high brain-to-plasma ratio (Table IV) shows that in the rat buprenorphine readily crosses the blood–brain barrier to exert a central action. The distribution within the brain is predominantly to the cerebrum [Manara et al., 1978].

Pontani et al. [1985] also described a high-affinity tightly bound buprenorphine component in rat brain that had a decay half-life of around 69 hr, an observation that

TABLE II. Relative Bioavailabilities of Buprenorphine in the Rat for Various Routes Over the Period 0–4 hr After Dosing[a,b]

Route	Area under blood concentration time curve (AUC_{0-4hr}, ng/ml/min)	Relative systemic availability (%) over the period 0–4 hr[c]
Intraarterial	1,852 ± 189	100
Intravenous	1,807 ± 242	98 ± 13
Intrarectal	1,000 ± 267	54 ± 14
Intrahepatoportal	900 ± 161	49 ± 9
Sublingual[d]	249 ± 39	13 ± 2
Intraduodenal	180 ± 71	9.7 ± 4

[a]Reproduced from Brewster et al. [1981b], with permission of the publisher.
[b]Values are the mean for four animals ± SEM.
[c]Intraarterial route assigned to represent complete availability.
[d]The slow absorption profile for this route results in a considerable underestimate of the sublingual availability.

they believe is consistent with the known high-affinity, slow-dissociation binding to opiate receptors.

Distribution

Tissue distribution studies have been carried out in the rat. Pontani et al. [1985] studied the distribution of a 200-μg/kg intravenous dose of [^{3}H]buprenorphine (Table IV) and found a distribution similar to that in another study, carried out as part of the development program, following a 20-μg/kg intramuscular dose (Table V). At early times after dosing, high tissue levels were found in lung, heart, kidney, and liver. Brain levels were higher than plasma levels in both studies, and separate studies showed that most of the brain radioactivity was unchanged drug. A study of

TABLE III. Peak Brain Concentrations of Buprenorphine After Administration of [^{3}H]Buprenorphine to Rat and Rhesus Monkey

Species	Route	Dose (μg/kg)	T_{max} (min)	C_{max} (ng/g)
Rat	IM	20	40	6[a]
	IM	20	60	14
	PO	100	80	2[a]
	PO	20,000	120	249
	SL	20	120	5
Rhesus monkey	IM	2	60	4[a]
	PO	15	60	0.3[a]

[a]Value based on total radioactivity.

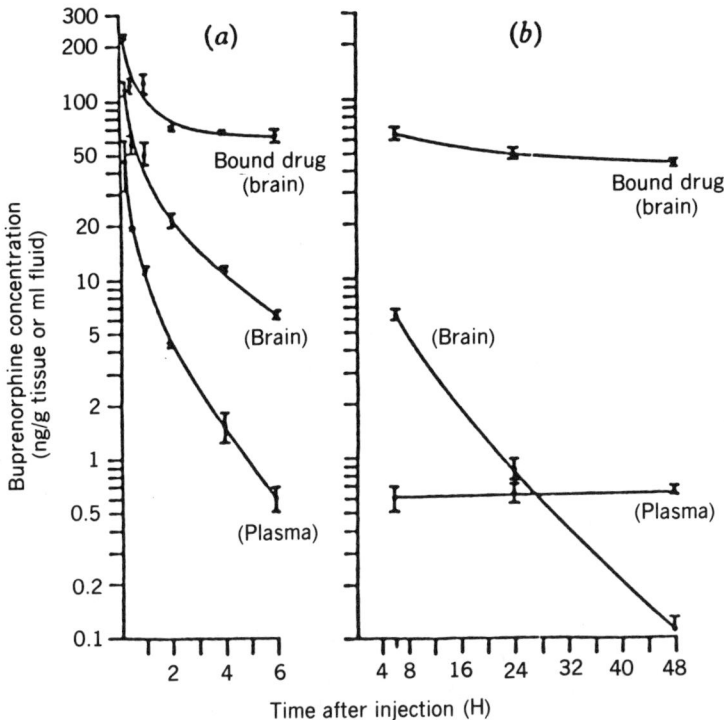

Fig. 3. Decay of extractable buprenorphine from brain and plasma and of bound drug in brain of rats injected with a single intravenous bolus dose of [^{3}H] buprenorphine (0.2 mg/kg). Data represent means ± SEM from three animals at each time up to 6 hr (*a*), and for times later than 6 hr (*b*). Reproduced from Pontani et al. [1985], with permission of the publisher.

the distribution of [^{3}H]buprenorphine in rats following a 20-mg/kg oral dose showed that maximum tissue radioactivity levels generally occurred within 1 hr of dosage. The highest levels were found in the excretory organs, liver and kidney. Again, the radioactivity in brain consisted almost entirely of unchanged drug, showing that such drug as survived first-pass metabolism was able to cross the blood–brain barrier. Chromatography studies showed that after intramuscular dosing of 20 μg/kg only 10%–36% of the radioactivity in liver was unchanged buprenorphine, whereas in other tissues, up to 30 min after dosing, most of the radiolabel (70%–97%) was the unchanged drug.

The distribution of buprenorphine was also compared in pregnant and nonpregnant rats, as were the levels of drug and metabolites measured in 11- and 21-day fetal tissues following a 5-mg/kg intramuscular dose of [^{3}H] buprenorphine. Levels of radioactivity in plasma and selected tissues were similar in pregnant and nonpregnant female rats, and the proportions of free drug in plasma were also similar, indicating that they had a similar metabolic capacity. Radioactivity was present in

TABLE IV. Distribution of Buprenorphine in Tissues of Male Wistar Rats Injected Intravenously With a Single-Bolus Dose of [15,16(n)-³H]Buprenorphine (0.2 mg/kg)[a,b]

Time after injection (hr)	Buprenorphine distribution (ng/g tissue or ml fluid)							
	0.25	0.5	1.0	2.0	4.0	6.0	24	48
Plasma	46 ± 15	19 ± 1	11 ± 0.4	4.3 ± 0.2	1.5 ± 0.3	0.6 ± 0.1	0.6 ± 0.1	0.6 ± 0.03
Brain	117 ± 12	57 ± 8	51 ± 8	21 ± 2	11 ± 0.3	6.3 ± 0.4	0.9 ± 0.1	0.1 ± 0.02
Liver	80 ± 7	41 ± 10	25 ± 5	12 ± 3	7.7 ± 0.8	6 ± 1	3.2 ± 0.2	1.8 ± 0.6
Heart	126 ± 11	63 ± 6	38 ± 4	13 ± 1	4.5 ± 0.3	1.8 ± 0.2	1.1 ± 0.1	1.1 ± 0.4
Lung	225 ± 24	123 ± 10	85 ± 10	27 ± 0.2	11 ± 1	6 ± 1	6.5 ± 0.4	3.5 ± 0.3
Kidney	182 ± 16	98 ± 9	66 ± 6	25 ± 2	9 ± 1	4 ± 1	3.3 ± 0.3	2.5 ± 0.6
Spleen	124 ± 8	66 ± 8	42 ± 7	17 ± 2	5.2 ± 0.2	3 ± 0.7	6.3 ± 0.2	4.0 ± 0.3
Testes	61 ± 4	60 ± 3	45 ± 0.4	20 ± 3	5.8 ± 0.4	2.5 ± 0.8	1.0 ± 0.1	0.4 ± 0.1
Skeletal muscle	77 ± 10	41 ± 8	31 ± 4	17 ± 5	5 ± 3	0.8 ± 0.3	1.3 ± 0.1	1.2 ± 0.3
Fat	239 ± 25	241 ± 8	262 ± 43	206 ± 22	89 ± 21	39 ± 15	19.3 ± 0.3	13.7 ± 0.9

[a]Reproduced from Pontani et al. [1985], with permission of the publisher.
[b]Data represent mean ± SEM from three animals at each time.

TABLE V. Distribution of Radioactivity in Tissues of Male Sprague-Dawley Rats Injected Intramuscularly With a Single Bolus Dose of [³H]Buprenorphine (0.02 mg/kg)[a]

Time after injection (hr)	Distribution of radioactivity (ng/g equivalent)						
	0.25	0.5	1.0	2.0	4.0	6.0	24
Plasma	5.5	3.0	2.1	1.3	0.4	0.6	0.1
Brain	9.4	11.9	8.7	6.6	5.3	3.5	0.4
Liver	25.8	24.5	17.8	13.7	8.2	9.8	3.4
Heart	21.1	12.7	6.4	4.3	1.6	1.6	0.3
Lung	34.6	21.4	11.0	7.8	3.4	3.5	0.8
Kidney	29.3	19.4	10.6	7.7	3.3	4.2	2.3
Spleen	5.7	12.3	7.1	4.1	2.9	3.3	0.9
Testes	2.7	3.8	4.0	3.1	0.8	0.7	0.1
Muscle (diaphragm)	15.3	10.4	5.7	3.9	1.3	1.2	0.2
Fat	12.3	24.5	28.9	29.5	33.0	15.7	1.6

[a]Data represent mean from two male animals at each time. The distribution in female rats carried out at the same time was similar.

the fetus at both 11 and 21 days, showing that drug-related material passed the placental barrier. In the 21-day rat fetus most of the radioactivity in the fetal gastrointestinal tract was as polar conjugates. Other studies showed that buprenorphine injected subcutaneously into female rats at the same high dose of 5 mg/kg/day for 14 days prior to mating, during mating, and throughout the gestation period caused no effect on fertility or gestation indices [Sutton et al., 1986].

Metabolism

Metabolite identification was carried out on the excretion products from rat, rabbit, dog, baboon, and rhesus monkey and the metabolic pattern was found to be similar in all species studied. Two pathways—conjugation with glucuronic acid and N-dealkylation—are well documented, leading to at least three metabolites: buprenorphine conjugate, N-dealkyl buprenorphine, and N-dealkyl buprenorphine conjugate (Fig. 4). In the rat and dog other nonhydrolyzable polar metabolites have been

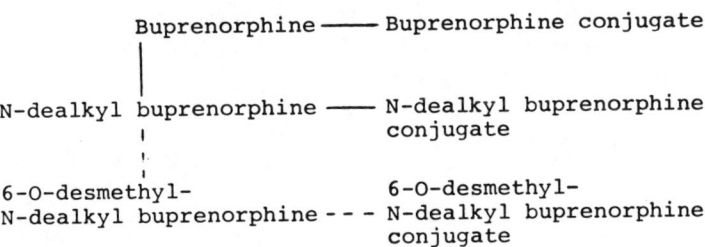

Fig. 4. Known and possible metabolic pathways for buprenorphine. ———, Known animal and human metabolic pathways; -----, metabolic pathway in Wistar rats.

TABLE VI. Proportion of Radioactivity Attributable to Buprenorphine and Its Metabolites in Rat Feces Following 4.5 mg/kg i.v., 4.5 mg/kg i.m., and 80 mg/kg p.o. of [³H]Buprenorphine

	Percentage of radioactivity		
	i.v.	p.o.	i.m.
Buprenorphine	34	61	36
N-Dealkyl buprenorphine	63	23	43
Nonhydrolyzable polar metabolites	4	15	20

observed by thin-layer chromatography (Tables VI and VII). One of these may be 6-O-desmethyl norbuprenorphine, a polar metabolite tentatively identified by Pontani et al. [1985] in Wistar rat urine; a conjugate of this metabolite was also observed (Table VIII).

The proportions of buprenorphine and its metabolites in rat and dog feces after intravenous, oral, and intramuscular administration are shown in Tables VI and VII. In rat, more N-dealkylation occurred following intravenous and intramuscular dosing and, conversely, more unchanged drug appeared in feces after oral dosing. There was also a higher proportion of the unknown polar metabolites in rat feces following oral and intramuscular dosing compared with intravenous dosing. In contrast, the metabolite patterns in dog feces after intravenous, oral, and intramuscular dosing were very similar (Table VII).

Rat bile samples contained over 94% of polar metabolites (Table IX). After enzyme hydrolysis around 70% of the radioactivity was hydrolyzed to buprenorphine and N-dealkyl buprenorphine, but around 20% was made up of nonhydrolyzable metabolites. There was a sex difference in the amount of N-dealkylation that was apparently greater in male rats [Brewster et al., 1981a].

Effects on Drug-Metabolizing Enzymes

Male and female rats were dosed intramuscularly with buprenorphine (0.1 or 4 mg/kg) or morphine (3 mg/kg) twice daily for 4 days to study the effects on hepatic

TABLE VII. Proportion of Radioactivity Attributable to Buprenorphine and Its Metabolites in Dog Feces Following 1.5 mg/kg i.v., 1.5 mg/kg i.m., and 15 mg/kg p.o. of [³H]Buprenorphine

	Percentage of radioactivity		
	i.v.	p.o.	i.m.
Buprenorphine	81	80	93
N-Dealkyl buprenorphine	4	6	4
Nonhydrolyzable polar metabolites	14	11	3

TABLE VIII. Proportions of Total Radioactivity as Metabolites in Unhydrolyzed Rat Urine (1-Week Sample) Following Dosing With 0.2 mg/kg i.v. of [^{3}H]Buprenorphine[a]

Compound	%
Buprenorphine	1.9
Buprenorphine conjugate	0.5–0.9
N-Dealkyl buprenorphine	9.4
N-Dealkyl buprenorphine conjugate	5.2
6-O-Desmethyl N-dealkyl buprenorphine	5.4
6-O-Desmethyl N-dealkyl buprenorphine conjugate	15.9

[a]Reproduced from Pontani et al. [1985], with permission of the publisher.

microsomal enzyme activity. There was a difference between the sexes; male rats showed an increase in microsomal enzyme activity following buprenorphine and morphine, whereas the activity in female rats was unaffected.

Excretion

As previously discussed, the elimination of buprenorphine-related material is mainly via the feces following biliary excretion of conjugated unchanged drug and conjugated phase I metabolite(s). A small amount of drug-related material is excreted in the urine. Balance study data from rat, dog, and rhesus monkey following intramuscular administration are presented in Table X. Similar results have been obtained following intravenous and oral administration in rat and dog and following sublingual administration to rats: In the latter a mean 79.6% of a 0.1-mg/kg sublingual dose of [^{3}H]buprenorphine was excreted in feces and a mean of 3.6% of dose in urine (three rats) over a 48-hr collection period.

TABLE IX. Proportion of Bile Radioactivity (0- to 24-hr Sample) in Male and Female Rats Following 100 μg/kg i.v. of [^{3}H]Buprenorphine[a,b]

	Male	Female
Before enzyme hydrolysis		
Buprenorphine	1.5 ± 0.8	0.8 ± 0.4
N-Dealkyl buprenorphine	0	0
Conjugates + nonhydrolyzable polar metabolites	94.2 ± 1.0	94.5 ± 1.4
After enzyme hydrolysis		
Buprenorphine	55.2 ± 2.2	72.5 ± 0.4
N-Dealkyl buprenorphine	15.3 ± 1.8	1.4 ± 0.7
Nonhydrolyzable polar metabolites	23.0 ± 2.5	18.9 ± 1.0

[a]Reproduced from Brewster et al. [1981a], with permission of the publisher.
[b]Values represent means ± SEM of three animals.

TABLE X. Cumulative Excretion of Radioactivity by Various Species Following a Single Intramuscular Dose of [³H]Buprenorphine[a,b]

Species	Sample	Collection period (hr)		
		0–24	0–72	0–144
Rat	Urine	5.1	6.9	7.1
	Feces	19.9	82.6	83.6
	Total	25.0	89.5	90.7[c]
Dog	Urine	3.0	5.1	5.8
	Feces	0.0	85.6	94.5
	Total	3.0	90.7	100.3
Rhesus monkey	Urine	7.4	12.7	18.5
	Feces	22.6	52.9	64.1
	Total	30.0	65.6	82.6

[a]Reproduced from Brewster et al. [1981a], with permission of the publisher.
[b]Values given for each animal species represent the mean result for two rats (female), two dogs (1 male and 1 female), and two rhesus monkeys (female) at doses of 20, 20, and 2 μg/kg, respectively.
[c]Collection period 0–96 hr only.

In all species it is likely that there is substantial enterohepatic cycling of drug and metabolites that gives rise to a slow excretion rate. In the rat, Brewster et al. [1981a] showed that the proportion of N-dealkyl buprenorphine in the bile was increased with each enterohepatic cycle (Table XI), indicating that the slow excretion of buprenorphine is linked with further biotransformation. The significance of these observations in the interpretation of human data is discussed later.

Excretion of Buprenorphine in the Milk of Lactating Female Rats

The transfer of buprenorphine and its metabolites to neonates via milk has been studied in lactating female rats following a single intramuscular injection of

TABLE XI. Change in Proportion of Metabolites Following Enterohepatic Cycling of Radiolabeled Material in Two Male Rats[a]

	% of bile radioactivity (hydrolyzed samples)	
	1st cycle	2nd cycle
Buprenorphine	54.4	36.8
N-Dealkyl buprenorphine	25.0	46.1
Nonhydrolyzable polar metabolites	17.1	16.8

[a]Reproduced from Brewster et al. [1981a], with permission of the publisher.

TABLE XII. Concentrations of Free Buprenorphine and Metabolites (μg or μg equiv. per 1 ml fluid or 1 g tissue) in Male Wistar Rats Each Implanted With a [15,16(n)-[³H]Buprenorphine Pellet for Various Times[a,b]

	Time (weeks)				
	2	4	6	10	12
Plasma	0.006 (—)	0.098 (29)	0.044 (—)	0.007 (16.7)	0.018 (9.7)
Liver	0.043 (8.8)	0.055 (5.1)	0.017 (6.2)	0.020 (4.0)	0.016 (4.3)
Heart	0.065 (6.5)	0.041 (4.2)	0.063 (2.6)	0.063 (1.4)	0.050 (1.3)
Lung	0.054 (5.4)	0.031 (0.7)	0.045 (0.6)	0.013 (1.1)	0.014 (0.6)
Kidney	0.070 (4.8)	0.042 (2.3)	0.057 (1.9)	0.016 (0.9)	0.022 (0.1)
Spleen	0.208 (7.7)	0.096 (6.7)	0.137 (4.5)	0.055 (2.4)	0.048 (0.7)
Testes	0.016 (6.5)	0.008 (4.2)	0.008 (2.4)	0.010 (2.7)	0.008 (1.7)
Skeletal muscle	0.350 (11.1)	0.204 (7.5)	0.362 (6.3)	0.067 (3.5)	0.109 (2.4)
Fat	0.175 (33.7)	0.187 (33.8)	0.140 (21.4)	0.261 (19.3)	0.366 (26.7)

[a]Reproduced from Pontani et al. [1985], with permission of the publisher.
[b]Data represent mean values from two animals. The concentrations of metabolites are given in parentheses. Plasma samples at 2 and 6 weeks were lost during determination of metabolite concentrations. Total radioactivity values were obtained by combustion of aliquots of tissue homogenates in a biological tissue oxidizer, and metabolite concentrations were determined by subtraction of the concentration of free buprenorphine from total radioactivity values. The dose of [³H] Buprenorphine was 10 mg.

[³H]buprenorphine (5 mg/kg). The studies showed that drug-related material is excreted in the milk of rats and that concentrations of unchanged buprenorphine in milk can equal or exceed that in plasma. From these data it seems likely that buprenorphine would be excreted in human breast milk.

Chronic Dosing of Buprenorphine

Pontani et al. [1985] studied the distribution of [³H]buprenorphine in the rat following its continuous release from a subcutaneously implanted pellet. The dose contained in the pellet was 10 mg of buprenorphine and the release was monitored over 12 weeks. The levels of unchanged buprenorphine and total metabolites in a number of tissues are given in Table XII. The results from chronic dosing show that metabolites form the major radioactive component of rat plasma, as would be expected because of the slow elimination of drug metabolites owing to enterohepatic circulation.

ADME STUDIES IN HUMANS

The absorption, plasma levels, metabolism, and excretion of buprenorphine in humans have been examined in a number of studies, some as part of the Reckitt & Colman drug development program. The human pharmacokinetics of buprenorphine are discussed by McQuay and Moore in this volume and are briefly considered here for comparison with animal results.

Four routes of administration have been examined: intravenous, intramuscular, sublingual, and oral. A major problem with all the pharmacokinetic studies has been the lack of a robust, reproducible assay that is specific for buprenorphine (particularly against the N-dealkyl metabolite) and retains sufficient sensitivity to measure low levels of drug in plasma especially at late times after dosing. HPLC methods were able to provide the specificity but none of the detection systems provided the required sensitivity. GC/MS methods provided both the sensitivity and specificity but were not capable of handling a large throughput of samples. The GC/MS method of Blom et al. [1985] served to describe the elimination of buprenorphine after intravenous administration of the drug and to measure the N-dealkyl metabolite for a few hours after dosing. However, application of this method for routine analysis was unsuccessful. Similar results with GC/MS were obtained more recently by Ohtani et al. [1989], who measured buprenorphine and N-dealkyl buprenorphine plasma profiles in one volunteer after a sublingual dose of the drug (Fig. 5). Again, the N-dealkyl metabolite was measurable only between two and three hours after dosing. There is no information about the robustness of this new method in routine analysis.

In most of the single-dose absorption studies, human plasma buprenorphine levels have been obtained by the best available assay for the drug, which is the radioimmunoassay method developed by Bartlett et al. [1980]. In early studies, [^{3}H]buprenorphine was used as the radioligand but this method was later modified to allow the use of [^{125}I]-labeled buprenorphine [Hand et al., 1986]. Although this assay does not distinguish between buprenorphine and N-dealkyl buprenorphine, the results of Blom et al. [1985] and Ohtani et al. [1989] have shown that this metabolite is measurable only at early times after dosing. However, it is likely that after a single dose the plasma immunoreactivity profiles will be made up in part from N-dealkyl buprenorphine and mostly from buprenorphine.

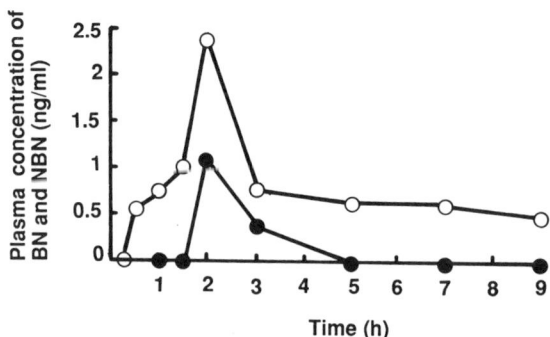

Fig. 5. Time courses of plasma concentrations of buprenorphine (BN) and norbuprenorphine (NBN) after administration of two sublingual tablets of buprenorphine to a healthy volunteer. (○) Buprenorphine; (●) norbuprenorphine. Adapted from Ohtani et al. [1989].

In chronic dosing studies, because of the slow excretion of drug-related material that is due to enterohepatic circulation of buprenorphine and metabolites, the simple radioimmunoassay method will not provide a useful picture of the levels of buprenorphine in view of the high levels of metabolites. A more suitable method has been described by Hand et al. [1986] in which immunoreactivity levels of diethyl ether-extracted buprenorphine are subtracted from total immunoreactivity levels to provide information about the levels of N-dealkyl buprenorphine and buprenorphine glucuronide, the latter two being distinguished from each other by using the two antibodies with selective metabolite cross-reactivity.

Early excretion balance studies were carried out with [^{3}H]buprenorphine. A later study by Cone et al. [1984] examined the excretion and metabolism of high sublingual doses of buprenorphine by applying GC/MS techniques to analyze the samples.

Single-Dose Absorption Studies

Early human pharmacology studies showed that oral doses of buprenorphine at least 10 times higher than intramuscular doses were needed to exert equivalent pharmacological effects. Excretion studies following an oral dose of 20 μg/kg of [^{3}H]buprenorphine showed that radioactivity was absorbed but the identity of the radioactive species in plasma was not determined. Although these data suggest that, as in animals, it is possible to overcome a marked first-pass metabolic effect by giving large oral doses of buprenorphine, this approach was seen as commercially impractical, and an oral dosage form of buprenorphine was not developed.

Plasma absorption profiles from 0 to 180 min in 10–24 patients following intravenous (0.3 mg), intramuscular (0.3 mg), and sublingual (0.4 mg) administration of buprenorphine have been reported by Bullingham et al. [1981] and are shown in Figure 1 of McQuay and Moore, this volume. The intravenous and intramuscular profiles are similar to those described by Bartlett et al. [1980] for three patients from 0 to 120 min, and the intravenous profile is similar to that described by Blom et al. (1985), who used GC/MS. Bullingham et al. [1982] extended the evaluation of sublingually administered buprenorphine by measuring bioavailability relative to an intravenous dose based on 0 to 10-hr plasma level data (see Table II in McQuay and Moore, this volume). The results obtained are very similar to a recent sublingual bioavailability study carried out with 24 subjects who received 0.3 mg of buprenorphine i.v. and 0.2, 0.4, and 0.8 mg of the drug sublingually in a randomized crossover design [Reckitt & Colman; data on file]. In this study, plasma buprenorphine was estimated by radioimmunoassay with the [^{125}I]radiolabel (Table XIII).

The elimination of buprenorphine immunoreactivity from plasma follows a multiexponential curve [Bullingham et al., 1980], and estimates of terminal elimination rates appear to be dependent on the time period of sampling. In the sublingual crossover study, plasma immunoreactivity was measured for times up to 24 hr and results showed that over the period from 8 hr to 24 hr the elimination of immunoreactivity was very slow, with an apparent half-life of the order of 18 hr. A similar terminal elimination rate was observed in the single patient studied by Ohtani et al.

TABLE XIII. Absorption/Bioavailability Results Following Sublingual (s.l.) Administration of Buprenorphine to 24 Men

Parameter	0.2 mg s.l.	0.4 mg s.l.	0.8 mg s.l.
C_{max} (ng/ml)	0.28 ± 0.02	0.47 ± 0.02	0.83 ± 0.03
AUC (0–24 hr) (ng/ml × min)	149 ± 16	206 ± 13	343 ± 20
T_{max} (min)	96 ± 11	91 ± 11	91 ± 9
Percentage bioavailability	52 ± 7	38 ± 4	31 ± 2

[1989], who used a specific buprenorphine GC/MS assay. This apparent slow rate of elimination may, however, be linked to reabsorption of immunoreactive components: buprenorphine, N-dealkyl buprenorphine, or both, following the first enterohepatic cycle. If this was the case, it is likely that the "terminal" elimination rate of buprenorphine is considerably faster, of the order of 3–5 hr, as described by Bullingham et al. [1982].

Bullingham et al. [1980] showed that buprenorphine, given intramuscularly, has a very rapid absorption. The peak plasma level was usually at 5 min after dosing, and in some patients it had occurred by 2 min. Mean plasma levels of intramuscular and intravenous buprenorphine differed little beyond 5 min, and systemic availability was generally close to 100%.

Absorption following sublingual dosing was much slower, the time to peak concentration being around 90–360 min (Table II in McQuay and Moore, this volume) or around 90 min (Table XIII). The first stage of the sublingual absorption process, the dissolution of the tablet and the uptake of drug into the sublingual mucosa, is quite rapid, however. In an unpublished study carried out by Reckitt & Colman volunteers, the amount of drug absorbed from the tablet after it was held for various times in the mouth was assessed by measuring the amount recoverable from the mouth contents. It was found that 27%–52%, 30%–55% and 26%–81% of the dose was transferred into the sublingual mucosa after the tablet had been retained in the mouth for 2, 4, and 8 min, respectively. Retaining the tablet for 16 min did not result in further absorption. Very little of the drug absorbed could be recovered by an acid (pH 4.67) rinse of the mouth. The transfer from sublingual mucosa to blood stream is thus a much slower process. Calculation of the overall absorption rate of buprenorphine for 0.4 mg and 0.8 mg sublingual doses from the cross-over study gave mean first-order rate constants of 0.041 per 1 min and 0.043 per 1 min, respectively, which equate to absorption half-lives of 0.50 hr (range 0.13–1.27 hr) and 0.46 hr (range 0.098–1.32 hr), respectively. Results from this study showed that over the dose range of 0.2–0.8 mg, the sublingual absorption of buprenorphine was proportional to the dose, and the mean bioavailability was between 31% and 52% (Table XIII). These parameters are of an order similar to those shown earlier by Bullingham et al. (Table II in McQuay and Moore, this volume).

In a separate study in normal volunteers, Weinberg et al. [1988] determined the percentage absorption of selected opioid analgesics from the oral cavity under

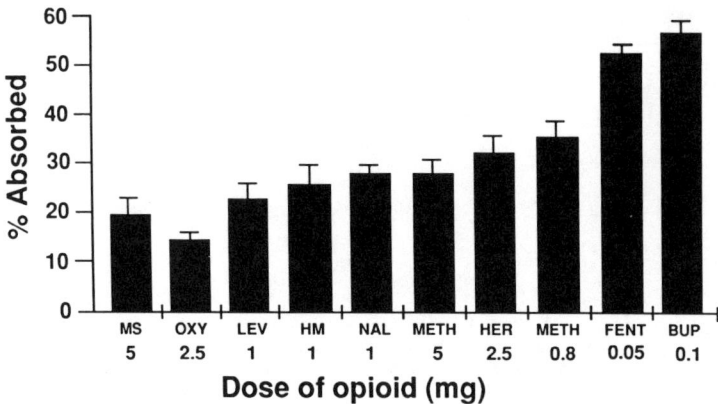

Fig. 6. Mean absorption (+SEM) of test opioids after 10 min in the oral cavity of normal subjects ($n = 10$ for each test condition). The pH of the dosing solution was 6.5. MS, morphine sulfate; OXY, oxycodone; LEV, levorphanol; HM, hydromorphone; NAL, naloxone; METH, methadone; HER, heroin; FENT, fentanyl; BUP, buprenorphine. Buprenorphine, fentanyl, and methadone were absorbed to a significantly greater extent ($p < 0.05$) compared with morphine, while the other opioids were not. Adapted from Weinberg et al. [1988].

conditions of controlled pH and swallowing when a 1.0-ml aliquot of the test drug was placed under the tongue for a 10-min period. Compared with morphine sulfate at pH 6.5 (18% absorption), buprenorphine (55% absorption) was absorbed to a greater extent (Fig. 6). Lipophilic opioids were better absorbed than hydrophilic opioids. Again, the mean absorption of buprenorphine was not significantly increased when the time in the oral cavity was increased from 2.5 to 10 min. Thus, the sublingual absorption of buprenorphine, whether in tablet or solution dosage forms, is relatively high.

Protein Binding

Buprenorphine binds highly to plasma proteins in the rat (85%–87%), dog (78%–81%), and human (96%). In human plasma the drug was bound primarily to α- and β-globulin fractions.

Metabolism and Excretion

Two radiolabel studies were carried out using [^{3}H]buprenorphine, with dosing by the intramuscular (2 μg/kg) and oral (15 μg/kg) routes in the first study, and by the oral route (20 μg/kg) in the second study. In a further study [Cone et al., 1984], excreted metabolites were monitored by gas–liquid chromatography following subcutaneous (1 and 2 mg), sublingual (2 and 4 mg), and oral (20 and 40 mg) administration of buprenorphine.

TABLE XIV. Excretion of Buprenorphine-Related Material in Humans

Route	Dose	Method	Collection period, days	% Dose Urine	Feces	Total
IM	2 µg/kg	³H	7	27.0	67.5	94.5
PO	15 µg/kg	³H	7	15.3	70.7	86.0
PO	20 µg/kg	³H	15	16.5	62.7	79.2
PO	20 µg/kg	³H	15	10.3	49.6	59.9
PO	20 mg[a]	GLC	4	11.9		
PO	40 mg[a]	GLC	4	12.3		
SL	2 mg[a]	GLC	4	12.5		
SL	4 mg[a]	GLC	4	14.3		
SC	1 mg[a]	GLC	4	1.9		
SC	2 mg[a]	GLC	4	28.6		

[a]Data from Cone et al. [1984], with permission of the publisher.

After oral or sublingual administration of buprenorphine, 10%–17% of the dose was excreted in the urine (Table XIV). A slightly greater amount (27%) was found in the urine following intramuscular administration and variable amounts after subcutaneous dosing. The variation in the level of urinary excretion is probably a consequence of the low number of subjects examined rather than a clear route difference.

In the radiolabel study, all radioactivity in the urine of both volunteers on Day 1 was present in the form of glucuronide conjugates of buprenorphine and N-dealkyl buprenorphine. In the urine of the volunteer dosed intramuscularly, the major aglycone was buprenorphine, whereas after oral dosing the major aglycone was N-dealkyl buprenorphine. In the study by Cone et al. [1984], no free buprenorphine was detected in the urine after administration by any of the routes studied and the amount of N-dealkyl metabolite (free plus conjugated) generally exceeded that of conjugated buprenorphine (Table XV).

In contrast, in the GC/MS study by Blom et al. [1985] both free and conjugated buprenorphine and free and conjugated N-dealkyl buprenorphine were excreted in the urine of one volunteer after an intravenous dose of 0.6 mg of buprenorphine (Fig. 7). The Blom et al. study also showed that the patterns of excretion of buprenorphine and N-dealkyl buprenorphine were quite different. The majority of buprenorphine conjugate was excreted in the first 24 hr after dosing, and thereafter the amounts excreted were very small. In contrast, the rate of excretion of both free and conjugated N-dealkyl buprenorphine remained reasonably constant over the 4-day period at about 2–3 µg/day of free and 11–19 µg/day of conjugated N-dealkyl buprenorphine (Fig. 7). The slow excretion of drug-related material was also shown by radioimmunoassay in an unpublished study with two buprenorphine antibodies, one that cross-reacts with buprenorphine glucuronide and the other with

TABLE XV. Urinary Excretion of Buprenorphine and Metabolites in a Human Subject[a]

Route	Dose (mg)	Buprenorphine		N-Dealkyl buprenorphine	
		Free	Conjugated	Free	Conjugated
PO	20	0	1.9	1.9	8.1
PO	40	0	1.5	3.4	7.4
SL	2	0	8.2	0.8	3.5
SL	4	0	6.5	0.8	7.0
SC	1	0	0	0.2	1.7
SC	2	0	8.4	5.0	15.2

[a]Reproduced from Cone et al. [1984], with permission of the publisher.

N-dealkyl buprenorphine. The mean excretion of immunoreactivity over 4 days following single buprenorphine doses of 0.3 mg intramuscularly or 0.4 mg sublingually is given in Table XVI.

The slow excretion of drug-related material and cross-reactivity of buprenorphine antibodies to metabolites have a bearing on the interpretation of results of

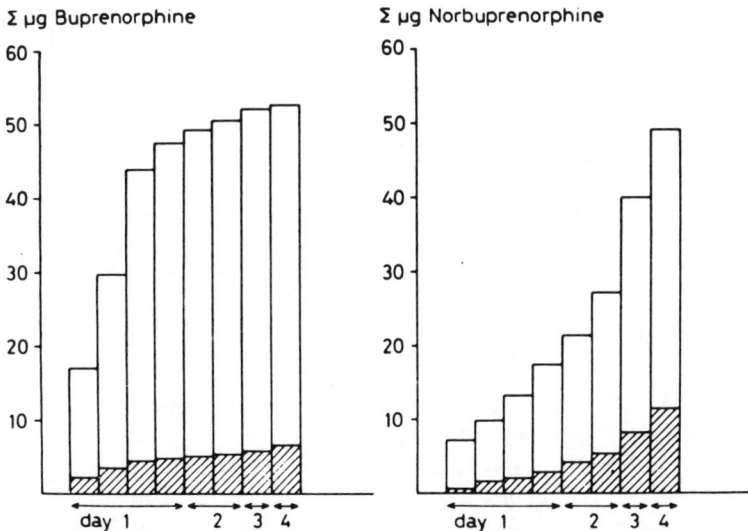

Fig. 7. Cumulative amount of buprenorphine and norbuprenorphine excreted in urine after intravenous administration of 0.6 mg of buprenorphine. (□) Conjugated drug or metabolite; (▨) free drug or metabolite. Reproduced from Blom et al. [1985], with permission of the publisher.

TABLE XVI. Urinary Excretion of Drug-Related Material Using a Buprenorphine Antibody That Cross Reacts With Buprenorphine Glucuronide (L160) or With N-Dealkyl Buprenorphine (L164)

Dose/route	Sampling period (hr)	Mean urinary immunoreactivity (μg) ($N = 3$) using antibody	
		L160	L164
0.3 mg i.m.	0–9/10	9.88	1.40
	9/10–24	3.59	0.68
	24–36	1.64	1.53
	36–48	2.53	2.26
	48–72	3.06	1.98
	72–96	2.36	3.39
0.4 mg s.l.	0–9	6.64	1.66
	9–24	3.76	3.25
	24–36	2.17	2.31
	36–48	2.04	2.66
	48–72	3.79	7.15
	72–96	2.02	4.11

buprenorphine immunoreactivity measurements in urine from opiate addicts. A study by Hand et al. [1989] of the use of an assay kit for screening urine for buprenorphine showed that there was extensive cross-reactivity to N-dealkyl buprenorphine and to glucuronides and they recommended a cutoff concentration of immunoreactivity. However, if this is set too low, it may simply show that the subject had taken a dose at some time in the previous 2–3 weeks but will not provide evidence of current usage.

Most of the buprenorphine dose (50%–71%) in the radiolabel studies was eliminated in the feces over the first 4 days (Table XIV). In the radiolabel study, chromatography of the methanolic extract of feces showed that the major radioactive component after both intramuscular and oral routes of administration was unconjugated buprenorphine; traces of N-dealkyl buprenorphine were also found. In the study by Cone et al. [1984], free and conjugated buprenorphine were present in fecal samples, the free buprenorphine generally exceeding that of the conjugate. N-dealkyl buprenorphine was found in all but one of the fecal samples and, following sublingual dosing, buprenorphine and N-dealkyl buprenorphine were excreted in nearly equal amounts. The total percentage of the dose eliminated in the feces was not determined in that study.

These results of the urinary excretion and fecal elimination of buprenorphine-related material, and the evidence from plasma levels of metabolites, show that the metabolism of buprenorphine in humans is similar to that found in animals, comprising phase I and phase II reactions to N-dealkyl buprenorphine and conjugates of

TABLE XVII. Buprenorphine and Metabolite Concentrations in Human Plasma (nmol/l)[a,b]

Sample time	N	Buprenorphine	Buprenorphine-3-O-glucuronide	N-Dealkyl buprenorphine
Chronic dosing				
Day 1 10:00	3	0.97 ± 0.38	1.53 ± 0.56	0.90 ± 0.78
Day 1 18:00	2	0.73 ± 0.59	1.41 ± 0.30	0.62 ± 0.41
Day 2 10:00	4	0.78 ± 0.40	1.86 ± 0.44	0.83 ± 0.68
Drug withdrawn				
Day 2 14:00	4	0.43 ± 0.17	1.52 ± 0.61	0.58 ± 0.56
Day 2 18:00	3	0.37 ± 0.14	1.24 ± 0.02	0.59 ± 0.31
Day 3 10:00	3	0.20 ± 0.07	0.85 ± 0.37	0.66 ± 0.06

[a]Reproduced from Hand et al. [1986], with permission of the publisher.
[b]Four patients who had been taking buprenorphine for at least 1 month were studied over three consecutive days. The sublingual buprenorphine dose was 1–2 mg daily in three divided doses and the drug was discontinued on the morning of the second day. Values for plasma buprenorphine and metabolite concentrations are the mean ± SD.

buprenorphine and N-dealkyl buprenorphine. No other metabolites have been reported in humans.

Chronic Dosing of Buprenorphine

Only low levels of metabolites are measurable in plasma after a single dose of buprenorphine, but this changes on chronic dosing of the drug. Hand et al. [1986] have reported levels of buprenorphine, buprenorphine 3-O-glucuronide, and N-dealkyl buprenorphine in four patients who had been taking buprenorphine for at least 1 month (Table XVII and McQuay and Moore, this volume). They made three measurements of buprenorphine and the two metabolites at steady state and then at 4 hr, 8 hr, and 24 hr after withdrawing the drug. When dosing ceased, both buprenorphine and buprenorphine-3-O-glucuronide declined in concentration over a period of 24 hr. During this period there was no discernible reduction in concentrations of N-dealkyl buprenorphine.

More recently, Hand et al. [1990] have measured plasma levels of buprenorphine and the two metabolites in patients with renal failure and have shown that levels of the metabolites were increased compared with patients with no renal impairment, whereas buprenorphine levels were similar in both groups. The results support a biliary excretion route for buprenorphine but point to the importance of the renal excretion route for the metabolites.

The pharmacological profile of the conjugates is unknown, whereas N-dealkyl buprenorphine profiles as a weak μ-opiate agonist of low intrinsic activity [Iizuka et al., 1981]. Distribution studies in animals have shown that buprenorphine metabolites do not pass the blood brain barrier and that buprenorphine is the only drug-related substance in the brain.

CONCLUSION

This chapter has reviewed studies of the ADME of buprenorphine in animals and humans and shown it to be similar in all species. The review has concentrated on the routes of administration (parenteral and sublingual) that are currently licensed in most countries. In Japan, the drug is also available as a suppository but ADME data following this route of administration were not available for inclusion in this review.

ACKNOWLEDGMENTS

The following current and former employees of Reckitt & Colman Products are responsible for the unpublished data reviewed here: JD Nichols, ME Havler, JM Clifford, Mrs. M Price, PT Bevan, JG Lloyd-Jones, BJ Jordan, KE Godfrey, MJ Humphrey, D Brewster, MA McLeavy, and PS Sherer. The authors thank Reckitt & Colman Products for their permission to use unpublished data and acknowledge the publishers of previously published material for permission to use the results. Dr. Inturrisi's research is supported in part by NCI Grant CA-32897 and NIDA Grants DA-01457 and DA-05130.

REFERENCES

Bartlett AJ, Lloyd-Jones JG, Rance MJ, Flockhart IR, Dockray GJ, Bennett MRD, Moore RA (1980): The radioimmunoassay of buprenorphine. Eur J Clin Pharmacol 18:339–345.

Blom Y, Bondesson U, Änggård E (1985): Analysis of buprenorphine and its N-dealkylated metabolite in plasma and urine by selected-ion monitoring. J Chromatogr 338:89–98.

Brewster D, Humphrey MJ, McLeavy MA (1981a): Biliary excretion, metabolism and enterohepatic circulation of buprenorphine. Xenobiotica 11:189–196.

Brewster D, Humphrey MJ, McLeavy MA (1981b): The systemic bioavailability of buprenorphine by various routes of administration. J Pharm Pharmacol 33:500–506.

Bullingham RES, McQuay HJ, Moore A, Bennett MRD (1980): Buprenorphine kinetics. Clin Pharmacol Ther 28:667–672.

Bullingham RES, McQuay HJ, Dwyer D, Allen MC, Moore RA (1981): Sublingual buprenorphine used postoperatively: Clinical observations and preliminary pharmacokinetic analysis. Br J Clin Pharmacol 12:117–122.

Bullingham RES, McQuay HJ, Porter EJB, Allen MC, Moore RA (1982): Sublingual buprenorphine used postoperatively: Ten hour plasma drug concentration analysis. Br J Clin Pharmacol 13:665–673.

Castle SJ, Tucker GT, Woods HF, Underwood JCE, Nicholson CM, Havler ME, Lewis CJ, Flockhart IR, Lloyd-Jones JG (1985): Assessment of an *in situ* rat intestine preparation with perfused vascular bed for studying the absorption and first pass metabolism of drugs. J Pharmacol Methods 14:255–274.

Cone EJ, Gorodetzky CW, Yousefnejad D, Buchwald WF, Johnson RE (1984): The metabolism and excretion of buprenorphine in humans. Drug Metab Dispos 12:577–581.

Hand CW, Ryan KE, Dutt SK, Moore RA, O'Connor J, Talbot D, McQuay HJ (1989). Radioimmunoassay of buprenorphine in urine: Studies in patients in a drug clinic. J Anal Toxicol 13:100–104.

Hand CW, Sear JW, Uppington J, Ball MJ, McQuay HJ, Moore RA (1990): Buprenorphine disposition in patients with renal impairment: Single and continuous dosing, with special reference to metabolites. Br J Anaesth 64:276–282.

Heel RC, Brogden RN, Speight TM, Avery GS (1979): Buprenorphine: A review of its pharmacological properties and therapeutic efficacy. Drugs 17:81–110.

Iizuka H, Shimada A, Yanagita T (1981): Pharmacological studies of buprenorphine and its major metabolite RX2007M. Jitchuken Zenrinsho Kenkyuho 7:279–321.

Lloyd-Jones JG, Robinson P, Henson R, Biggs SR, Taylor T (1980): Plasma concentration and disposition of buprenorphine after intravenous and intramuscular doses to baboons. Eur J Drug Metab Pharmacokinet 5:233–239.

Manara L, Cerletti C, Luini A, Tavani A (1978): Rat brain levels and subcellular distribution of *in vivo* administered buprenorphine: Effects of naloxone. In Van Ree JM, Terenius L (eds): "Characteristics and Function of Opioids." Amsterdam: Elsevier/North-Holland, pp 225–226.

Numata H, Tsuda T, Atai H, Tanaka M, Yanagita T (1981): Pharmacokinetics of buprenorphine in rats and monkeys. Jitchuken Zenrinsho Kenkyuho 7:347–357.

Ohtani M, Shibuya F, Kotaki H, Uchino K, Saitoh Y, Nakagawa F, Nishitateno K (1989): Quantitative determination of buprenorphine and its active metabolite, norbuprenorphine, in human plasma by gas chromatography–chemical ionization mass spectrometry. J Chromatogr 487:469–475.

Pontani RB, Vadlamani NL, Misra AL (1985): Disposition in the rat of buprenorphine administered parenterally and as a subcutaneous implant. Xenobiotica 15:287–297.

Rance MJ, Shillingford JS (1976): The role of the gut in the metabolism of strong analgesics. Biochem Pharmacol 25:735–741.

Rance MJ, Shillingford JS (1977): The metabolism of phenolic opiates by rat intestine. Xenobiotica 7:529–536.

Rance MJ, Robinson JD, Taylor KT (1976): The preparation of ^{2}H and ^{3}H-tertiary amines. Catalytic hydrogenation in the presence of the N-cyclopropylmethyl group. J Labelled Compd Radiopharm 12:467–472.

Sutton ML, Graham TC, Van Petten LE (1986): General reproduction study of Buprenex in Sprague-Dawley rats. Toxicologist 6:239.

Weinberg DS, Inturrisi CE, Reidenberg B, Moulin DE, Nip TJ, Wallenstein S, Houde RW, Foley KM (1988): Sublingual absorption of selected opioid analgesics. Clin Pharmacol Ther 44:335–342.

BUPRENORPHINE KINETICS IN HUMANS

H. J. McQUAY
Oxford Regional Pain Relief Unit, Churchill Hospital, Oxford OX3 7LJ, and Nuffield Department of Anaesthetics, Radcliffe Infirmary, Oxford, UK

R. A. MOORE
Oxford Regional Pain Relief Unit, Churchill Hospital, Oxford OX3 7LJ, Nuffield Department of Anaesthetics, Radcliffe Infirmary, Oxford, and Euro/DPC Ltd, Glyn Rhonwy, Llanberis, Caernarfon, Wales, UK

INTRODUCTION

Buprenorphine is a synthetic opiate analgesic with partial agonist properties [Heel et al., 1979]. Its slow dissociation from the receptor [Hambrook and Rance, 1976] and its long (6–10 hr) duration of clinical effect [McQuay et al., 1980] lead to particular interest in its kinetics.

This chapter reviews kinetic data for single-dose parenteral and sublingual buprenorphine and data on single-dose 0.6-mg intravenous (i.v.) buprenorphine, single-dose 0.3-mg i.v. buprenorphine in end-stage renal failure, single-dose 0.4-mg oral buprenorphine, chronic sublingual kinetics after dosing with 0.4 mg three times daily, i.v. infusion kinetics in patients with renal dysfunction, and cerebrospinal (CSF) and plasma concentrations after intrathecal and extradural doses of buprenorphine. In all of these studies, plasma buprenorphine concentrations were measured by a sensitive and specific radioimmunoassay [Bartlett et al., 1980] and single-dose studies used arterial samples exclusively.

INTRAVENOUS KINETICS

Single-dose pharmacokinetic data for intravenous buprenorphine at 0.3-mg and 0.6-mg doses are summarized in Table I and Figure 1. Sufficient samples, including early

Buprenorphine: Combatting Drug Abuse With a Unique Opioid, pages 137–147
© 1995 Wiley-Liss, Inc.

TABLE 1. Intravenous and Intramuscular Buprenorphine Kinetics[a]

Dose route (mg)	No. of pts	Study duration (hr)	Weight (kg)	A (ng/ml)	B (ng/ml)	C (ng/ml)	$t_{1/2a}$ (min)	$t_{1/2b}$ (min)	$t_{1/2c}$ (min)	Cl (ml/min)
Anesthetized[b]										
0.3, i.v.	24	3	68 ± 2	26.5 ± 3.2	5.8 ± 0.6	1.1 ± 0.1	2.1 ± 0.1	11.2 ± 0.7	139.6 ± 14	901.3 ± 39.7
Anesthetized, end-stage renal failure[c]										
0.3, i.v.	5			148.8 ± 68	8.9 ± 1.8	1.3 ± 0.4	0.5 ± 0.1	4.8 ± 0.5	126 ± 29.5	1127 ± 269
Postoperative[d]										
0.3, i.v.	10	3	68 ± 2	18.1 ± 3.1	2.8 ± 0.6	0.7 ± 0.1	2.2 ± 0.3	18.7 ± 3.2	183.6 ± 37.0	1275 ± 89
0.3, i.v.	5	13	65 ± 4			0.4 ± 0.02			310.9 ± 32.8	
0.3, i.v.	10	3	63 ± 4	66 ± 33	4.4 ± 1.1	1.0 ± 0.2	1.3 ± 0.3	0.7 ± 2.1	112.6 ± 27	1402 ± 211
0.6, i.v.	10	3	72 ± 3	142 ± 27	20.8 ± 6.3	3.0 ± 0.5	0.9 ± 0.1	6.2 ± 1.3	77.9 ± 13.3	1073 ± 68
0.3, i.m.[e]	7	3	67 ± 4		4.7 ± 0.8	1.4 ± 0.3		16.5 ± 2.5	138.5 ± 41.8	993 ± 70

[a] Values are mean ± SEM. A, B, and C were derived from the formula: plasma concentration = $A(\exp - t) + B(\exp - t) + C(\exp - t)$. Plasma clearance (Cl) values were obtained by dividing the dose by the area under the concentration–time curve (AUC), and bioavailability by comparing the AUC_{im} with the AUC_{iv}. Half-lives were calculated as $\log_e^{-2}$ (rate constant).
[b] Bullingham et al. [1980].
[c] Summerfield et al. [1985].
[d] Bullingham et al. [1980]; Watson et al. [1982].
[e] Bioavailability 40%–90%.

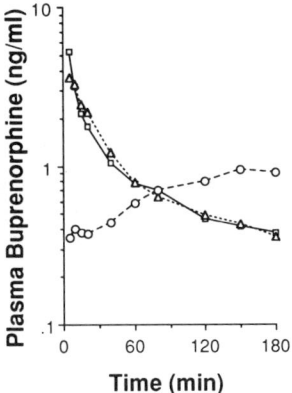

Fig. 1. Mean plasma buprenorphine concentrations for 10 patients given a postoperative dose of 0.3 mg intravenous buprenorphine [Bullingham et al., 1980] (□); 10 patients given a postoperative dose of 0.3 mg intramuscular buprenorphine [Bullingham et al., 1980] (△); and 10 patients given a postoperative dose of 0.4 mg sublingual buprenorphine [Bullingham et al., 1981a] (○). Patients had previously been given an intraoperative 0.3-mg intravenous dose.

points, were taken to allow tri-exponential fitting to 3-hr data [Bullingham et al., 1980; Watson et al., 1982]. These 3-hr studies gave a terminal half-life estimate of 2–3 hr; this was increased to 5 hr when the study duration was extended to 13 hr [Bullingham et al., 1982a].

Plasma drug clearance was high, with a mean value of about 1,000 ml/min or more. The clearance values for anaesthetized patients and postoperative clearance values in the same patients were significantly different; the approximately one-third lower clearance in the anaesthetized patient was attributed to the reduction in hepatic blood flow caused by the intraoperative use of halothane [Bullingham et al., 1980]. A fixed (not weight-related) dose was given in these studies and no correlation between weight and plasma concentrations was found.

Plasma buprenorphine concentrations in postoperative patients given 0.6 mg i.v. were approximately twice those in similar patients given 0.3 mg i.v. [Watson et al., 1982], and the kinetic analysis summarized in Table I reflects this linearity.

Single-Dose Intravenous Kinetics in Renal Failure

Plasma buprenorphine concentrations were measured after a 0.3-mg i.v. dose in five patients with end-stage renal failure who were undergoing surgery [Summerfield et al., 1985]. Sample times were the same as in previous studies [Bullingham et al., 1980]. The kinetic analysis (Table I) shows that renal failure caused no change in the kinetic fate of 0.3 mg i.v. buprenorphine; this is in sharp contrast with morphine studied in a similar renal patient group when the morphine clearance fell markedly [Moore et al., 1984].

INTRAMUSCULAR KINETICS

Plasma buprenorphine concentrations after a postoperative 0.3-mg intramuscular (i.m.) dose showed very rapid absorption [Bullingham et al., 1980]. Average peak plasma concentrations of 3.6 ng/ml occurred 2–5 min after the dose was given; by ten minutes there was no significant difference from the plasma concentrations seen after the same dose given intravenously. The mean kinetic data for the i.m. dose are summarized in Table I and are shown in Figure 1. Even though 2- and 5-min samples were taken, absorption was too fast to allow calculation of an absorption rate constant. Systemic availability of the 0.3-mg i.m. dose was greater than 90% in 7 of the 11 patients; in the other 4 it was between 40% and 60%.

SUBLINGUAL KINETICS

Single Dose

The mean peak plasma buprenorphine concentration, time to peak concentration, and systemic availability for 0.4-mg and 0.8-mg single-dose sublingual buprenorphine are shown in Table 2, and for the 0.4-mg dose are shown in Figure 1. After sublingual doses there was no significant rise in mean plasma buprenorphine concentration for 20 min [Bullingham et al., 1981a; Bullingham et al., 1982a]. The pattern of the rise varied greatly between patients; peak plasma buprenorphine concentration occurred anywhere between 20 and 360 min, but generally at about 3 hr. From about 2 hr, the plasma concentrations were sustained for some hours at a higher level than seen after comparable parenteral doses (Fig. 1). Similar plasma concentrations were reported by an enzyme immunoassay method [Tiong and Olley, 1988].

The average maximum increase in plasma concentration of 0.50 ng/ml after a

TABLE II. Pharmacokinetic Data for Sublingual and Oral Buprenorphine[a]

Route dose (mg)	No. of pts	Study duration (hr)	Weight (kg)	C_{max} (ng/ml)	T_{max} (min)	Systemic availability (%)
Sublingual[b]						
0.4	5	10	66 ± 3	0.50 ± 0.06	210 ± 40	57.7 ± 6
0.8	5	10	65 ± 4	1.04 ± 0.27	192 ± 49	54.1 ± 12.7
0.4	10	3	69 ± 3	0.74 ± 0.16		31
Oral						
0.4	6	3	70 ± 4	0.25 (<0.1–0.62)	20 (?–180)	14 (0–33)

[a]Values are mean ± SEM. No T_{max} quoted for 3-hr sublingual study because concentrations were still rising for some patients. Values (with ranges) shown for the oral study because two patients had no measured absorption.
[b]Bullingham et al. [1981a, 1982a].

0.4-mg single sublingual dose was doubled to 1.04 ng/ml after a 0.8-mg dose (Table II). The increase in dose roughly doubled the average plasma concentrations and hence the area under the plasma concentration–time curve, but made no difference to the systemic availability or to the time to reach peak plasma concentration. Average systemic availability, calculated as 30% by 3 hr [Bullingham et al., 1981a], rose to 55% by increasing study duration to 10 hr; in individual patients this varied from 16%–94% [Bullingham et al., 1982a]. The absorption half-life was estimated at 76 min based on a single-compartment model, first-order absorption, and mean plasma drug concentrations [Bullingham et al., 1982a].

Chronic Dosing

Four patients with chronic pain, who had been taking 0.4 mg sublingual buprenorphine three times a day for at least 4 weeks (median 2 months, range 1–12 months), were studied when they stopped taking the drug to obtain decay kinetics from steady-state chronic dosing (Hand et al., 1986]. The mean plasma buprenorphine level on this 0.4-mg three times daily dose was 0.5 ± 0.2 ng/mL (SEM) 2 hr after dose (i.e., close to peak). Venous samples were taken 2, 6, 10, 26, 30, 50, 54, 74. 98, and 122 hr after the final dose. Kinetic analysis of the mean plasma concentrations revealed two components of this decay. The first had a half-life of 3.7 hr (similar to the terminal half-life reported in the 3-hr parenteral studies; Table I); the second had a much longer half-life of 45 hr, but only became evident when plasma buprenorphine concentrations were below 0.2 ng/ml.

ORAL KINETICS

Buprenorphine 0.4 mg was given orally to postoperative patients at the same sample times (over 3 hr) as a previous sublingual study [Bullingham et al., 1981a]. From 1 to 3 hr, oral buprenorphine resulted in significantly lower plasma buprenorphine concentrations than after the same dose given sublingually (Fig. 2). The median systemic availability (Table II) was 16% (range 0%–36%), consistent with the mean availability of 15% calculated from the intravenous bolus kinetics [Bullingham et al., 1980].

SPINAL KINETICS

There is little published information on the kinetics of intrathecal or extradural administration of buprenorphine. When buprenorphine 0.03 mg (diluted with saline to 1 ml injectate) was given intrathecally at the lumbar level for treatment of painful spasm in 6 paraplegic patients [Glynn et al., 1984], there was a rapid fall in the mean CSF concentrations (Fig. 3).

After an extradural dose of 0.3 mg of buprenorphine to three patients with painful paraplegic spasm a peak CSF concentration of 9.5 ng/ml was reached after

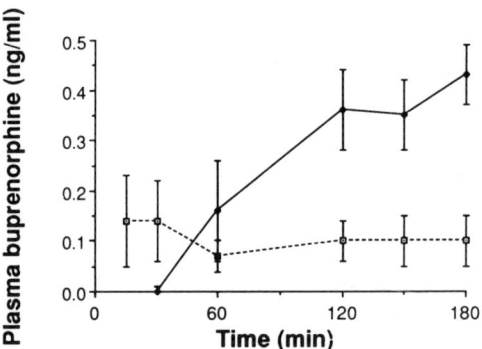

Fig. 2. Mean (± SEM) plasma buprenorphine concentrations against time for 6 patients given 0.4 mg oral buprenorphine (□) and 5 patients given 0.4 mg sublingually (●) [Bullingham et al., 1982a]. Patients had previously been given an intraoperative 0.3-mg intravenous dose. The computed contribution from that dose was subtracted from the measured concentrations to derive the data used in the figure.

45 min, with a mean concentration of 0.2 ng/ml at 24 hr (Fig. 4) [Jamous, 1987]. The CSF absorption half-life was calculated at 3.5 min, with an elimination half-life of 84 min, a mean time to peak concentration of 16 min, and a mean maximum concentration of 13.2 ng/ml [Jamous, 1987]. The peak plasma buprenorphine concentration was reached much earlier than in CSF (Fig. 4).

CSF concentrations were measured after an intravenous dose of 0.6 mg of buprenorphine in three patients undergoing neurosurgery [Jamous, 1987]. CSF buprenorphine concentrations were very low, with 0.17 ng/ml measurable at the first sample time (10 min), and the highest mean value (0.3 ng/ml) occurring at 3 hr.

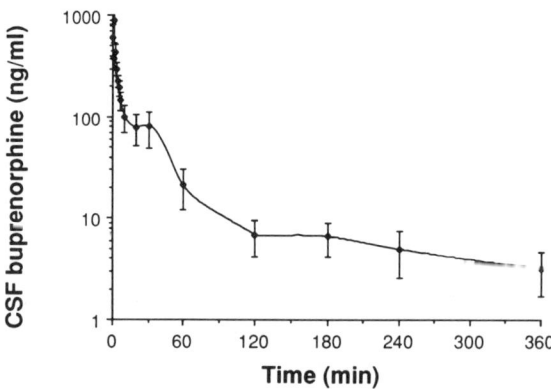

Fig. 3. Mean (± SEM) CSF buprenorphine concentrations against time for six patients given 0.03 mg intrathecal buprenorphine in 1 ml at the lumbar level [Jamous, 1987].

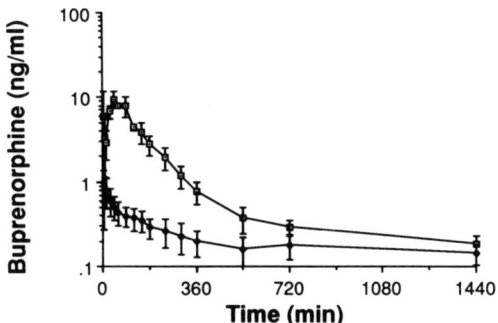

Fig. 4. Mean (± SEM) CSF (□) and plasma (♦) buprenorphine concentrations against time for three patients given 0.3 mg extradural buprenorphine in 5 ml at the lumbar level [Jamous, 1987].

METABOLISM

Buprenorphine is metabolized in humans by N-dealkylation and conjugation of N-dealkyl buprenorphine and the parent drug; about 70% of a radioactive oral or intravenous dose is recovered in feces and most of the remainder appears in the urine [Heel et al., 1979]. The N-dealkyl metabolite has even slower elimination than the parent compound [Hand et al., 1986]. The red blood cell:plasma buprenorphine ratio is reported to be close to unity [Bullingham et al., 1980]. Buprenorphine and its glucuronide appear in urine in 1–2 days, and the N-dealkyl metabolite and its glucuronide in 1–4 days [Cone et al., 1984; Blom et al., 1985].

Impaired Renal Function

Kinetic analysis with 3-hr sampling after a single i.v. dose showed no change in the fate of buprenorphine in patients with impaired renal function [Summerfield et al., 1985]. In 9 dialysis-dependent patients there were no differences in buprenorphine kinetics with 24-hr sampling after a single i.v. 0.3-mg dose compared with 6 patients with normal renal function [Hand et al., 1990]. Continuous i.v. infusion in 8 patients with renal failure showed similar plasma buprenorphine concentrations and clearance values compared with values for infused patients with normal renal function [Hand et al., 1990]. Plasma concentrations of norbuprenorphine, however, were increased by a median of 4 times compared with normal renal function patients, and buprenorphine-3-glucuronide concentrations were increased by a median of 15 times [Hand et al., 1990].

IMPLICATIONS FOR CLINICAL USE

The high systemic clearance of buprenorphine (> 1,000 ml/min) approaches hepatic blood flow and leads to the prediction [Wilkinson and Schenker, 1976] that drugs or interventions that lower hepatic blood flow should decrease buprenorphine clearance. The intraoperative fall in buprenorphine clearance [Bullingham et al., 1980] when halothane was used intraoperatively agrees with this. Another consequence of high clearance is the prediction of a substantial first-pass effect and low oral bioavailability. The median oral bioavailability for buprenorphine of 16% reported here is consistent with the prediction, as was the analgesic effect of oral doses of 1-2 mg [Bullingham et al., 1981b].

Sublingual administration provides a way of avoiding the first-pass effect, and the systemic availability for sublingual doses up to 0.8 mg of 55% is borne out by the effectiveness of analgesia obtained by this route, which is equivalent to that seen with parenteral routes [Bullingham et al., 1981a]. Both oral and sublingual buprenorphine resulted in large variation in calculated systemic availability. Low availability after sublingual doses may result if the dose is swallowed rather than kept under the tongue; unexpectedly high availability after an oral dose might result if the dose were held up in the oesophagus and mucosal absorption, which is presumably like sublingual, should occur there.

The long terminal half-life revealed by the chronic dosing kinetics suggests that it may take up to 10 days for steady-state plasma drug concentrations to be achieved by sublingual dosing alone. Equally, such a long half-life supports the use of buprenorphine in suppressing the use of heroin by addicts [Mello and Mendelson, 1980]. The lack of withdrawal symptoms until Day 14 after 8-mg total daily dose may again reflect this long half-life [Jasinski et al., 1978].

While the spinal route for opiates is capable of producing remarkable analgesia [Bullingham et al., 1982b; Bullingham et al., 1983], there is little spinal kinetic information available for buprenorphine. In vivo experiments showed that very high intrathecal doses of buprenorphine would be required to produce analgesic effect [Dickenson et al., 1990], and this was attributed to the high lipophilicity. The high doses necessary would be subject to systemic absorption, and it is doubtful whether there would be any measurable clinical gain from giving such high doses intrathecally as opposed to systemically. By analogy with methadone it might also be unsafe [Jacobson et al., 1990]. In vitro experiments demonstrated that the permeability of buprenorphine for both lumbar and cranial dura was the lowest of any opioid studied [Moore et al., 1982], and this was attributed to the relatively high molecular weight. Calculations indicated that only 0.2% of an extradural dose of buprenorphine would cross the dura, compared with 20% for extradural morphine [Moore et al., 1982]. The limited information in patients (Fig. 4) [Jamous, 1987] shows that CSF concentrations are considerably higher after an extradural dose than they are after the equivalent dose given i.v. The CSF concentration required for a spinal as opposed to a central effect is probably much greater again [Dickenson et al., 1990].

The relative availabilities for intramuscular and sublingual buprenorphine quoted above agree well with the relative potency ratio for intramuscular and sublingual analgesia of 2:1 in postoperative or cancer patients [Wallenstein et al., 1982]. Such a direct relation between dose, plasma concentration, and effect may not always be so apparent with this drug.

Doubling the intravenous dose from 0.3 to 0.6 mg produced a dose-related increase in analgesia and increased neuroendocrine effects without significant change in blood gases [Watson et al., 1982]. The same doses given as infusions over 20 min to volunteers for rebreathing studies also produced no dose response for respiratory depression [De Klerk et al., 1981]. This suggests a ceiling effect in the therapeutic dose range for respiratory depression but not for analgesia. Although plasma concentrations after intravenous or intramuscular doses did not differ after 10 min [Bullingham et al., 1980], the analgesia and the neuroendocrine effects were significantly greater after intramuscular use [McQuay et al., 1980] without greater respiratory effects.

Men and women may respond differently to the same plasma buprenorphine concentration. Different analgesic and neuroendocrine effects were obtained in men and women who were given the drug by the same route; in men buprenorphine had less analgesic effect but produced a significantly greater suppressive effect on neuroendocrine stress responses to surgery [McQuay et al., 1980; Moore et al., 1981]. These differences could not be explained by any difference in plasma drug concentration or kinetic fate.

Buprenorphine is an intriguing and complicated drug; it is a potent partial agonist, is highly lipophilic, and dissociates slowly from the receptor. It has proved to be an effective, strong analgesic, and the apparent ceiling effect on respiration and unchanged kinetics in renal failure are two specific advantages for the drug. These properties make sublingual buprenorphine an important alternative for patients in both acute and chronic pain.

ACKNOWLEDGMENTS

We wish to thank Mike Allen, Dene Baldwin, Martin Bennett, Roy Bullingham, Dawn Carroll, Peter Evans, Chris Glynn, Chris Hand, Ali Jamous, John Lloyd, Geraldine O'Sullivan, Patsy Poppleton, Brian Porter, John Sear, Richard Summerfield, and Jane Watson for their help in this work.

REFERENCES

Bartlett AJ, Lloyd-Jones JG, Rance MJ, Flockhart IR, Dockray G, Bennett MRD, Moore RA (1980): The radioimmunoassay of buprenorphine. Eur J Clin Pharmaco 18:339–345.

Blom Y, Bondesson U, Änggård E (1985): Analysis of buprenorphine and its N-dealkylated metabolite in plasma and urine by selected-ion monitoring. J Chromatogr 338:89–98.

Bullingham RES, McQuay HJ, Moore RA, Bennett MRD (1980): Buprenorphine kinetics. Clin Pharmacol Ther 28:667–672.

Bullingham RES, McQuay HJ, Dwyer D, Allen MC, Moore RA (1981a): Sublingual buprenorphine used postoperatively: Clinical observations and preliminary pharmacokinetic analysis. Br J Clin Pharmacol 12:117–122.

Bullingham RES, McQuay HJ, Moore RA, Weir L (1981b): An oral buprenorphine and paracetamol combination compared with paracetamol alone: A single dose double-blind postoperative study. Br J Clin Pharmacol 12:863–867.

Bullingham RES, McQuay HJ, Porter EJB, Allen MC, Moore RA (1982a): Sublingual buprenorphine used postoperatively: Ten hour plasma drug concentration analysis. Br J Clin Pharmacol 13:665–673.

Bullingham RES, McQuay HJ, Moore RA (1982b): Intrathecal and extradural narcotics. In Atkinson RS, Langton Hewer C (eds): "Recent Advances in Anaesthesia and Analgesia, 14." London: Churchill Livingstone, pp 141–156.

Bullingham RES, McQuay HJ, Moore RA (1983): Principles of use of intrathecal and extradural narcotics. In Kaufman L (ed): "Anaesthesia Review 2." London: Churchill Livingstone, pp 137–147.

Cone EJ, Gorodetzky CW, Yousefnejad D, Buchwald WF, Johnson RE (1984): The metabolism and excretion of buprenorphine in humans. Drug Metab Dispos 12:577–581.

De Klerk G, Mattie H, Spierdijk J (1981): Comparative study on the circulatory and respiratory effects of buprenorphine and methadone. Acta Anaesthesiol Belg 32:131–139.

Dickenson AH, Sullivan AF, McQuay HJ (1990): Intrathecal etorphine, fentanyl and buprenorphine on spinal nociceptive neurones in the rat. Pain 42:227–234.

Glynn CJ, McQuay HJ, Lloyd JW, Moore RA, Teddy PJ (1984): Intrathecal buprenorphine for painful muscle spasms in paraplegic patients. Pain [Suppl 2], S341.

Hambrook JM, Rance MJ (1976): The interaction of buprenorphine with the opiate receptor. In Kosterlitz HW (ed): "Opiates and Endogenous Opioid Peptides." Amsterdam: Elsevier/North-Holland, pp 295–301.

Hand CW, Baldwin D, Moore RA, Allen MC, McQuay HJ (1986): Radioimmunoassay of buprenorphine with iodine label: Analysis of buprenorphine and metabolites in human plasma. Ann Clin Biochem 23:47–53.

Hand CW, Sear JW, Uppington J, Ball MJ, McQuay HJ, Moore RA (1990): Buprenorphine disposition in patients with renal impairment: Single and continuous dosing, with special reference to metabolites. Br J Anaesth 64:276–282.

Heel RC, Brogden RN, Speight TM, Avery GS (1979): Buprenorphine; a review of its pharmacological properties and therapeutic efficacy. Drugs 17:81–110.

Jacobson L, Chabal C, Brody MC, Ward RJ, Wasse L (1990): Intrathecal methadone: A dose-response study and comparison with intrathecal morphine 0.5 mg. Pain 43:141–148.

Jamous MA (1987): Master of science thesis, University of Oxford.

Jasinski DR, Pevnick JS, Griffith JD (1978): Human pharmacology and abuse potential of the analgesic buprenorphine. Arch Gen Psychiatry 35:501–516.

McQuay HJ, Bullingham RES, Paterson GMC, Moore RA (1980): Clinical effects of buprenorphine during and after surgery. Br J Anaesth 52:1013–1019.

Mello N, Mendelson JH (1980): Buprenorphine suppresses heroin use by heroin addicts. Science 207:657–659.

Moore RA, Smith RF, McQuay HJ, Bullingham RES (1981): Sex and surgical stress. Anaesthesia 36:263–267.

Moore RA, Bullingham RES, McQuay HJ, Hand CW, Aspel JB, Allen MC, Thomas D (1982): Dural permeability to narcotics: In vitro determination and application to extradural administration. Br J Anaesth 54:1117–1128.

Moore RA, Sear JW, Baldwin D, Allen MC, Hunniset A, Bullingham R, McQuay HJ (1984):

Morphine kinetics during and after renal transplantation in man. Clin Pharmacol Ther 35:641–648.

Summerfield RJ, Allen MC, Moore RA, Sear JW, McQuay HJ (1985): Buprenorphine in end stage renal failure. Anaesthesia 40:914.

Tiong GKL, Olley JE (1988): Enzyme immunoassay of buprenorphine. Naunyn-Schmeidebergs Arch Pharmakol 338:202–206.

Wallenstein SL, Kaiko RF, Rogers AG, Houde RW (1982): Clinical analgesic assays of buprenorphine and morphine (abstract). Clin Pharmacol Ther 31:278.

Watson PJQ, McQuay HJ, Bullingham RES, Allen MC, Moore RA (1982): Single-dose comparison of buprenorphine 0.3 and 0.6 mg iv given after operation: Clinical effects and plasma concentrations. Br J Anaesth 54:37–43.

Wilkinson GR, Schenker S (1976): Effects of liver diseases on drug disposition in man. Biochem Pharmacol 25:2675–2681.

CLINICAL PHARMACOLOGY AND EVALUATION

CLINICAL PHARMACOLOGY OF BUPRENORPHINE IN RELATION TO ITS USE AS AN ANALGESIC

JOHN W. LEWIS
School of Chemistry, University of Bristol, Bristol BS8 1TS, UK

INTRODUCTION

One of the important landmarks in the search for an analgesic able to treat severe pain, but lacking the abuse potential and other undesirable effects of morphine, was the discovery that nalorphine, an antagonist of morphine, had analgesic activity in humans [Lasagna and Beecher, 1954]. This discovery led to the era of the so-called mixed agonist/antagonist analgesics, a designation based on their characterization as antagonists at the morphine (μ) receptor and agonists at a second receptor, later called kappa after the prototype full agonist ketazocine. More sophisticated profiling of these opioids has led to the recognition that they are partial agonists at both μ and κ receptors with insufficient intrinsic activity at μ receptors to cause significant abuse liability.

The profiles of pentazocine, butorphanol, and nalbuphine fall within this class. Buprenorphine has a different profile, being a partial agonist only at the μ receptor. At κ it has very high affinity but low intrinsic activity so that it is effectively a κ antagonist. There is no evidence that κ antagonism contributes to the clinical pharmacological profile of buprenorphine but lack of κ agonism accounts for its freedom from dysphoric and psychotomimetic effects. That buprenorphine is a μ partial agonist and not a full agonist gives rise to significant pharmacological differences between it and morphine or other μ agonists. In this chapter the clinical pharmacology data relating to buprenorphine's μ partial agonist profile will be discussed as well as other factors that affect its utility as a treatment for both acute and chronic pain.

One of the important factors determining the extent of utilization of an analgesic is its ability to be administered by mouth, by far the most convenient route.

Buprenorphine: Combatting Drug Abuse With a Unique Opioid, pages 151–163
© 1995 Wiley-Liss, Inc.

Buprenorphine's oral bioavailability is poor but it is very well absorbed sublingually, which offers the potential for more rapid absorption and onset of action relative to the oral route, since sublingual absorption delivers the drug directly into the systemic circulation, avoiding hepatic first-pass clearance. Furthermore, sublingual administration may be particularly beneficial in cancer patients who are unable to tolerate oral administration because of nausea and vomiting, or dysphagia resulting from the disease, and are unable to tolerate parenteral administration because of decreased venous access, emaciation, or coagulation defects. Weinberg et al. [1988] showed that, among nine therapeutically important opioids, buprenorphine had the highest sublingual bioavailability, being greater than 50%, which confirmed the results of other studies (see Walter and Inturrisi, and McQuay and Moore, this volume).

CLINICAL EFFICACY

The important characteristics of a μ partial agonist that can affect its clinical efficacy as an analgesic are (1) the ceiling to its agonist (analgesic) effect; (2) the potential for antagonizing the acute effects of a full agonist if given after the agonist or for blocking the effects of the agonist if given before; and (3) the possibility of precipitating abstinence when the partial agonist is administered to patients previously treated for lengthy periods with opioid agonists, as well as the possibility of interfering with the effect of a full agonist administered after treatment with the partial agonist over a period of time.

Ceiling Effect to Buprenorphine's μ Agonist Actions

Only anecdotal data from patients in pain exist to indicate at what dose a ceiling to the analgesic effect of buprenorphine may be encountered. In order to prove that there is a ceiling effect it would be necessary to carry out single-dose controlled studies in which the effects of several individual doses were studied to show that at some point further dose increase produces no increase in analgesic effect. Such studies in the dose range likely to demonstrate the ceiling have not been reported.

Relevant Data From Single-Dose Studies. Budd [1981] studied the analgesic and respiratory depressant effects of buprenorphine in 50 female patients immediately following cesarian section. Successive increments of buprenorphine were provided intravenously until pain disappeared. Most patients received several milligrams of drug up to a maximum of 7 mg and all were successfully treated without any clinically significant respiratory depression, which was directly monitored in ten patients. The data suggest that buprenorphine's ceiling effect dose not occur below 7 mg.

There had been suggestions from clinical pharmacology studies not involving pain patients that the ceiling to buprenorphine's μ agonist effect could occur at parenteral doses substantially lower than 7 mg. From single-dose studies in detoxified opiate addicts Jasinski et al. [1978] concluded that the maximum euphorigenic dose of buprenorphine was in the range of 0.8–1.2 mg subcutaneously, and that this dose was equivalent to 20–30 mg morphine. In contrast, in a preliminary report of a study of ascending single doses of sublingual buprenorphine in a similar group of subjects, Walsh et al. [1992] showed that the ceiling for mood-elevating effects was reached at 16 mg and for no measure was the effect of 32 mg greater than that of 16 mg. Indeed, there was the suggestion that for some measures 32 mg produced less effect than 16 mg (Walsh, private communication). If this is confirmed, it will be the first evidence of a bell-shaped dose-response curve for a clinical effect of buprenorphine. Bell-shaped dose-response curves have been reported for rodent studies of buprenorphine's antinociceptive activity [Cowan et al., 1977a], respiratory depressant effect; and inhibition of gastrointestinal transit [Cowan et al., 1977b]. In rat tail flick and electrical stimulation of tail tests, the bell-shaped dose-response curve was totally shifted to the right in the presence of naloxone [Rance et al., 1980] or naltrexone [Dum and Herz, 1981] so that very high doses of buprenorphine that alone produced very little antinociception were very much more effective in the presence of the antagonist.

In a clinical study in which high doses of buprenorphine (30 and 40 μg/kg i.v.) were used as the sole anesthetic in 12 patients undergoing cholecystectomy, it was found that in the immediate postoperative period only half the patients were free from pain [Schmidt et al., 1985]. The rest were in severe pain, which could be totally abolished by the administration of low doses (0.08–0.4 mg) of naloxone. Using this procedure, these patients obtained long-lasting pain relief. It was suggested that the high doses of buprenorphine in the naloxone-treated group were on the descending arm of a bell-shaped dose-response curve so that the effect of the antagonist was to restore buprenorphine's analgesic effect as in the rodent studies [Rance et al., 1980; Dum and Herz, 1981]. In light of the extreme difficulty and high doses of naloxone required to reverse the effects of buprenorphine once established [Cowan et al., 1977 a,b; Heel et al., 1979], it is surprising that low doses of naloxone given about 2 hr after buprenorphine could have such a dramatic effect. An alternative explanation might be that the low dose of naloxone was exerting an independent, possibly non-opioid, analgesic action. Such an effect has been demonstrated both in animals [Ueda et al., 1986] and in pain patients [Lasagna, 1965].

Relevant Data From Chronic Studies. There are no hard data from clinical analgesic studies in chronic pain patients to indicate at what daily dose of buprenorphine maximum analgesic effect is achieved. Since the sublingual tablet is the usual form for chronic pain and until recently it has been available only at 0.2-mg strength (now also 0.4-mg), a working maximum daily dose of about 4 mg has been fairly arbitrarily set.

Evidence from studies in opiate addicts suggests that buprenorphine's maximum agonist effect in chronic dosing is shown at daily sublingual doses about 4 mg. In a rising-dosage study in which subjects were maintained for 2-week periods on increasing daily doses of sublingual buprenorphine, a hydromorphone challenge (cumulative 6 mg + 12 mg subcutaneously) was given nearly 24 hr after the maintenance dose [Bickel et al., 1988a]. In this situation the effect of the challenge drug depends on the extent of its cross-tolerance with the maintenance level of buprenorphine, which relates to the agonist effect of the latter. Over the buprenorphine maintenance dose range 2 mg, 4 mg and 8 mg, there was dose-related attenuation of hydromorphone's agonist effects, whereas there was no significant difference between 8 mg and 16 mg. In their primary dependence study in post addicts, Jasinski et al. [1978] used a maintenance dose of 8 mg subcutaneously per day, roughly equivalent to the sublingual dose giving maximum hydromorphone-blocking effect. The effects of a morphine challenge (30 mg s.c.) given 29 hr after buprenorphine were almost completely blocked.

Thus, it appears that buprenorphine's maximum agonist effect on repeated dosing is shown by daily sublingual doses in the range 8–16 mg. However, it must be recognized that the higher doses will have a longer duration of action—giving greater area under the time-action curve and greater total analgesic effect.

Morphine Dosage Equivalents of the "Ceiling" Dosages of Buprenorphine

Since the dose-response curves for a full agonist and a partial agonist are not parallel in the region of the partial agonist's ceiling effect, extrapolation from dosage equivalents in the lower parts of the dose-response curves is not valid. At these lower doses the curves for a partial agonist and a full agonist may be parallel. If the analgesic effect level required in a particular pain situation is in the lower range, the partial agonist and the full agonist will be equally effective. This is the case in most pain situations. Only when the pain requires very high doses of a full agonist will the effect level be in the shallower part of the partial agonist's dose-response curve. Since duration of analgesia is extended in the higher dose range, quantum of analgesia as opposed to peak effect is increased at higher doses. The inherently longer duration of buprenorphine's interaction with the receptor compared with morphine also contributes to the apparent potency ratio.

Reliable relative potency estimates of buprenorphine's clinical analgesic effect compared with morphine's are available only for single doses and in the effect range defined by 4 mg and 16 mg of intramuscular morphine [Wallenstein et al., 1986]. In these studies intramuscular buprenorphine was approximately 25 times more potent and sublingual buprenorphine 15 times more potent than intramuscular morphine. From their studies in opiate addicts, Jasinski et al. [1978] concluded that the maximum μ agonist effect of buprenorphine in single doses subcutaneously was equivalent to 20–30 mg morphine based on the persistence of agonist effects (as measured by "chronic liking" scores) in the primary dependence study; these workers estimated the effects of an 8-mg/day dose of buprenorphine to be equivalent to a dose of about 30 mg morphine administered four times daily. Since the 8-mg subcutaneous daily dose of buprenorphine appears to give maximal agonism, it can be concluded that the morphine daily dose equivalents are in the range 90–150 mg s.c. or 270–450 mg p.o. This range of morphine doses is very much higher than that suggested in a recent review by Hoskin and Hanks [1991], who commented on the apparently narrow effective dose range of buprenorphine for cancer pain and suggested that it may be a possible alternative to morphine only in a lower dose range (10–20 mg p.o. morphine every 4 hr).

Antagonism by Buprenorphine of the Effects of μ Agonists

Acute Effects. A partial agonist may antagonize the effects of a previously administered agonist depending on the proportion of receptors occupied by the agonist and the time interval between the administration of the two drugs. A high level of receptor occupancy by the agonist is required for the antagonist effect of the partial agonist to be manifested. This situation can prevail when fentanyl is used in anesthesia, and buprenorphine administered immediately after surgery can reverse it and leave the patient with postoperative analgesia. Boysen et al. [1988] compared buprenorphine with naloxone in 32 hysterectomy patients for the reversal of postoperative respiratory depression caused by preoperative fentanyl (25 μg/kg). The two agents were equally effective except at the earliest recording time (15 min), when naloxone (median dose 2.5 μg/kg) was more effective than buprenorphine (median dose 8 μg/kg).

There is no evidence that in the normal analgesic dose range buprenorphine antagonizes or blocks the effects of morphine or other μ agonists. In single-dose crossover studies comparing buprenorphine with morphine in postoperative pain there was no evidence of an effect related to the order of drug administration, which might be expected to occur if buprenorphine were exerting an antagonist or blocking effect [e.g., Wallenstein et al., 1986].

Effects After Chronic Administration of Opioids. When chronic pain patients who have received large quantities of opiates so that they are physically dependent are switched to buprenorphine, abstinence effects may be elicited. There can seldom be justification for such a prescribing decision, since at the point when substantial quantities of opiates are required a partial agonist cannot provide greater pain relief. However, it is important to understand what risk of discomfort there is to the patient should such a situation arise.

There are very few reports in the literature of precipitated abstinence caused by buprenorphine. In their single-dose study of intramuscular buprenorphine versus morphine in postoperative cancer patients, Wallenstein et al. [1986] reported that two patients who had had substantial exposure to opiates prior to operation and had received 0.8 mg and 1.6 mg buprenorphine experienced mild and moderate symptoms of withdrawal, respectively. In a case study reported by Hughes et al. [1991], a male patient with a history of opioid and other drug abuse and severe Crohn's disease, previously treated with morphine sulfate (MS) 3.6 mg i.v. every 2 hr (shown to be physically dependent by i.v. naloxone challenge), was then stabilized on MS 9 mg i.v. every 6 hr for 3 days. With an overnight interval of 12 hr since the last MS dose, buprenorphine 0.6 mg i.v. every 6 hr was substituted with no significant withdrawal effects or loss of pain control. In an earlier case study report, Breivik [1981] cited the experience of a female cancer patient with extensive skeletal metastases who was maintained on a daily regime including 120 mg p.o. methadone and was switched to 3 mg i.v. buprenorphine. It gave neither pain relief nor signs of abstinence.

There is substantial literature on the induction of opiate addicts on to a maintenance regime with sublingual buprenorphine (see Fudala and Johnson, this volume). These studies confirm that the risk of precipitating abstinence is small when transferring to buprenorphine from long-term treatment with opiates. Most of the studies in which subjects have had a controlled level of physical dependence have involved methadone maintenance. Since morphine is the most used opioid for severe chronic pain, investigation of the transfer to buprenorphine of addicts with a controlled level of morphine dependence would be more relevant to the situation of chronic pain patients. Only one such study has been reported (Jasinski and Preston, this volume). This used a small number of addicts who were stabilized on 15 mg morphine given s.c. four times a day. In these subjects, 2 mg buprenorphine both subcutaneously and sublingually produced minimal withdrawal effects. It can be deduced that pain patients receiving up to 200 mg/day oral morphine should be exposed to very little hazard in switching to an equivalent dose ($\sim$4 mg) of sublingual buprenorphine. Though the risk of precipitating abstinence is small in transferring patients from moderate doses of oral opiates to buprenorphine, such a move should not be encouraged beyond the earlier phases of pain management.

Transfer From Chronic Buprenorphine to μ Agonist Analgesics. Whereas transfer of chronic pain patients from μ agonist analgesics to buprenorphine should be unusual, the reverse is a normal occurrence when pain becomes so severe that buprenorphine fails to provide relief. There is no evidence to suggest that transferring a patient from buprenorphine to morphine or an alternative μ agonist is any more difficult than switching from one μ agonist to another. In the treatment of chronic pain with opioids over long periods, the patient may develop a degree of tolerance that reduces the analgesic effect. One solution is to change the opioid. If the two opioids are of similar intrinsic activity, and particularly if the second has lower intrinsic activity than the first, cross-tolerance will prevent the realization of an improved analgesic response. Switching will be most effective if the intrinsic activity of the second opioid is substantially higher than that of the first. This is the situation in replacing a partial agonist like buprenorphine with a full agonist.

This relationship between cross-tolerance and intrinsic activity of opioids has been investigated in rodents [Neil, 1982; Stevens and Yaksh, 1989; Adams et al., 1990]. The explanation suggested is that the prolonged administration of an agonist down-regulates receptors, which causes tolerance, to an extent that is determined by the agonist's intrinsic activity. The extent of cross-tolerance is determined by the intrinsic activity of the second opioid, since an agent of high intrinsic activity requires activation of only a small number of receptors. Since the receptor system is down-regulated in the tolerant state, a partial agonist requiring activation of a larger proportion of receptors will be ineffective.

A comparative study of buprenorphine versus methadone, a μ agonist of higher intrinsic activity than morphine, in opiate addicts supports these observations. Bickel et al. [1988b] studied addicts maintained on buprenorphine (2 mg/day s.c.) or methadone (30 mg/day p.o.) and challenged them with hydromorphone (6 mg i.m.)

approximately 24 hr after they had received the treatment medication. The subjective effects of the challenge drug were significantly higher in the buprenorphine group than in the group maintained on methadone, which would be predicted from the relative intrinsic activities of the two maintenance drugs.

That a transfer from sublingual buprenorphine to oral morphine in cancer patients can be achieved with no loss of analgesic effect was demonstrated in a study by Atkinson et al. [1990]. In this study, switching from buprenorphine (0.2 or 0.4 mg s.c. every 6 hr for 14 days) to morphine (10 or 20 mg p.o. every 4 hr for 14 days) was successful in all 16 patients in whom the transfer was made.

Summary of Effects of μ Partial Agonist Profile on Buprenorphine's Clinical Efficacy and Safety

The designation of buprenorphine as a μ partial agonist implies that there is a ceiling to its analgesic effect. The available evidence suggests that the ceiling effect occurs at a level equivalent to that of single doses of 30 mg i.m. morphine in acute pain and greater than 400 mg/day p.o. morphine for chronic pain. These limits allow buprenorphine to be an effective analgesic in most patients with moderate to severe pain.

In the normal analgesic dose range there is no attenuation of the acute effects of μ agonists administered before, after, or with buprenorphine. Patients can be transferred from prolonged treatment with μ agonists to buprenorphine with no risk of precipitating abstinence unless they are at a very high level of physical dependence. However, in this situation when a μ agonist has ceased to provide relief, there would be no justification for switching to a partial agonist. When it is required to transfer chronic pain patients maintained on buprenorphine to morphine, it can be accomplished without complications.

In regard to respiratory depression, buprenorphine—as a partial agonist—has a much greater margin of acute safety than μ agonist analgesics. However, it must be appreciated that in the therapeutic dose range the respiratory depressant effect of the partial agonist and the full agonist will not differ markedly (see Reversal of Buprenorphine's Opioid Actions, below). Other unwanted effects, notably constipation, are also limited by buprenorphine's partial agonism, but again there appears to be no evidence of substantial benefit in the normal therapeutic dose range.

INFLUENCE OF RECEPTOR KINETICS

A most important factor in the pharmacology of buprenorphine is the nature of its association with, and dissociation from, opioid receptors. Receptor kinetics were investigated in the electrically stimulated guinea pig ileum preparation by Schulz and Herz [1976]. Depression of twitch height, which is virtually instantaneous when μ agonists such as normorphine are introduced into the bath and is equally rapidly reversed by naloxone, is achieved only very gradually with buprenorphine, which takes very nearly an hour to reach peak effect and is not antagonized by naloxone.

These unusual receptor kinetics have important influences on buprenorphine's clinical profile: (1) its opioid effects are long lasting; (2) it is difficult to antagonize the acute effects of buprenorphine once the drug is established on the receptors; and (3) it is associated with a low level of physical dependence as manifested by the intensity and time course of abstinence effects when administration of buprenorphine is abruptly stopped. The biochemical equilibria that are established at new levels following repeated dosing return to normal levels so slowly when dosing is stopped that the imbalances that are the cause of abstinence effects do not occur. The partial agonism also contributes to the lower level of physical dependence by limiting the effect on the resetting of biochemical equilibria when high doses are chronically administered.

Duration of Clinical Actions

The literature contains apparently conflicting reports of buprenorphine's duration of analgesic action, especially when intramuscular buprenorphine is compared with morphine. Downing et al. [1977] found 0.6 mg i.m. buprenorphine gave longer relief than 15 mg morphine. Dobkin et al. [1977] reported that doses of up to 0.4 mg i.m. buprenorphine provided postoperative analgesia for up to 6.3 hr, compared with only 4.8 hr for 10 mg i.m. morphine. However, in a 3-day multiple-dose study of equianalgesic doses of morphine and buprenorphine, Ouellette [1982] reported similar time–effect curves for the two drugs, and this was also found in a single-dose twin crossover study by Wallenstein et al. [1986]. In the latter study, sublingual buprenorphine was also compared with i.m. morphine and was found to have a substantially longer duration of action.

In studies in which the analgesic was administered before pain became established, that is, pre- or perioperatively, or in the recovery room following operation, the extent of buprenorphine's analgesic action is very clearly shown [Kamel and Geddes, 1978; Carl and Crawford, 1984; Piepenbrock and Zenz, 1984]. In such a study, the superiority of buprenorphine's duration of action over that of morphine was also clearly demonstrated [Kay, 1978].

In a controlled study in recovered addicts of the time course of single subcutaneous doses of buprenorphine (1 mg), morphine (30 mg), and methadone (30 mg), the duration of action of buprenorphine was either longer than, or equivalent to, that of morphine for different measures [Jasinski et al., 1978]. In particular, the durations of euphorigenic effects were similar, whereas pupillary constriction due to buprenorphine far outlasted that associated with morphine. The duration of methadone's effects was similar to that of buprenorphine's for all measures.

Is there an explanation for these apparently conflicting results? The time course of effects that can be objectively measured, for example, pupillary constriction and respiratory depression (see later) clearly establish that buprenorphine has a protracted duration of action. It is when subjective effects (euphoria, analgesia) are considered that buprenorphine's duration of action appears sometimes only to be equivalent to that of morphine. Differences in receptor output required for an effect to be measurable may be important when drugs of different intrinsic activity are compared. Thus, the level of pain in a group of patients in a study may be such that a μ partial agonist can provide relief over only a narrow period around its time of peak effect. This will not be the case if the opioid-treatable pain is of much lower intensity, when a longer duration of action will be observed. In the studies of Wallenstein et al. [1986], a dose of 12 mg of intramuscular (i.m.) morphine was equivalent in duration and effect to 0.4 mg i.m. buprenorphine. In the comparison of i.m. morphine with sublingual buprenorphine, 4 mg morphine was of shorter duration than 0.8 mg buprenorphine. This indicates that the levels of pain in the two studies may have been different; when pain was of lower intensity, the duration of buprenorphine's effect relative to that of morphine was greater.

Reversal of Buprenorphine's Opioid Actions

The difficulty with which buprenorphine's opioid actions are reversed with specific opioid antagonists was demonstrated in early animal experiments [Cowan et al., 1977a]. In studies in healthy human volunteers, Orwin et al. [1976] showed that only very high doses of naloxone (8–12 mg; cf, reversal of morphine by 0.4 mg) could bring about substantial reversal of the respiratory depressant effect of buprenorphine given 30 min before. The lack of instant and complete reversal of buprenorphine's respiratory depressant effects is a major defect in its profile as a clinically useful analgesic for acute use. Though the depressant effects are only rarely of clinical significance, their long duration and relative difficulty of reversal puts buprenorphine at a competitive disadvantage against morphine and other reversible μ agonists. Reversal of buprenorphine's effects by doxapram, a specific respiratory stimulant (Orwin, 1977; Gibbs et al., 1982; Pederson et al., 1986; Sekar and Mimpriss, 1987), though having the advantage that analgesia is undiminished, does not give the same level of confidence to anesthetists as they have with other μ agonists that are rapidly and completely reversed with low doses of naloxone.

The circumstances in which respiratory depression has been significant with buprenorphine are when it is used together with other depressants, particularly benzodiazepines, in surgery [Faroqui et al., 1983; Papworth, 1983; Nishioka and Amaha, 1984; Schmidt et al., 1985]. In contrast to these reports, Budd [1981] showed no clinically significant respiratory depression when large doses of buprenorphine (1–7 mg i.v.) were given to young women who had undergone cesarean section.

An even clearer demonstration of the safety of large doses of buprenorphine was the report of Walsh et al. [1992] of the good acceptance without respiratory depression of sublingual buprenorphine in doses up to 32 mg in nontolerant recovered addicts. This lack of significant respiratory depression with high doses of buprenorphine is in accord with results in rats showing that the dose-response curve is bell-shaped, with very high doses causing less respiratory depression than intermediate doses [Doxey et al., 1982].

Since sublingual buprenorphine is usually administered at a substantial time interval after operation when other depressants have cleared, reports of respiratory depression in these circumstances have been extremely rare. This also applies to the use of buprenorphine in the treatment of chronic pain.

Physical Dependence

The favorable physical dependence profile of buprenorphine is described and discussed in Negus and Woods, and Jasinski and Preston, this volume. When high doses of buprenorphine are administered over a substantial period of time, stopping the drug produces very mild acute abstinence effects and this is also the case with naloxone challenge. The effects of withdrawal are delayed (up to 2 weeks) and are then of only moderate intensity [Jasinski et al., 1978]. Not surprisingly, reports of signs of physical dependence in pain patients treated with buprenorphine have been extremely rare.

Tolerance

In the clinical context, tolerance to individual μ agonist effects develops at different rates, apparently more rapidly for respiratory depression and sedation but more slowly for constipation. Fortunately, development of tolerance to analgesia is usually fairly slow. These differential rates can be related to the receptor reserve involved. Effects that require activation of a high proportion of receptors will be associated with more rapid development of tolerance. Similar reasoning predicts that tolerance to a partial agonist's effects will develop more rapidly than those to a full agonist. Tolerance to the analgesic effect could limit buprenorphine's efficacy in chronic pain. In practice, for moderate to severe chronic pain there is no published evidence of a rapid development of tolerance to buprenorphine's analgesic action. When buprenorphine's efficacy becomes insufficient for whatever reason in treating chronic pain, the switch to a full agonist can be made very easily (see Transfer From Chronic Buprenorphine to μ Agonist Analgesics, above).

CONCLUSIONS

Buprenorphine is used for the indications for which opioids are usually prescribed. These are the "opioid-sensitive" pains, particularly acute postoperative pain, cancer pain, and certain nonmalignant pain conditions. As a μ partial agonist, it has advantages over the full agonists but these are balanced by some disadvantages. In both acute and chronic pain, the ceiling to buprenorphine's clinical analgesic action is sufficiently high for it to be a constraint only rarely in the most severe pain.

The main advantages of the partial agonist profile are the very favorable safety from overdosage and the very mild physical dependence that follows extended dosage. The main weakness in buprenorphine's pharmacological profile is the difficulty in reversing its acute effects, which puts it at a competitive disadvantage in acute use. However, the slowness of receptor dissociation that is responsible for this effect also confers prolonged duration of analgesic action and contributes to the low level of physical dependence.

In short, buprenorphine is a *unique* opioid that offers viable alternative therapy to the agonist opiates for the treatment of moderate to severe pain.

REFERENCES

Adams JU, Paronis CA, Holtzman SG (1990): Assessment of relative intrinsic activity of mu opioid analgesics in vivo by using β-funaltrexamine. J Pharmacol Exp Ther 255:1027–1032.

Atkinson RE, Schofield P, Mellor P (1990): Buprenorphine to morphine transfer: An assessment in patients with advanced cancer pain. Pain (Suppl 5):S369.

Bickel WK, Stitzer ML, Bigelow GE, Liebson IA, Jasinski DR, Johnson RE (1988a): A clinical trial of buprenorphine: Comparison with methadone in the detoxification of heroin addicts. Clin Pharmacol Ther 43:72–78.

Bickel WK, Stitzer ML, Bigelow GE, Liebson IA, Jasinski DR, Johnson RE (1988b): Buprenorphine: Dose-related blockade of opioid challenge effects in opioid dependent humans. J Pharmacol Exp Ther 247:47–53.

Boysen K, Hertel S, Chraemmer-Jørgensen B, Risbo A, Poulsen NJ (1988): Buprenorphine antagonism of ventilatory depression following fentanyl anaesthesia. Acta Anaesth Scand 32:490–492.

Breivik H. (1981): Experience with parenteral and sublingual buprenorphine in the treatment of chronic pain in cancer patients. In Valentin N, Crawford M, Volquardsen CR (eds): "Nordic Symposium on Temgesic." Copenhagen: Meda, pp 104–111.

Budd K (1981): High dose buprenorphine for postoperative analgesia. Anaesthesia 36:900–903.

Carl P, Crawford ME (1984): A comparison between buprenorphine, fentanyl and pentazocine in analgesic-supplemented flunitrazepam–nitrous oxide anaesthesia. Proc R Soc Med Int Cong Ser 65:71–76.

Cowan A, Lewis JW, Macfarlane IR (1977a): Agonist and antagonist properties of buprenorphine: A new antinociceptive agent. Br J Pharmacol 60:537–545.

Cowan A, Doxey JC, Harry EJR (1977b): The animal pharmacology of buprenorphine: An oripavine analgesic agent. Br J Pharmacol 60:547–554.

Dobkin AB, Esposito B, Philbin C (1977): Double-blind evaluation of buprenorphine hydrochloride for post-operative pain. Can Anaesth Soc J 24:195–202.

Downing JW, Leary WP, White ES (1977): Buprenorphine: A new potent long-acting synthetic analgesic. Comparison with morphine. Br J Anaesth 49:251–255.

Doxey JC, Everitt JE, Frank LW, MacKenzie JE (1982): A comparison of the effects of buprenorphine and morphine on the blood gases of conscious rats. Br J Pharmacol 75(Suppl):118P.

Dum JE, Herz A (1981): In vivo receptor binding of the opiate partial agonist, buprenorphine, correlated with its agonistic and antagonistic actions. Br J Pharmacol 74:627–633.

Faroqui MH, Cole M, Curran J (1983): Buprenorphine, benzodiazepines and respiratory depression. Anaesthesia 38:1002–1003.

Gibbs JM, Johnson H, Davis FM (1982): Patient administration of i.v. buprenorphine for post-operative pain relief using the "Cardiff" demand analgesia apparatus. Br J Anaesth 54:279–284.

Heel RC, Brogden RN, Speight TM, Avery GS (1979): Buprenorphine: A review of its pharmacological properties and therapeutic efficacy. Drugs 17:81–110.

Hoskin PJ, Hanks GW (1991): Opioid agonist–antagonist drugs in acute and chronic pain states. Drugs 41:326–344.

Hughes JR, Bickel WK, Higgins ST (1991): Buprenorphine for pain relief in a patient with drug abuse. Am J Drug Alcohol Abuse 17:451–455.

Jasinski DR, Pevnick JS, Griffith JD (1978): Human pharmacology and abuse potential of the analgesic buprenorphine. Arch Gen Psychiatry 35:501–516.

Kamel MM, Geddes IC (1978): A comparison of buprenorphine and pethidine for immediate post-operative pain relief by the i.v. route. Br J Anaesth 50:599–603.

Kay B (1978): A double-blind comparison of morphine and buprenorphine in the prevention of pain after operation. Br J Anaesth 50:605–609.

Lasagna L (1965): Drug interaction in the field of analgesic drugs. Proc R Soc Med 58:978–983.

Lasagna L, Beecher HK (1954): The analgesic effectiveness of nalorphine–morphine combinations in man. J Pharmacol Exp Ther 112:356–363.

Neil A (1982): Morphine and methadone-tolerant mice differ in cross tolerance to other opiates. Naunyn-Schmidebergs Arch Pharmacol 320:50–53.

Nishioka K, Amaha K (1984): Evaluation of analgesic, sedative and respiratory depressant effects of intravenous buprenorphine in surgical and gynaecological patients during the immediate post-operative period. Proc R Soc Med Int Cong Ser 65:107–113.

Orwin JM (1977): The effect of doxapram on buprenorphine-induced respiratory depression. Acta Anaesthesiol Belg 28:93–106.

Orwin JM, Robson PJ, Orwin J, Price M (1976): Antagonist action of naloxone on the acute effects of buprenorphine. In "Proceedings of the Sixth World Congress on Anaesthesiology, Mexico City," pp 189–206.

Ouellette RD (1982): Buprenorphine and morphine efficacy in postoperative pain: A double-blind multiple-dose study. J Clin Pharmacol 22:165–172.

Papworth DP (1983): High dose buprenorphine for postoperative pain. Anaesthesia 38:163.

Pedersen JE, Chraemmer-Jørgensen B, Schmidt JF, Risbo A (1986): Peroperative buprenorphine: Do high dosages shorten analgesia postoperatively? Acta Anaesthesiol Scand 30:660–663.

Piepenbrock S, Zenz M (1984): Postoperative pain relief with intravenous opiates: Buprenorphine vs pentazocine. Proc R Soc Med Int Cong Ser 65:95–105.

Rance MJ, Lord JAH, Robinson T (1980): Biphasic dose response to buprenorphine in the rat tail flick assay: Effect of naloxone pretreatment. In Way EL (ed): "Endogenous and Exogenous Opiate Agonists and Antagonists." New York: Pergamon, pp 387–390.

Schmidt JF, Chraemmer-Jørgensen B, Pedersen JE, Risbo A (1985): Postoperative pain relief with naloxone. Severe respiratory depression and pain after high dose buprenorphine. Anaesthesia 40:583–586.

Schulz R, Herz A (1976): The guinea-pig ileum as an In vitro model to analyse dependence liability of narcotic drugs. In Kosterlitz HW (ed): "Opiates and Endogenous Opioid Peptides." Amsterdam: Elsevier, pp 319–326.

Sekar M, Mimpriss TJ (1987): Buprenorphine, benzodiazepines and prolonged respiratory depression. Anaesthesia 42:567–568.

Stevens CW, Yaksh TL (1989): Potency of infused spinal antinociceptive agents is inversely related to magnitude of tolerance after continuous infusion. J Pharmacol Exp Ther 250: 1–8.

Ueda H, Fukushima N, Kitao T, Ge M, Takagi H (1986): Low doses of naloxone produce analgesia in the mouse brain by blocking presynaptic autoinhibition of enkephalin release. Neurosci Lett 65:247–252.

Wallenstein SL, Kaiko RF, Rogers AG, Houde RW (1986): Crossover trials in clinical analgesic assays: Studies of buprenorphine and morphine. Pharmacotherapy 6:228–235.

Walsh SL, Preston KL, Stitzer ML, Bigelow GE, Liebson IA (1992): The acute effects of high dose buprenorphine in non-dependent humans. NIDA Res Monogr 119:245.

Weinberg DS, Inturrisi CE, Reidenberg B, Moulin DE, Nip TJ, Wallenstein S, Houde RW, Foley KM (1988): Sublingual absorption of selected opioid analgesics. Clin Pharmacol Ther 44:335–342.

BUPRENORPHINE: EPIDURAL AND INTRATHECAL USE

CARL E. ROSOW

Department of Anesthesia, Massachusetts General Hospital, Boston, MA 02114

GENERAL CONSIDERATIONS

Pharmacodynamics

Spinal (i.e., intrathecal and epidural) administration of opioids can produce very intense, segmental analgesia for acute or chronic pain states that involve the legs or torso. Epidural administration is the more common mode of administration, because placement of an epidural catheter is a safer way to perform repeated dosing. Intrathecal administration is usually reserved for situations in which a lumbar puncture is already planned, and a single injection of the opioid will suffice.

Epidural opioids are highly effective for most types of acute and chronic pain. They are frequently used postoperatively when an epidural catheter has already been inserted for administration of local anesthetics. Small doses of most opioids produce marked prolongation of local anesthetic effect, so they are often combined with drugs like bupivacaine. In cancer pain and other chronic pain states they may be given by intermittent injection through a tunneled catheter and by continuous infusion with external or implantable pump systems.

Virtually all clinically available opioids have been given by spinal injection, although only morphine is FDA-approved for administration by this route. The advantages of spinally injected morphine have been demonstrated in numerous clinical trials: Relatively small doses achieve CSF concentrations 100–1,000 times those seen after systemic administration. A 2- to 5-mg dose of morphine given epidurally can produce analgesia that frequently lasts for 12–24 hr. For potent, lipophilic opioids like fentanyl and buprenorphine the advantages are not quite so dramatic (see below).

The basis for so-called selective spinal analgesia is now fairly well understood:

Buprenorphine: Combatting Drug Abuse With a Unique Opioid, pages 165–174
© 1995 Wiley-Liss, Inc.

Opioids suppress the activity of nociceptive afferent neurons in laminae I and V of the spinal dorsal horn. The activity of other neurons (e.g., those in lamina IV, which process light touch and proprioception) are generally unaffected. Unlike local anesthetics, opioids do not cause vasodilation or hypotension by blocking sympathetic preganglionic nerves. Opioids do not block somatic sensory or motor function, although weak local anesthetic properties have been demonstrated for meperidine. Spinal opioid analgesia is probably mediated by actions on μ, δ, and κ opioid receptor systems.

PHARMACOKINETICS

The pharmacokinetics of spinally administered opioids are determined primarily by lipid solubility. After epidural administration, a portion of the opioid is taken up by the epidural venous plexus or trapped in epidural fat. The onset of analgesia depends upon the rate at which the remaining opioid penetrates the dura and the lipid-rich environment of the dorsal horn. Molecular size and shape may also influence opioid passage through the dura [Moore et al., 1982]. The termination of opioid effect depends mainly upon vascular absorption and redistribution away from the site of action.

Morphine is a highly ionized, hydrophilic opioid that has an octanol–water partition coefficient (PC) of 1.42. It penetrates both dura and cord slowly, so effective analgesia may not occur for as much as 1 hr. There is not much uptake of morphine by epidural fat or by the spinal cord, so a relatively small epidural dose eventually produces a high concentration of drug in the CSF. Morphine has an extraordinarily long duration of action, because its vascular absorption and removal from the cord is slow. In addition, morphine remaining in the CSF acts as a depot to prolong the analgesic effect.

The situation is quite different with lipophilic opioids like fentanyl (PC = 860) or buprenorphine (PC = 2320). Penetration of the cord is rapid, and useful analgesia may be seen within 10–15 min. Some of the early analgesic effect is actually due to systemic absorption, because these drugs undergo rapid uptake into the vertebral venous plexus. Significant plasma concentrations of fentanyl are measurable within minutes of epidural administration. Large amounts of fentanyl are also sequestered in epidural fat, so concentrations of the opioid in CSF remain relatively low. Fentanyl is rapidly removed from the cord by redistribution, so the duration of analgesic effect is short (2–6 hr).

These differences in distribution influence epidural analgesic potency. Unlike morphine, the dose of fentanyl or buprenorphine required for effective epidural analgesia is very close to that used for systemic administration. The inverse relationship between lipid solubility and epidural analgesic potency has been documented for a number of opioids. Van den Hoogen and Colpaert [1987] showed that the ratio of subcutaneous to epidural potency decreases as the octanol–water partition coefficient increases. Similar relationships have been described for intrathecal potency (Fig. 1) [Dickenson et al., 1990].

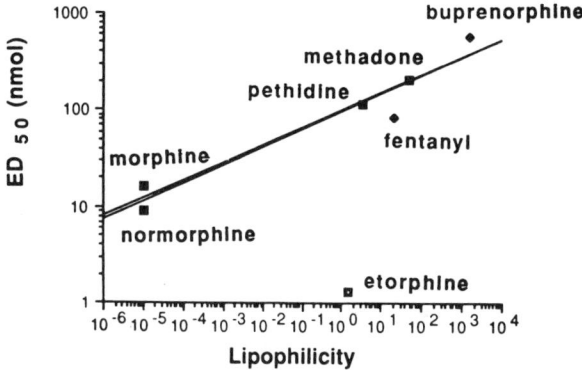

Fig. 1. Relationship between lipophilicity (heptane–water partition coefficient) and potency for opioids administered intrathecally to halothane-anesthetized rats. Electrical activity in dorsal horn nociceptive neurons was evoked by C-fiber stimulation. Potency is expressed as the ED_{50} for suppression of this activity. Results from two series of experiments are shown. For all opioids except etorphine potency is inversely related to lipophilicity ($r^2 = 0.93$) [Dickenson et al., 1990].

The influence of lipophilicity on duration of epidural opioid analgesia was demonstrated in a clinical study by Torda and Pybus [1982] (Table I).

Given epidurally, methadone has a *shorter* duration than morphine. It is slightly longer-acting than meperidine despite being more lipid-soluble, and this may be a reflection of its high receptor affinity. Similarly, the duration of epidural buprenorphine is in the range of 6–12 hr [Lanz et al., 1984], substantially longer than would be predicted on the basis of lipid solubility alone.

Since a decrease in the dose and an increase in duration are major advantages of spinally administered morphine, some have questioned whether there is any advantage to giving lipophilic opioids by this route. Loper et al. [1990] compared epidural and intravenous administration of fentanyl in surgical patients. They were unable to show a significant difference in either pain scores or plasma concentrations of fentanyl. Although most clinicians still feel that epidural and parenteral fentanyl are different, the issues are germane when we consider a highly lipophilic opioid like buprenorphine.

TABLE I. Comparison of Four Opioids Given Epidurally

Drug	Dose (mg)	Lipid solubility (oct:H_2O)	Duration (hr)
Morphine	6	1.42	12.26
Meperidine	60	39	6.61
Methadone	6	116	8.71
Fentanyl	0.06	860	5.71

Toxicity

The toxicology of spinal opioids raises several issues that are not pertinent to systemic administration.

1. High concentrations of the opioid are applied directly to nerve tissue, so local neurotoxicity must be ruled out. The pH and preservatives (e.g., metabisulfites, chlorbutanol) also become important considerations. Histological studies in humans and animals have established that even chronic administration of preservative-free morphine does not cause local damage. Lesser amounts of pathological data are available for fentanyl, sufentanil, and meperidine. No clinically used opioid has definitely been shown to cause direct neural damage.

2. Respiratory depression, sedation, nausea, and other typical opioid effects can occur soon after epidural injection. This is most common following lipophilic opioids and is due to systemic absorption. Rarely, clinically significant respiratory depression and sedation can occur 8–12 hr later. This phenomenon is more likely with a hydrophilic opioid like morphine, because the high concentration of drug in the CSF undergoes slow rostral migration. (Water-soluble radioopaque dye has been shown to take about 6 hr to travel from the lumbar cistern to the level of medullary respiratory centers.) Lipophilic opioids tend to bind at or near the level of injection, so rostral spread is much less likely.

3. Urinary retention requiring bladder catheterization occurs commonly, especially in males. It is due to a dramatic reduction in tone of the detrusor muscle and bladder wall. Some studies suggest that low doses of naloxone may be used to reverse this effect without eliminating all analgesia.

4. Generalized pruritus that is nonsegmental in distribution can be produced by all μ agonists, particularly morphine. It is much less common with agonist–antagonists like butorphanol [Abboud et al., 1987]. The effect is not due to histamine release, but it appears to be a dysesthesia mediated at the level of the cord and brainstem. The pruritus is rapidly reversible by naloxone in doses that may preserve adequate analgesia. This side effect is particularly frequent and intense when spinal opioids are used during labor and delivery—the itch can be more distressing than the pain being treated.

SPINAL BUPRENORPHINE

Preclinical Studies

Hassan et al. [1989] demonstrated in the rat that perineural injection of buprenorphine (20 μg/ml) did not produce measurable local anesthesia.

There is good reason to doubt that buprenorphine produces analgesia by a specific spinal action. Dickenson et al. [1990] measured evoked electrical activity of dorsal horn nociceptive neurons in the halothane-anesthetized rat. In this preparation, intrathecal fentanyl and etorphine suppressed the responses to C-fiber stimulation in a dose-related fashion. Buprenorphine was inactive unless very high doses were used. At these doses Aβ fibers were inhibited as well. Unlike fentanyl and etorphine, the effects of buprenorphine were not reversed by naloxone.

There have been several animal studies using behavioral endpoints to compare the analgesic effects of intrathecal buprenorphine and morphine. One study showed that intrathecal buprenorphine was substantially more potent than morphine in the rat tail flick assay [Bernatsky and Jurna, 1986]. In two other studies using tail-flick, tail-immersion, and paw-pressure tests, the two drugs appeared equipotent [Bryant et al., 1983; Ossipov et al., 1988]. By way of contrast, buprenorphine is 1,000 times more potent than morphine when the drugs are injected directly into the median raphe nucleus [Bryant et al., 1982]. This is further evidence for a predominantly supraspinal site of action.

Clinical Studies

Intrathecal. There have been only three published studies of intrathecal buprenorphine in patients, and in all cases the opioid was combined with a local anesthetic. Lipp et al. [1987] performed a double-blind comparison of isobaric tetracaine anesthesia with and without 0.15 mg buprenorphine. The opioid increased the level of tetracaine block and prolonged the duration of postoperative analgesia from 9 to 13 hr. Capogna et al. [1988] carried out a blind evaluation in 90 elderly males undergoing suprapubic prostatectomy. Each received hyperbaric bupivacaine with either 0, 0.03, or 0.045 mg of buprenorphine. The opioid significantly increased the intensity and duration of postoperative analgesia. Buprenorphine also increased nausea and vomiting, but there were few other side effects. The same protocol was used to study the effects of intrathecal buprenorphine in 45 women undergoing elective cesarean section [Celleno and Capogna, 1989]. Once again, the duration of analgesia was prolonged in a dose-dependent fashion. There was a significant increase in nausea and vomiting, but only 3/30 buprenorphine patients experienced pruritus.

Epidural. In most cases, epidural buprenorphine has been studied together with, or following, local anesthetics. The available clinical studies are summarized in Table II. The first and largest dose-response study was that of Lanz et al. [1984], who tested 0.15 and 0.3 mg (with mepivacaine or bupivacaine) in postoperative pain. The lower dose proved to be insufficient, since 40% of the patients complained of moderate to severe pain within the first 3 hr. A dose of 0.3 mg was satisfactory, and this has been the dose used in the majority of studies. The analgesic effect was measurable for 6–12 hr after injection. It is virtually certain that systemic effects were occurring with these doses: Volunteer studies have shown that 0.15 and 0.3 mg of buprenorphine given epidurally cause miosis and significant depression of CO_2 response for up to 8 hr [Molke-Jensen et al., 1987; Ravnborg et al., 1987].

The range of effective doses is still debatable, since good effects have been claimed with doses as low as 60 or 90 µg [Murphy and MacEvilly, 1984; Murphy et al., 1984; Simpson et al., 1988]. Doses as high as 0.9 mg have been safely administered by the lumbar epidural route, but it is not clear whether they are any more effective [Gundersen et al., 1986].

Subsequent studies have generally confirmed these findings. Given without local anesthetics, epidural buprenorphine can sometimes produce analgesia equivalent to

TABLE II. Epidural Buprenorphine in Acute Pain

Study[a]	N	Local anesth[b]	Dose (mg)	Comparison drugs[c]	Clinical setting
Rondomanska et al. [1982]	72	L	0.3	±lidocaine	Abdominal surgery
Cahill et al. [1983]	24	—	0.06	mor	Abdominal surgery
Lanz et al. [1984]	158	B,M	0.15, 0.3		Orthopedic surgery
Murphy and MacEvilly	20	B	0.06	mor (i.m.)	Spine surgery
Murphy et al. [1984]	30	—	0.06	mor (i.m.)	Hip surgery
Bilsback et al. [1985]	60	L,B	0.3	lof, pla	Orthopedic surgery
Gundersen et al. [1986]	45	B	0.3, 0.9	mor	Abdominal thoracic surgery
Wolff et al. [1986]	40	—	0.3	mor	Orthopedic surgery
Chrubasik et al. [1987]	34	—	0.15 + infusion	mor	Abdominal surgery
Petit et al. [1988]	24	—	6.6 µg/kg	mor	Thoracic surgery
Simpson et al. [1988]	57	B	0.09, 0.18	mor	Cesarean
Ackerman et al. [1989]	60	B	0.3	mor, fent, but	Cesarean
Celleno et al. [1991]	90	L	0.3	mor, fent, suf, oxy, pla	Cesarean
Lehmann et al. [1992]	80	B	0.3	pla	Labor

[a]Bold print indicates double-blind protocol.
[b]B, bupivacaine; L, lidocaine; M, Mepivacaine.
[c]mor, morphine; fent, fentanyl; but, butorphanol; suf, sufentanil; lof, lofentanil; oxy, oxymorphone; pla, placebo.

that produced by morphine [Wolff et al., 1986; Chrubasik et al., 1987; Pasqualucci et al., 1987; Petit et al., 1988]. When it is given during local anesthetic blockade, buprenorphine can produce significant prolongation of analgesia, although at the price of additional side effects. In most published studies buprenorphine analgesia is rated no better than or inferior to that of morphine and other agonists [Bilsback et al., 1985; Celleno et al., 1991; Lehmann et al., 1992], but the frequency and severity of side effects is usually better.

There are only a few studies dealing with long-term administration of epidural buprenorphine in chronic pain. Carl et al. [1986] published a retrospective analysis of 61 patients treated an average of 53 days (range, 7–262). Both benign and malignant pain syndromes were included. The mean daily dose was 1.3 mg given by an average of 2.6 injections. Buprenorphine appeared to be well-tolerated and effective: Satisfactory pain relief was obtained in 67% of patients, and the frequency of side effects was significantly better than that of epidural morphine.

Toxicity

Direct Tissue Toxicity. Like so many other opioids, buprenorphine was not tested for possible spinal application prior to approval and marketing. As a result, the issue of neurotoxicity was never addressed in a systematic fashion. Thus far, thousands of patients have received epidural and intrathecal buprenorphine and there have been no reports of tissue toxicity. One in vitro study by Börner et al. [1982] showed that buprenorphine is physically compatible with CSF at a concentration of 0.015 mg/ml. As mentioned previously, Carl et al. [1986] reported on a series of patients receiving epidural injections of buprenorphine up to 262 days. Postmortem examination of dura and surrounding tissue was carried out in one patient (length of treatment unspecified), and no abnormalities were found. To my knowledge, there have been no other histologic studies of neural tissue exposed to buprenorphine either acutely or chronically.

Respiratory Depression. We have seen that normal epidural doses of buprenorphine are sufficient to depress CO_2 response in normal volunteers [Molke-Jensen et al., 1987]. Knape [1986] reported two cases of total apnea 12–17 min after injection of 0.3 mg. In both cases the drug was administered in the setting of an established block with lidocaine–epinephrine. Naloxone (0.8–1.0 mg) was ineffective in reversing the respiratory depression. There have been no subsequent reports of such severe reactions.

Pruritus. The incidence of pruritus is consistently low in comparison with morphine reaction. Ackermann et al. [1989] conducted a randomized double-blind comparison of four epidural opioids in 60 patients undergoing cesarean delivery. Morphine, fentanyl, butorphanol, and buprenorphine caused itching in 60%, 47%, 7%, and 0%, respectively. Itching has been a minor or infrequent complication in all but two studies [Lipp et al., 1987; Simpson et al., 1988]. Keaveny and Harper [1989] report a case in which epidural buprenorphine was used to treat severe pruritus caused by epidural morphine.

Other Side Effects. Nausea and sedation are relatively common occurrences and probably reflect systemic absorption of the drug.

Urinary retention requiring catheterization probably occurs in 10% or more of patients receiving epidural morphine [Gustafsson et al., 1982]. None of the published studies mentions this as a significant side effect of buprenorphine. Drenger and Magora [1989] compared the urodynamic effects of intrathecal fentanyl and buprenorphine in the dog. Fentanyl produced dramatic decreases in bladder tone, while buprenorphine caused small and inconsistent changes.

Hallucinations occur in only 0.1% of patients given parenteral buprenorphine [Harcus et al., 1980]. MacEvilly and O'Carroll (1989) reported five cases of hallucinations when repeated epidural doses of buprenorphine were used for postoperative analgesia. The total doses of buprenorphine ranged from 750 to 1,200 μg. There are no other reports to suggest that hallucinations are more frequent when the epidural route is used.

Finally, one must consider the possible consequences of the antagonist effects of buprenorphine. Christensen and Andersen [1982] reported treating two women with terminal cancer who became tolerant to epidural morphine (requiring 12–15 mg twice daily). Each received 0.3 mg of buprenorphine epidurally 12 hr after the last morphine dose, and within 2 hr both had developed pallor, hypotension, and tachycardia. It is certainly possible that these reactions represented some variant of precipitated abstinence.

CONCLUSIONS

Spinally administered buprenorphine produces a significant analgesic effect but only at doses that have systemic activity. It has a moderate duration of action (6–12 hr), and it can significantly prolong the analgesia produced by local anesthetics. Compared with morphine, buprenorphine has a lower incidence of troublesome side effects such as pruritus and urinary retention. It remains to be established whether epidural or intrathecal buprenorphine is superior to parenteral buprenorphine in terms of analgesic efficacy or side effect profile.

REFERENCES

Abboud TK, Moore M, Zhu J, Murakawa K, Minehart M, Longhitano M, Terrasi J, Klepper ID, Choi Y, Kimball S, Chu G (1987): Epidural butorphanol or morphine for the relief of post-cesarean section pain: Ventilatory responses to carbon dioxide. Anesth Analg 66:887–893.

Ackerman WE, Juneja MM, Kaczorowski DM, Colclough GW (1989): A comparison of the incidence of pruritus following epidural opioid administration in the parturient. Can J Anaesth 36:388–391.

Bernatzky G, Jurna I (1986): Intrathecal injection of codeine, buprenorphine, tilidine, tramadol and nefopam depresses the tail-flick response in rats. Eur J Pharmacol 120:75–80.

Bilsback P, Rolly G, Tampubolon O (1985): Efficacy of the extradural administration of lofentanil, buprenorphine or saline in the management of postoperative pain: A double-blind study. Br J Anaesth 57:943–948.

Börner U, Müller H, Stoyanov M, Hempelmann G (1982): Epidural opiate analgesia (EOA): Compatibility of opiates with tissue and CSF. In Yaksh TL, Müller H (eds): "Spinal Opiate Analgesia: Experimental and Clinical Studies." Berlin: Springer-Verlag, pp 33–36.

Bryant RM, Olley JE, Tyers MB (1982): Involvement of the median raphe nucleus in antinociception induced by morphine, buprenorphine and tilidine in the rat. Br J Pharmacol 77:615–624.

Bryant RM, Olley JE, Tyers MB (1983): Antinociceptive actions of morphine and buprenorphine given intrathecally in the conscious rat. Br J Pharmacol 78:659–663.

Cahill J, Murphy D, O'Brien D, Mulhall J, Fitzpatrick G (1983): Epidural buprenorphine for pain relief after major abdominal surgery: A controlled comparison with epidural morphine. Anaesthesia 38:760–764.

Capogna G, Celleno D, Tagariello V, Loffreda-Mancinelli C (1988): Intrathecal buprenorphine for postoperative analgesia in the elderly patient. Anaesthesia 43:128–130.

Carl P, Crawford ME, Ravlo O, Bach V (1986): Longterm treatment with epidural opioids: A retrospective study comprising 150 patients treated with morphine chloride and buprenorphine. Anaesthesia 41:32–38.

Celleno D, Capogna G (1989): Spinal buprenorphine for postoperative analgesia after caesarean section. Acta Anaesthesiol Scand 33:236–238.

Celleno D, Capogna G, Sebastiani M, Costantino P, Muratori F, Cipriani G, Emanuelli M (1991): Epidural analgesia during and after cesarean delivery: Comparison of five opioids. Reg Anesth 16:79–83.

Christensen FR, Andersen LW (1982): Adverse reaction to extradural buprenorphine. Br J Anaesth 54:476.

Chrubasik J, Vogel W, Trötschler H, Farthmann EH (1987): Continuous-plus-on-demand epidural infusion of buprenorphine versus morphine in postoperative treatment of pain. Arzneim-Forsch 37:361–363.

Dickenson AH, Sullivan AF, McQuay HJ (1990): Intrathecal etorphine, fentanyl and buprenorphine on spinal nociceptive neurones in the rat. Pain 42:227–234.

Drenger B, Magora F (1989): Urodynamic studies after intrathecal fentanyl and buprenorphine in the dog. Anesth Analg 69:348–353.

Gundersen RY, Andersen R, Narverud G (1986): Postoperative pain relief with high-dose epidural buprenorphine: A double-blind study. Acta Anaesth Scand 30:664–667.

Gustafsson LL, Schildt B, Jacobsen K (1982): Adverse effects of extradural and intrathecal opiates: Report of a nationwide survey in Sweden. Br J Anaesth 54:479–486.

Harcus AH, Ward AE, Smith DW (1980): Buprenorphine in postoperative pain: Results in 7500 patients. Anaesthesia 33:382–386.

Hassan HG, Åkerman B, Pilcher CWT, Renck H (1989): Antinociceptive effects of localized administration of opioids compared with lidocaine. Reg Anesth 14:138–144.

Keaveny JP, Harper NJN (1989): Treatment of epidural morphine-induced pruritus with buprenorphine. Anaesthesia 44:691.

Knape JTA (1986): Early respiratory depression resistant to naloxone following epidural buprenorphine. Anesthesiology 64:382–384.

Lanz E, Simko G, Theiss D, Glocke MH (1984): Epidural buprenorphine—A double-blind study of postoperative analgesia and side effects. Anesth Analg 63:593–598.

Lehmann KA, Stern S, Breuker K-H (1992): Geburtshilfliche peridural-anästhesie mit bupivacain und buprenorphin. Anaesthesist 41:414–422.

Lipp M, Daubländer M, Lanz E (1987): Buprenorphin 0,15 mg intrathekal zur postoperativen Analgesie. Anaesthesist 36:233–238.

Loper KA, Ready LB, Downey M, Sandler AN, Nessly M, Rapp S, Badner N (1990): Epidural and intravenous fentanyl infusions are clinically equivalent after knee surgery. Anesth Analg 70:72–75.

MacEvilly M, O'Carroll C (1989): Hallucinations after epidural buprenorphine. Br Med J 298:928–929.

Molke-Jensen F, Jensen N-H, Holk IK, Ravnborg M (1987): Prolonged and biphasic respiratory depression following epidural buprenorphine. Anaesthesia 42:470–475.

Moore RA, Bullingham RES, McQuay HJ, Hand CW, Aspel JB, Allen MC, Thomas D (1982): Dural permeability to narcotics: In vitro determination and application to extradural administration. Br J Anaesth 54:1117–1128.

Murphy DF, MacEvilly M (1984): Pain relief with epidural buprenorphine after spinal fusion: A comparison with intramuscular morphine. Acta Anaesth Scand 28:144–146.

Murphy DF, MacGrath P, Stritch M (1984): Postoperative analgesia in hip surgery: A controlled comparison of epidural buprenorphine with intramuscular morphine. Anaesthesia 39:181–183.

Ossipov MH, Suarez LJ, Spaulding TC (1988): A comparison of the antinociceptive and behavioral effects of intrathecally administered opiates, α-2-adrenergic agonists, and local anesthetics in mice and rats. Anesth Analg 67:616–624.

Pasqualucci V, Tantucci C, Paoletti F, Dottorini ML, Bifarini G, Belfiori R, Berioli MB, Grassi V, Sorbini CA (1987): Buprenorphine vs. morphine via the epidural route: A controlled comparative clinical study of respiratory effects and analgesic activity. Pain 29:273–286.

Petit J, Comar D, Pigot B, Eustache ML, Oksenhendler G, Winckler C (1988): Analgésie péridurale après chirurgie thoracique: Morphine versus buprénorphine. Ann Fr Anesth Réanim 7:464–470.

Ravnborg M, Jensen FM, Jensen N-H, Holk IK (1987): Pupillary diameter and ventilatory CO_2 sensitivity after epidural morphine and buprenorphine in volunteers. Anesth Analg 66:847–851.

Rondomanska M, de Castro J, Lecron L (1982): The use of epidural buprenorphine for the treatment of postoperative pain. In Yaksh TL and Müller H (eds): "Spinal Opiate Analgesia: Experimental and Clinical Studies." Berlin: Springer-Verlag, pp 91–94.

Simpson KH, Madej TH, McDowell JM, MacDonald R, Lyons G (1988): Comparison of extradural buprenorphine and extradural morphine after caesarean section. Br J Anaesth 60:627–631.

Torda TA, Pybus DA (1982): Comparison of four narcotic analgesics for extradural analgesia. Br J Anaesth 54:291–294.

Van den Hoogen R, Colpaert FC (1987): Epidural and subcutaneous morphine, meperidine, fentanyl, and sufentanil in the rat: Analgesia and other in vivo pharmacologic effects. Anesthesiology 66:186–194.

Wolff J, Carl P, Crawford ME (1986): Epidural buprenorphine for postoperative analgesia: A controlled comparison with epidural morphine. Anaesthesia 46:77–79.

BUPRENORPHINE IN PSYCHIATRIC DISORDERS

DAVID NUTT, SIMON GROVES, NICK COUPLAND, and PAUL GLUE
Psychopharmacology Unit, School of Medical Sciences, University of Bristol, Bristol BS8 1TD, UK

INTRODUCTION

The use of opioid treatment in psychiatry has a long if somewhat controversial history [McDiarmid, 1876; Comfort, 1977; Emrich et al., 1988]. In the last century various formulations of opiates became popular remedies for minor complaints. The risks of overdose and dependence began to be reevaluated at about the turn of the century and to a great extent opioids have disappeared from the pharmacotherapeutic armamentarium, excepting their use in analgesia and more recently in drug-dependence therapy. Why should buprenorphine be different?

There are two reasons why buprenorphine might be of value in psychiatric illness: its μ partial agonism and its κ antagonism. There is little doubt that μ agonists are powerful anxiolytic and calming agents whose use is curtailed by their addictive liability. Moreover, there is a body of evidence that implicates opioids and/or their receptors in depression and schizophrenia [Fink et al., 1970; Kline et al., 1977; Angst et al., 1979; Pickar et al., 1981; Pfeiffer et al., 1986; Schmauss and Emrich, 1988; Frecksa et al., 1989; Matussek and Hoehe, 1989]. A μ agonist that did not cause dependence would provoke a rethinking of the use of this class of drugs and a reappraisal of their therapeutic utility. The low intrinsic activity of buprenorphine, particularly its limited capacity to cause respiratory depression, makes it a candidate for evaluation, provided intravenous administration can be avoided by a suitable pharmaceutical preparation.

The κ antagonism of buprenorphine also offers a new approach to the investigation, if not treatment, of psychiatric illness. A number of lines of evidence point towards involvement of endogenous κ agonists in these disorders, originating in the disappointing revelation that the clinical potential of κ agonists as analgesics was

---

Buprenorphine: Combatting Drug Abuse With a Unique Opioid, pages 175–186
© 1995 Wiley-Liss, Inc.

nullified by their psychotomimetic and dysphorogenic propensities [Pfeiffer et al., 1986; Millan, 1990]. Subsequently, theories of depression, schizophrenia, and various anxiety disorders have developed that invoke overproduction or underremoval of the presumed endogenous κ agonist, dynorphin [Chavkin et al., 1981; Corbett et al., 1982]. Since buprenorphine is the most potent κ antagonist that it is currently possible to use clinically, it offers the best means of testing these theories. Indeed, one might speculate that the admixture of κ antagonism and μ partial agonism would give the best of both worlds in this line of pharmacotherapy.

So far there have been few attempts to use buprenorphine to evaluate these theories. This chapter reviews those that have been published and presents some of our own findings on its use in schizophrenia.

DEPRESSION

The major protagonists of the use of buprenorphine in depression have been the group from the Max Planck Institute in Munich. Their rationale derived from several lines of evidence. First was the belief that endogenous opioid tone was abnormal in these patients, and that it might be rectified by treatments such as electroconvulsive therapy (ECT). Second was the observation of Fink et al. [1970] that the mixed agonist–antagonist cyclazocine was antidepressant despite having a psychotomimetic tendency. Third were the suggestions that infusions of the presumed endogenous μ agonist β-endorphin had antidepressant actions [Kline et al., 1977; Angst et al., 1979; Gerner et al., 1980]. Finally, the fact that buprenorphine had beneficial actions on mood in postoperative patients [Harcus et al., 1980] and in drug addicts [Mello and Mendelson, 1980] argued for a trial in another condition of altered mood: depression.

A study by Emrich et al. [1982], still the only double-blind one carried out in this condition with this drug, used 13 patients who met Research Diagnostic Criteria for major depression and who had shown little response to conventional therapy. The patients were washed out of other drugs for 4 days before receiving a placebo for a variable period of 1–7 days. They then entered the active phase, being given buprenorphine 0.2 mg sublingually twice daily for a period of 5–8 days. Mood ratings were determined with the Hamilton scale and showed a 30% reduction in the first couple of days that was sustained till the terminal placebo period, when they began to creep up again.

One interesting aspect of the study was the observation that the response showed great individual variation, as some patients improved greatly whereas others changed little. Since this group were relative nonresponders, a real benefit from buprenorphine in half of its members represents a potentially very significant clinical gain. Side effects were limited to slight dizziness, nausea, and sedation, with the exception of one patient who had to be withdrawn because of vomiting.

Two other recent reports have also found an antidepressant effect of buprenorphine in open trials. One was carried out in opioid abusers segregated into depressed and nondepressed groups on the basis of the Beck Depression Inventory and the Short Depression Scale [Kosten et al., 1990]. It revealed progressive improvement in mood ratings of the depressed group over the period of 4 weeks of treatment with buprenorphine (mean dose 3.2 mg/day), so that the final rating was half that at entry. The distribution of response was very similar to that found in the German study, that is, half the group did well and the other half showed little change in symptoms. The studies also concurred in revealing that the antidepressant activity of buprenorphine was rapid, with 75% of the responders having shown improvement in the first week. This observation is similar to that made in regard to schizophrenia (see below) and argues for a mode of action that differs from that of the tricyclic antidepressants, which normally take 3–5 weeks to work.

Similarly positive effects on mood were reported by Mongan and Callaway [1990] in a group of former drug abusers with depressive symptoms. A rapid antidepressant effect was observed in five subjects with major depression, and no response in one. The buprenorphine was administered sublingually in solution, at an initial dose of 150 μg, which was doubled if no response was seen within 45 min. Subjective improvements were seen for periods that varied from 2 min to 3 hr, a range that probably reflects a mixture of placebo and drug effects, since the onset of analgesic action of buprenorphine is usually about 1 hr [Houde, 1979; Wallenstein et al., 1982]. Interestingly, white subjects showed a slower onset of action than blacks.

One final mention on the role of buprenorphine in depression is warranted. It has to do with an unsolicited letter sent to our group several years ago, an extract of which reads:

> I have been wanting to write to you for some time regarding your painkiller Temgesic (buprenorphine). I was first given it to relieve back pain which it did wonderfully. I have tried to speak to doctors about the effect these tablets have on pain in the *mind*— not just bodily pain! They have kept me out of hospital for depression for ten years. Just one or even half a tablet will help to lift the depression—taking it over a few days, allowing me to come out of this black hole. I do not understand how they work, I just know that *they do!* My husband says they are like magic—if I am unable to get out of bed, he will put one under my tongue and within an hour I am able to get on with my life as normal."

Perhaps the most likely explanation of the observed actions of buprenorphine in depression is that they reflect μ partial agonism. There is no real evidence that opioid antagonists are of value in depression [Terenius et al., 1977]. On the basis of the data presented here it would seem reasonable to conclude that further trials of buprenorphine in depressed patients are warranted. Such studies should best be carried out by randomized double-blind placebo-controlled procedures. Any ethical concerns would be minimized by the use of treatment-resistant patients.

SCHIZOPHRENIA

The rationale for the evaluation of buprenorphine in schizophrenia is based on a set of arguments similar to those used to justify assessing its use in depression. Of particular importance were the observations that the κ agonists cyclazocine and bremazocine were psychotomimetic in volunteers [Pfeiffer et al., 1986]. Moreover, there are data that show increased levels of endogenous opioids in patients with chronic psychosis [Terenius et al., 1976], and opioid antagonists have been reported to be effective in reducing symptoms (particularly hallucinations) in some patients [Pickar et al., 1982]. Thus, one possibility is that some aspects of schizophrenia may be due to the overproduction or overactivity of endogenous κ agonists such as the dynorphins.

An alternative hypothesis derives from observations that μ agonists have marked effects on dopamine-mediated behaviors [Joyce and Iversen, 1979], probably by releasing dopamine in both accumbens and striatum [Spanagel et al., 1990]. If excessive release of dopamine underlies schizophrenia, then it could reflect increased μ tone and would explain the actions of naloxone and naltrexone on hallucinatory experiences (see above). As buprenorphine is a μ partial agonist, it could block the actions of endogenous opioids if they were full agonists.

Finally, it is possible that μ agonism itself might prove beneficial. In past centuries opioids were used to quell the symptoms of madness, and more recently Brizer et al. [1985] have shown methadone to augment the therapeutic efficacy of neuroleptics. Mu agonists are excellent calming and anxiolytic agents whose use is limited by their dependence and abuse liabilities. It is somewhat surprising that drugs such as the opioids, which are thought to act by releasing dopamine [Spanagel et al., 1990], should be antipsychotic. However, they may also work downstream of the dopamine projections to limit some aspects of dopamine function [Joyce et al., 1981]. Moreover, a simplistic view that equates dopamine release with either schizophrenia or opioid addiction is likely to be wrong on at least one count. Thus, for whatever reason μ agonists are antipsychotic, the availability of a less dangerous and less abusable variant such as buprenorphine would strengthen the case for their reappraisal in schizophrenia.

STUDIES

The most controlled investigation so far is that of Schmauss and colleagues [1987], who performed a double-blind comparison of buprenorphine and placebo in ten acutely psychotic schizophrenics with predominantly paranoid or hallucinatory symptoms. This study was remarkable in that buprenorphine was the sole therapy in a 4-day double-blind crossover study against placebo. A sublingual dose of 0.2 mg buprenorphine was found to be significantly superior to placebo. The onset of action was fast, within 1 hr of drug administration, and it lasted for the next 4 hr. A specific action on the positive symptoms of delusions and hallucinations was detected, whereas the negative ones of apathy, retardation, and reduced social functioning were not altered. Additionally, no significant changes in anxiety or depression were recorded, which suggests that the changes seen in the schizophrenic symptoms were not secondary to a more general effect on psychological well-being.

Improvements in schizophrenic symptoms were also reported in one of two patients by Mongan and Callaway [1990]. The number of patients was too small to make any definite conclusions about subtype or symptom specificity, but a 50% response rate supports the findings of the Munich group that some patients gain benefit, if only for a limited time.

In our unit we have performed two studies on the effects of a single administration of buprenorphine on symptoms in schizophrenics who were actively psychotic at the time of testing. In both studies we examined the actions of buprenorphine by rating symptoms every hour for the 4 hr following a sublingual dose of 0.2 mg. The first eight patients were examined in an open trial design. We found a variable degree of responsiveness to buprenorphine, some patients becoming very much less symptomatic for a few hours after the dose [Groves and Nutt, 1991].

Because in several patients the preeminent symptoms of schizophrenia disappeared or markedly ameliorated after the administration of buprenorphine, we decided to repeat the study with a double-blind design. So far we have been able to study ten schizophrenics who have been given both buprenorphine and placebo in a double-blind crossover design. The subjects were stable schizophrenics who were tested 1 week apart in a balanced presentation of the two treatments. Neuroleptic medication was kept constant for the period between tests. The Brief Psychiatric Rating Scale (BPRS) was modified to be administered every hour, starting just before drug dosing and continuing for the next 4 hr. All ratings were performed by the same psychiatrist (S.G.). Two subjects dropped out during the study, one early in the first test (placebo) and the other between tests.

The effects of buprenorphine and placebo at each hour were computed as change scores from the baseline, and these were then totaled to give a total change score for the whole test. Figure 1 (a and b) shows the differences between the buprenorphine and placebo-induced changes for 8 of the 17 BPRS items that showed the greatest changes. It can be seen that there was a tendency for buprenorphine to produce a more marked improvement in symptoms than placebo, and this reached a significance of $p < 0.05$ (Wilcoxon, one-tailed) for two items: somatic symptoms and depression. For several other items some patients showed a very marked improvement on buprenorphine, although the majority experienced no clear effect. Overall BPRS change scores showed a significant effect of buprenorphine at $p < 0.02$ (Wilcoxon, two-tailed) (see Fig. 2). The interviewing doctor was able to correctly identify the drug and placebo condition in 6/8 cases. Side effects were not recorded, but in the three cases in which we were able to get the schizophrenics to attempt to identify the active condition they all guessed correctly.

Because of the occasional dramatic improvement following administration of buprenorphine, we attempted to treat a very ill schizophrenic with large doses of this drug. The findings are outlined below and in Figure 3.

CASE REPORT

The patient was a 38-year-old white male, with a 20-year history of schizophrenia. While initially his illness was intermittent, he had been continuously psychotic for the past 12 years, most of this time being spent in psychiatric hospitals. His illness

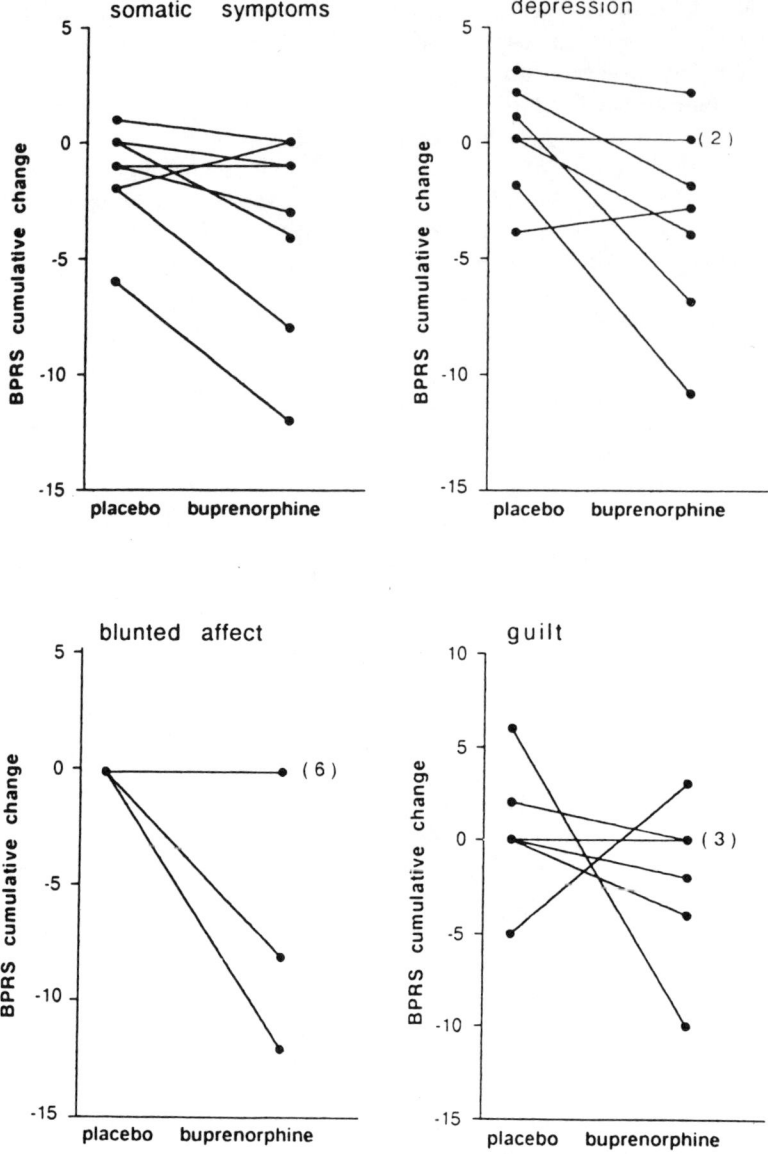

Fig. 1. Total change scores over the 4 hr after buprenorphine or placebo administration for eight items on the Brief Psychiatric Rating Scale (BPRS). Lines join scores for the same patient and negative scores represent improvement.

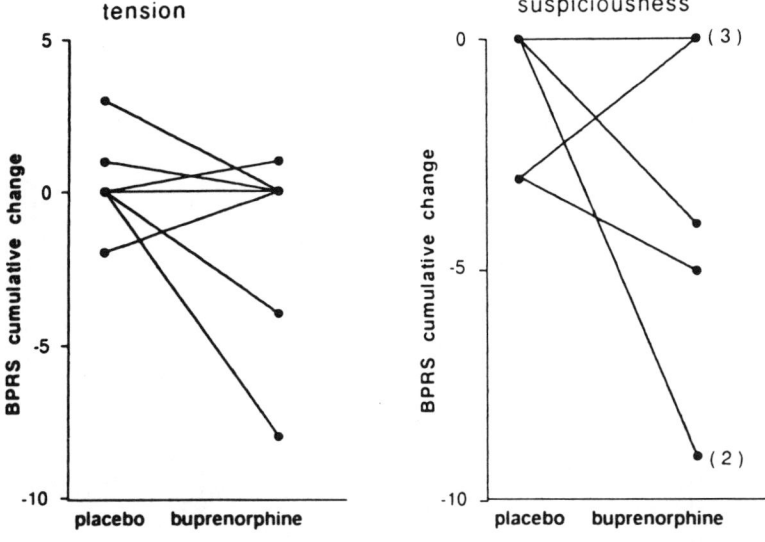

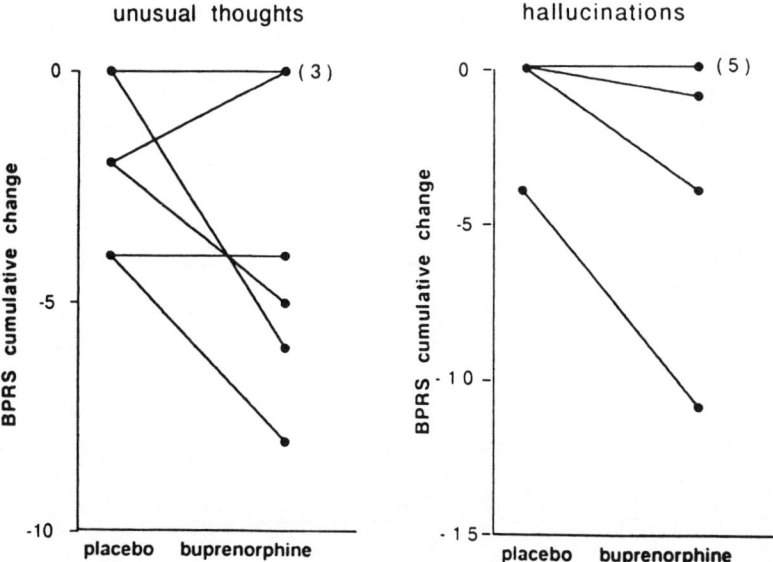

Fig. 1. *(Continued)*

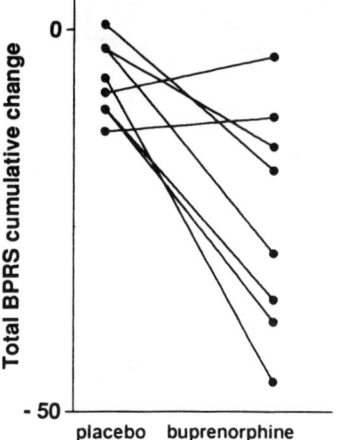

Fig. 2. Total change scores over the 4 hr after buprenorphine or placebo administration for the BPRS. Lines join scores for the same patient and negative scores represent improvement. Mean (± SEM) total BPRS change: buprenorphine = 24 ± 5 and placebo = 6 ± 2.

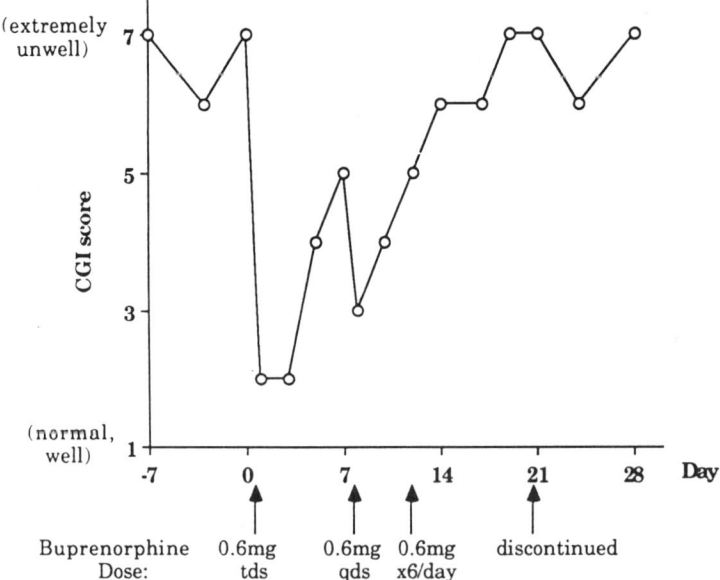

Fig. 3. Effects of buprenorphine on symptoms of schizophrenia in a single individual.

was characterized by prominent but fleeting delusional and hallucinatory symptoms (e.g., he was a crown prince or a famous World War I general; he heard voices commenting about his behavior or telling him about things he should do). His thoughts were occasionally confused, and he complained of people "messing with my thinking." He also had swings in his mood that were incongruous with his behavior or circumstances (e.g., laughing when being told of a relative's death; weeping or becoming angry while eating a meal). His behavior ranged from being slow and lethargic to episodes during which he would stamp around and scream, occasionally breaking windows. Comparison of his clinical state with descriptions in his clinical notes indicated that he had become more apathetic and underactive and displayed poverty in his range of thought, in addition to the positive symptoms described above. Previous extensive treatment with large doses of a range of oral and depot neuroleptics, lithium, and electroconvulsive therapy had not been effective in improving his symptoms, and he was too psychotic to take part in social or rehabilitation programs. At the time of assessment prior to starting buprenorphine he was maintained on chlorpromazine 200 mg/day, which he felt helped him to sleep.

Initially the patient was started on buprenorphine 0.6 mg sublingually tds. Within 2 hr of the first dose he commented spontaneously that his voices had gone. He spoke coherently to the staff, and did not talk about any delusional beliefs. His mood, while flattened, was not inappropriate. He became more active and asked to be taken for a walk. Over the next 48 hr, this change in his thought, behavior, and mood persisted. However, there was a gradual deterioration over the next 3 days which resulted in the buprenorphine dose being increased to 0.6 mg qds on Day 7. Following this, a further brief improvement occurred for 24 hr, with subsequent deterioration in his thoughts, mood, and behavior. On Day 10 his dose of buprenorphine was increased to 0.6 mg × 6/day, with no obvious effect on his symptoms. Within 2 weeks of starting the buprenorphine, his clinical state was similar to that noted prior to treatment. After a further 7 days of treatment the buprenorphine was discontinued, with no change in his symptoms.

Subsequent attempts to readminister buprenorphine to this patient have been unsuccessful, as he now believes that it was a poison and that he will die if given any more. He has had no further episodes of symptom amelioration.

This finding, albeit in a single subject, is remarkable for the massive improvement occurring in someone considered unresponsive to therapy. The speed of onset was also dramatic, although perhaps not unexpected, given the earlier findings.

OTHER CONDITIONS

There are few data on buprenorphine in other forms of mental illness. Some success in the treatment of personality disorders has been reported by Mongan and Callaway [1990]. Similarly Resnick (personal communication) has reported buprenorphine to be of use in borderline patients, producing a rapid normalization of dysphoric state.

However, as most of his patients also had a history of drug abuse, it would be interesting to attempt to confirm this observation in a drug-naive population.

CONCLUSIONS

There is a small but growing body of data that support the view that buprenorphine may have some utility in the treatment of some psychiatric disorders. Several interesting issues are raised by this possibility. One concerns the ethics of using a drug that has some tendency to engender abuse in an opiate-naive patient with the theoretical risk of producing dependence. In practice the pharmacokinetics of buprenorphine make it unlikely that withdrawal will be a significant problem if the drug is used for a long period. Moreover, if previous drug abusers were excluded from therapy, the risk of buprenorphine abuse would be very low, and could be further reduced by developing combination treatments to preclude i.v. use. Nevertheless, it may be very difficult to overcome the prejudice against μ-receptor-acting drugs that permeates the medical profession and society in general.

Another issue that is raised by the therapeutic actions of buprenorphine is the speed of response to treatment. Conventional antidepressants and antipsychotics work over weeks, whereas maximal benefit was apparent after a couple of hours following administration of buprenorphine. The fact that such a rapid alteration in mental state can be produced by a pharmacological agent belies the established theory that slow adaptive changes underly the therapeutic response in schizophrenia, unless the result of chronic therapy is to produce the same changes as acute buprenorphine. The buprenorphine findings give hope to those looking for other treatments with faster onsets. In addition, they suggest that schizophrenia and depression are ongoing processes that can be acutely interrupted. Could tonic activation of opioid receptors contribute to both conditions? Alternatively the μ agonism of buprenorphine may lead to a euphoriant-type of mood elevation that secondarily ameliorates the symptoms of schizophrenia. These theories could be distinguished by comparing the actions of full μ agonists with those of pure antagonists in the same subjects. We hope that the data presented in this chapter will encourage further work along these lines.

ACKNOWLEDGMENTS

We thank our colleagues in the Bristol and Weston and Bath Health Districts for helping in the recruitment of research patients and Jayne Bailey for editorial assistance.

REFERENCES

Angst J, Autenrieth V, Brem F, Koukkou M, Meyer H, Stassen HH, Storck U (1979): Preliminary results of treatment with β-endorphin in depression. In Usdin E, Bunney WE,

Kline NS (eds): "Endorphins in Mental Health Research." London: Macmillan, pp 518–528.

Brizer DA, Hartman N, Sweeney J, Millman RB (1985): Effect of methadone plus neuroleptics on treatment-resistant chronic paranoid schizophrenia. Am J Psychiatry 142:1106–1107.

Chavkin C, James IF, Goldstein A (1981): Dynorphin is a specific endogenous ligand of the κ-opioid receptor. Science 215:413–415.

Comfort A (1977): Morphine as an antipsychotic. Relevance of a 19th-century therapeutic fashion. Lancet 2:448–449.

Corbett AD, Paterson SJ, McKnight AT, Magnan J, Kosterlitz HW (1982): Dynorphin 1-8 and dynorphin 1-9 are ligands for the kappa-subtype of opiate receptors. Nature 299:79–81.

Emrich HM, Vogt P, Herz A (1982): Possible antidepressive effects of opioids: Action of buprenorphine. Ann NY Acad Sci 398:108–112.

Emrich HM, Schmauss C, Dose M, Lonati-Galligani M, Pirke KM, Weber M (1988): New concepts in the action of mood stabilisers and antidepressants. In Briley M, Fillion G (eds): "New Concepts in Depression." London: Macmillan, pp 320–333.

Fink M, Simeon J, Itil TM, & Freedman AH (1970): Clinical antidepressant activity of cyclazocine—A narcotic antagonist. Clin Pharmacol Ther 11:41–48.

Frecska E, Arato M, Banki CM, Mohari K, Perenyi A, Badgy G, Fekete MIK (1989): Prolactin response to fentanyl in depression. Biol Psychiatry 25:629–696.

Gerner RH, Catlin DH, Gorelick DA, Hui KK, Li CH (1980): β-Endorphin: Intravenous infusion causes behavioural change in psychiatric inpatients. Arch Gen Psychiatry 37:642–647.

Groves S, Nutt DJ (1991): Buprenorphine and schizophrenia. Hum Psychopharmacol 6:71–72.

Harcus AH, Ward AE, Smith DW (1980): Buprenorphine in post-operative pain: Results in 7500 patients. Anaesthesia 35:382–386.

Houde RW (1979): Analgesic effectiveness of the narcotic agonist–antagonists. Br J Clin Pharmacol 7:297S–308S.

Joyce EM, Iversen SD (1979): The effect of morphine applied locally to mesencephalic dopamine cell bodies on spontaneous motor activity in the rat. Neurosci Lett 14:207–212.

Joyce EM, Koob GF, Strecker R, Iversen SD, Bloom FE (1981): The behavioural effects of enkephalin analogues injected into the ventral tegmental area and globus pallidus. Brain Res 221:359–370.

Kline NS, Li CH, Lehmann HE, Lajtha A, Laski E, Cooper T (1977): β-Endorphin-induced changes in schizophrenic and depressed patients. Arch Gen Psychiatry 34:1111–1113.

Kosten TR, Morgan C, Kosten TA (1990): Depressive symptoms during buprenorphine treatment of opioid abusers. J Substance Abuse Treat 7:51–54.

Matussek N, Hoehe M (1989): Investigations with the specific μ-opiate receptor agonist fentanyl in depressive patients: Growth hormone, prolactin, cortisol, noradrenaline and euphoric responses. Neuropsychobiology 21:1–8.

McDiarmid J (1876): The hypodermic injection of morphia in insanity. J Med Sci 22:18–42.

Millan MJ (1990): Kappa opioid receptors and analgesia. Trends Pharmacol Sci 11:70–76.

Mello NK, Mendelson JH (1980): Buprenorphine suppresses heroin use by heroin addicts. Science 207:657–659.

Mongan L, Callaway E (1990): Buprenorphine responders. Biol Psychiatry 28:1065–1080.

Pfeiffer A, Brantl V, Herz A, Emrich HM (1986): Psychotomimesis mediated by κ opiate receptors. Science 233:774–776.

Pickar D, Davis GC, Schulz SC, Extein I, Wagner R, Naber D, Gold PW, van Kammen DP, Goodwin FK, Wyatt RJ, Li CH, Bunney WE (1981): Behavioral and biological effects of acute β-endorphin injection in schizophrenic and depressed patients. Am J Psychiatry 138:160–166.

Pickar D, Vartanian F, Bunney W, Maier HP, Gastpar MT, Prakash R, Sethi BB, Lideman R, Belyaev BS, Tsutsulkovskaja MVA, Jungkunz G, Nedopil M, Verhoeven W, van Praag H (1982): Short-term naloxone administration in schizophrenic and manic patients. Arch Gen Psychiatry 39:313–319.

Schmauss C, Emrich HM (1988): Narcotic antagonist and opioid treatment in psychiatry. In Rodgers RJ, Cooper SJ (eds): "Endorphins, Opiates and Behavioural Processes." Chichester: Wiley, pp 327–351.

Schmauss C, Yassouridis A, Emrich HM (1987): Antipsychotic effect of buprenorphine in schizophrenia. Am J Psychiatry 144:1340–1342.

Spanagel R, Herz A, Shippenberg TS (1990): The effects of opioid peptides on dopamine release in the nucleus accumbens: An in vivo microdialysis study. J Neurochem 55:1734–1740.

Terenius L, Wahlstrom A, Lindstrom L, Widerlov E (1976): Increased CSF levels of endorphins in chronic psychosis. Neurosci Lett 3:157–162.

Terenius L, Wahlstrom A, Agren H (1977): Naloxone (Narcan) treatment in depression: Clinical observations and effects on CSF endorphins and monoamine metabolites. Psychopharmacology 54:31–33.

Wallenstein MS, Kaiko RF, Rogers AG, Houde RW (1982): Clinical analgesic assays of buprenorphine and morphine. Clin Pharmacol Ther 31:278.

STUDIES RELATING TO TREATMENT OF SUBSTANCE ABUSE

LABORATORY STUDIES OF BUPRENORPHINE IN OPIOID ABUSERS

DONALD R. JASINSKI
Department of Medicine, Francis Scott Key Medical Center, Johns Hopkins University School of Medicine, Baltimore, MD 21224

KENZIE L. PRESTON
Addiction Research Center, National Institute on Drug Abuse, Baltimore, MD 21224

INTRODUCTION

To deal with the public health and social problems of opiate addiction in the United States, the Committee on Drug Addiction of the National Research Council, National Academy of Sciences was formed in 1929 [Eddy, 1973]. Their specific research project was to replace opiates with substitutes that have no addictive properties and thus reduce or eliminate opiate abuse.

The basis for their approach was a general acceptance of the clinical impression that codeine (methyl morphine) relative to morphine had little or no addiction liability, even though codeine was widely used for pain relief and cough suppression. It was suggested that addiction and analgesia would be disassociated through further chemical modifications of morphine. Additional support for this concept was the observation that the frequency of abuse of cocaine had waned simultaneously with the introduction of the synthetic substitute procaine, which had reduced therapeutic use of cocaine as a local anesthetic. As a consequence, large numbers of semisynthetic and synthetic opioids were produced and studied in the hope of doing for morphine what procaine had done for cocaine.

Of the thousands of opioids produced as a result of this project, a few have been introduced into therapeutics as analgesics of lesser abuse potential than morphine.

Buprenorphine: Combatting Drug Abuse With a Unique Opioid, pages 189–211
© 1995 Wiley-Liss, Inc.

In addition, other drugs in this class have been introduced as drugs to treat addiction, as antidotes for opiate poisoning, and as tools in neuropharmacology. Some two hundred of the opioids have been studied in human volunteers with histories of opioid addiction, primarily to assess abuse liability [Jasinski, 1977].

The formal program to identify opioids of low abuse liability waned during the 1970s. The last major compound introduced as a result of this effort was the opioid buprenorphine. This drug most closely achieved the objective of an opioid that demonstrates a disassociation of morphine-like analgesia from risk of physical dependence. In addition, buprenorphine has a pharmacological profile appropriate for use in the treatment of opiate addiction.

The objective of this chapter is to summarize the experimental studies of buprenorphine in human volunteers with histories of opiate addiction and to review those results that provide the basis for use of buprenorphine as an analgesic of low abuse potential, as a treatment drug for opiate addiction, and as a tool for understanding the mechanism of opioid action in the brain.

BACKGROUND AND METHODS

A discussion of the pharmacological effects of buprenorphine in opiate abusers necessitates relating these effects to current concepts of abuse potential, the addictive process, and the mode of action of opioids in humans. The early studies with opiate abusers were governed by prevailing scientific concepts and the initial situations in which patients could be studied. The major clinical phenomenon observed in patients presenting for treatment was the occurrence of a discomforting abstinence syndrome that was believed to be responsible for drug seeking. As a result, the addictive process was equated with the production of physical dependence, and early treatment efforts were directed towards detoxification or alleviating the condition of physical dependence. Abuse or addiction potential was consequently viewed in terms of the capacity to produce physical dependence. Furthermore, opioids were viewed as either acting like morphine or lacking morphine-like activity—an all-or-none phenomenon. The systematic study of opioids over the years led changes in these initial scientific concepts as well as changes in the methods for measuring drug effects. In recent years the role of their capacity to cause physical dependence both in the addictive process and in abuse potential have been deemphasized. In addition, opioids are no longer viewed as working through a single mechanism of action. The changes in concepts, the view of the addictive process, and the impact on methods for assessing abuse potential have been extensively reviewed and summarized. [Cami, 1991; Henningfield et al., 1987; Jasinski, 1977, 1979, 1981, 1984, 1988, 1990, 1991; Jasinski and Henningfield, 1988, 1989; Jasinski et al., 1984a; Martin, 1979; Preston and Jasinski, 1991].

Opioids produce diverse pharmacological effects that are believed to occur as a result of varying degrees of agonist activity at distinct types of receptors. In human pharmacology, drugs acting through μ receptors are viewed as producing typical morphine-like euphoria, physical dependence, and respiratory depression that can lead to apnea. Drugs acting through the κ receptor produce feelings of sedation, minimal respiratory depression not leading to apnea, dysphoric mood changes, and a type of physical dependence with little or no drug seeking. Less clearly defined in human pharmacology are opioids acting through the σ and δ receptors. In humans, the description of opioids in terms of receptor types is based upon the pharmacological profile observed, usually in studies in opioid abusers. The classification of opioids that arises from studies of these drugs in tissue-binding studies, studies of endogenous opioids, studies on isolated tissues, and studies in species other than humans has not systematically nor clearly been related to the clinical syndromes produced in humans.

Buprenorphine has been studied in opiate abusers under a number of experimental paradigms. Single doses and multiple doses of buprenorphine have been given to nondependent opiate abusers and to opioid abusers dependent on morphine or methadone. In nondependent opiate abusers, buprenorphine has been characterized in terms of its ability to alter mood, feeling, thinking, and perception (subjective effects); and discriminative stimulus effects, behavioral effects, and physiological effects. In opiate-dependent subjects its ability to serve as a substitute and to suppress or produce a withdrawal syndrome has been tested. Administration of multiple doses of buprenorphine has also been studied in regard to its impact on the production of physical dependence, changes in subjective effects, alteration in laboratory parameters, changes in vital signs, and effects on the self-administration of heroin and other drugs.

CLINICAL PHARMACOLOGY IN OPIATE ABUSERS

Nondependent Subjects

Acute Effects. Eight studies assessing the subjective profile, time course, and relative potency of buprenorphine have been conducted (Table I). The results of these studies have generally been consistent. Buprenorphine, given acutely, produced self-reported increases in intensity of effects that were described as opiate-like rather than nalorphine-like (Table II). Buprenorphine produced predominantly morphine-like responses on the Morphine–Benzedrine Group (MBG) Scale of the Addiction Research Center Inventory (ARCI), with little or no responses in the Pentobarbital–Chlorpromazine (PCAG) or LSD Scales (Table II). In general, liking scores increased in a dose-related manner except at high doses of buprenorphine. The effects of buprenorphine were different from morphine in terms of duration of action and potency, and the responses to increasing doses.

Studies that compared equally effective doses of buprenorphine (1 mg), morphine (30 mg), and methadone (30 mg)—all given subcutaneously—indicated that buprenorphine, morphine, and methadone had similar durations of action for subjective effects [Jasinski et al., 1978]. In contrast, the duration of the miotic effects of buprenorphine was greater than that for morphine, miosis being demonstrated at 72 hr for both methadone and buprenorphine, but not morphine.

TABLE I. Single-Dose Studies for Subjective Effect and Miosis

Study	Drug dosages route	Number of subjects	Design	Reference	Comments
1	Buprenorphine hydrochloride, 0.2, 0.4, 0.8 mg s.c. Morphine sulfate, 15, 30 mg s.c. Placebo s.c.	9	Randomized block	Jasinski et al. [1978]	Therapeutic dose range for analgesia
2	Buprenorphine hydrochloride, 0.6, 1.2 mg s.c. Morphine sulfate, 20, 40 mg s.c.	9	Randomized block	Jasinski et al. [1978]	
3	Buprenorphine hydrochloride, 1.0 mg s.c. Morphine sulfate, 30 mg s.c. Methadone, 30 mg s.c. Placebo s.c.	14	Randomized block	Jasinski et al. [1978]	Duration of action
4	Buprenorphine hydrochloride, 1.0, 1.4, 2.0 mg s.c. Morphine sulfate, 20, 40 mg s.c.	7	Latin square (5 subjects); randomized block (2 subjects)	Jasinski et al. [1978]	
5	Buprenorphine 1, 2, and 4 mg s.l. Buprenorphine 1, 2 mg s.c. Placebo s.c., s.l.	10	Latin square	Jasinski et al. [1989]	Route comparison
6	Buprenorphine 20 and 40 mg p.o. Buprenorphine 1, 2 mg s.c. Placebo p.o., s.c.	10	Latin square	Jasinski et al. [1982]	Route comparison
7	Buprenorphine 0, 1, 2, 4, 8, 16, 32 mg s.l.	4	Dose run-up	Walsh et al. [1992]	
8	Buprenorphine 0.5, 2, 8, 16, 32 mg, s.l. Methadone 3.75, 15, 60 mg, p.o. Placebo p.o., s.l.	9	Latin square	Walsh et al. [1993]	

TABLE II. Composite Summary of the Subjective and Miotic Effects Profile of Acute Administration of High Doses of Morphine and Buprenorphine[a]

	Morphine	Nalorphine	Buprenorphine
Drug identification	Opiate	Barbiturate	Opiate
Prototypic symptom	Skin itching, coasting, relaxed	Sleepy, drunken	Skin itching, relaxed, turning of the stomach
Liking	Dose-related to 30 mg	Present in low dose but not in larger dose	Dose-related except at high doses
ARCI scales:			
MBG (euphoria)	Dose-related increase	Little or no effect	Dose-related increase
PCAG (apathetic sedation)	Little or no effect	Dose-related increase	Little or no effect
LSD	Little or no effect	Dose-related increase	Little or no effect

[a]Data from Jasinski [1977], Jasinski et al. [1978], and Walsh et al. [1992].

With regard to potency, buprenorphine is generally regarded as being 25–50 times more potent than morphine in relieving pathological pain, such that the standard therapeutic dose (0.3 mg) given parenterally is equivalent to 10 mg of morphine [Jasinski et al., 1978]. When single subcutaneous doses of buprenorphine (0.2–0.8 mg) were compared with 15-mg and 30-mg doses of morphine for the ability to produce subjective effects in addict volunteers, it was found that buprenorphine was approximately 33 times more potent than morphine. These potencies were consistent with that for relief of pathological pain. Consequently, in therapeutic dose ranges there appeared to be no selective analgesic action of buprenorphine relative to morphine. Subsequent studies involved comparing doses of buprenorphine by different routes of administration. In general, it was found that when given sublingually buprenorphine was approximately two thirds as potent as when administered parenterally [Jasinski et al., 1989]. When given orally, buprenorphine was approximately one-twentieth as potent as parenteral morphine [Jasinski et al., 1982].

From calculations of relative potencies, doses of buprenorphine can be expressed in terms of morphine-equivalent units (Table III). These morphine equivalencies allow evaluation of the effects of a large range of doses of buprenorphine relative to those of other opioids. With increasing doses, buprenorphine becomes less potent relative to morphine.

As is consistent with the decreasing relative potency measures on subjective and physiological measures, high doses of buprenorphine did not produce any respiratory depression leading to apnea and caused no marked alterations in cardiovascular effects or excessive sedation or other signs of intoxication that would be expected

TABLE III. Morphine-Equivalent Doses of Buprenorphine Doses Used in Various Single-Dose Studies

Study	Maximum dose per route (mg)	Morphine-equivalent parenteral units (mg)	Reference
1	0.8 s.c.	24	Jasinski et al. [1978]
2	1.2 s.c.	36	Jasinski et al. [1978]
3	2.0 s.c.	60	Jasinski et al. [1978]
4	4.0 s.l.	80	Jasinski et al. [1989]
5	40 p.o.	60	Jasinski et al. [1982]
6	32 s.l.	639	Walsh et al. [1993]

from equi-analgesic doses. Nausea, vomiting, urinary retention, and constipation were, however, seen. In recent studies by Walsh et al. [1992, 1993], 32 mg of buprenorphine given sublingually, which is approximately 80 times the standard therapeutic dose of buprenorphine by this route (usually 0.4 mg), was administered without significant respiratory depression or serious side effects.

The recognition that opiate addicts could discriminate morphine from placebo and between morphine and other psychoactive drugs on the basis of their past experience of such drugs, and the hypothesis that these same drugs would produce interoceptive stimuli in laboratory animals, led to the development of behavioral procedures for drug discrimination in laboratory animals. Subsequently, these behavioral techniques in animals were adapted to human studies of agonist–antagonist opioids in opiate abusers [Preston et al., 1989, 1992; Bigelow et al., 1992]. In general, these studies involve the administration of an initial set of training doses of standard drugs, such as an opioid agonist, and placebo, which subjects have been trained to discriminate. Correct responses are reinforced with the payment of money. After meeting preset criteria for acquisition of the discrimination, subjects are tested in random order with placebo, or training doses of morphine, and with doses of other drugs of interest. The subjects are asked to compare the effects of the test drug to placebo or the training doses of the opioid agonist.

Three such studies have been conducted with buprenorphine. The first compared buprenorphine, hydromorphone, pentazocine, butorphanol, and nalbuphine in patients trained to discriminate hydromorphone from saline [Preston et al., 1992]. Buprenorphine fully substituted for hydromorphone, indicating an opiate-like profile of stimulus effects. A second study compared the same group of drugs in subjects trained to discriminate hydromorphone from both saline and pentazocine [Preston et al., 1989]. Under these conditions, buprenorphine fully substituted for neither hydromorphone nor pentazocine. A similar study was conducted in which the subjects were trained to discriminate butorphanol, hydromorphone, and saline; under these conditions buprenorphine again produced hydromorphone-like stimulus effects [Bigelow et al., 1992]. The investigators concluded that, at the doses used, buprenorphine produced effects consistent with μ partial agonist activity. Subjective effect data collected during these studies were consistent with profiles of effects reported in the studies summarized in Table I.

Only limited metabolic and disposition studies have been done in opiate abusers. Urine and feces samples were collected from subjects participating in studies of orally and sublingually given buprenorphine (Studies 5 and 6 in Table I). Analysis of these samples [Cone et al., 1984] showed that buprenorphine underwent an initial metabolism to norbuprenorphine with subsequent excretion of buprenorphine, norbuprenorphine and their congeners. Buprenorphine itself was not found in the urine. Relatively small amounts of norbuprenorphine congeners were excreted through the urine (in the range of 2%–13%). Buprenorphine was excreted primarily through the feces, with a maximum excretion occurring at 4–6 days after drug administration, indicating an enterohepatic circulation. Pharmacokinetic profiles from plasma concentrations of buprenorphine are not available for the doses used in experimental studies in opiate abusers.

Chronic Effects. Four experimental studies conducted in opiate abusers to evaluate the effects of repeated large doses of buprenorphine are described in Table IV.

The first, and most extensive, study was that of Jasinski et al. [1978]. Buprenorphine was given subcutaneously once a day to five subjects in increasing doses to 8 mg/day for a total administration period of 30–57 days. This 8-mg dose was equivalent to 240 mg of morphine. The data on the profile of effects in this study have been extracted and summarized in Tables V–VII. Chronic administration of buprenorphine significantly lowered systolic and diastolic blood pressure (Table V). In addition, the following effects were noted (but not statistically significant changes): constricted pupils, elevated rectal temperature, and reduced body weight and caloric intake (Table V). There was little evidence that buprenorphine has a significant effect on respiratory rate (Table V). Both subjects and observers reported that the effects were opiate-like. The profile of symptoms reported by subjects, and the signs reported by observers, have been extracted and tabulated in Table VI. The major symptoms were sleepy, tired, skin itchy, nodding, relaxed, nervous stomach, feel hot, and coasting. This profile of symptoms is consistent with an opiate-like drug. Similarly, the signs reported by observers were typical of an opiate-like drug (Table VI). During the period of chronic administration there were changes in some laboratory parameters, including decreases in red blood cell count, hemoglobin and

TABLE IV. Studies of Chronic Effects and Physical Dependence Capacity of Buprenorphine

Study	Maximum buprenorphine dose and route	Number of subjects	Duration (days)	References
1	8 mg s.c. q day	5	30–57	Jasinski et al. [1978]
2	8 mg s.c. q day	10	29	Mello et al. [1982]
3	16 mg s.l. q day	5	84	Bickel et al. [1988]
4	3 mg s.l. q day	15	30	Kosten et al. [1990]

TABLE V. Means and Standard Errors of Physiological Measures on Five Subjects for Addiction Cycle of Buprenorphine[a]

Measure	Control (C)	Stabilization (S)	t(S vs. C)	p
Pupils, mm	3.9 ± 0.3	3.5 ± 0.2	−2.26	<.10
Temperature, °C	36.7 ± 0.1	36.9 ± 0.2	+2.36	<.10
Respiratory rate, breaths per min	17.1 ± 0.7	16.6 ± 0.5	−1.16	<.40
Systolic blood pressure, mm Hg	116 ± 4	107 ± 3	3.09	<.05
Diastolic blood pressure, mm Hg	73 ± 3	65 ± 1	3.11	<.05
Pulse beats per min	70 ± 2	65 ± 2	2.64	<.10
Body weight, kg	78.3 ± 3.8	76.7 ± 3.2	2.17	<.10
Caloric intake	1,978 ± 154	1,395 ± 213	2.71	<.10

[a]Data from Jasinski et al. [1978].

hematocrit, and total protein (Table VII), that were consistent with similar effects seen in chronic administration of morphine and other opiates [Jasinski et al., 1978]. Additional changes not seen with other opioids included increased sedimentation rate and decreased alkaline phosphatase.

Subsequently, Mello et al. [1982] administered an identical dose of 8 mg once daily subcutaneously to ten opiate abusers residing in a research unit. The primary purpose of these studies was to evaluate the effects of buprenorphine on self-administration of heroin. As part of their study, they examined the effects of buprenorphine on subjective and physiological measures. Subject-reported side effects included constipation, sedation, nausea, vomiting, decreased libido, anxiety, headache, dizziness, tinnitus, hypotension, dry mouth, urinary hesitancy, decrease of appetite, and some early minimal insomnia. One subject reported severe nausea and vomiting and received his dose increases at a lesser rate than the other subjects. Another subject developed hypotension after 2 mg/day of buprenorphine (reported as 100 mm Hg systolic and 60 mm Hg diastolic) that was associated with feelings of panic. The investigators reported that buprenorphine was well tolerated in five of seven subjects and that all subjects reported opiate-like effects that were characterized by contentment.

Bickel et al. [1988] conducted a study in which buprenorphine was administered repeatedly by the sublingual route in doses of 2–16 mg once daily to opiate addicts to determine the ability of these doses of buprenorphine to block the effects of opiate agonists. In this study the investigators reported few observable dose-related effects from buprenorphine. The major effect noted was a decline in respiratory rate of two breaths per minute at the highest buprenorphine dose. They remarked that the effects seen were less than those reported by Jasinski et al. [1978] in their study of buprenorphine given in an 8-mg/day dose subcutaneously.

TABLE VI. Rank Order Comparison of the Profile of Symptoms and Signs Produced by Repeated Administration of 8 mg Buprenorphine s.c. Once Daily[a,b]

Cited by subjects		Rank	Cited by observers		Rank
Relaxed	52	5	Relaxed	136	1
Tired	71	2	Tired	51	8.5
Anxious	13	14	Anxious	22	16
Nervous	24	9	Nervous	27	14.5
Drunk	1	20	Drunk	2	22.5
Pins and needles	0	23	Depressed	4	20.5
Rush or flush	5	16	Disheveled	2	22.5
Cold or chills	3	18	Unfriendly	5	19
Joint pains	5	16	Primping	44	10
Soapboxing	20	11	Vomiting	4	20.5
Nausea	19	12	Soapboxing	95	2.5
Feel hot	40	7	Stays to himself	43	11
Driving	23	10	Driving	59	6
Coasting	29	8	Coasting	90	4
Nodding	53	4	Nodding	54	7
Sleepy	85	1	Sleepy	51	8.5
Skin itchy	58	3	Scratching	95	2.5
Stomach cramps	0	23	Uncooperative	0	24.5
Diarrhea	0	23	Rides the bed	32	12
Gooseflesh	1	20	Complaining	95	2.5
Can't sleep	1	20	Can't sleep	0	24.5
Irritable	17	13	Irritable	9	18
Weak	5	16	Moody	29	13
Nervous stomach	46	6	Stays close to ward	72	5
			Room disorderly	27	14.5

[a]Reproduced from Jasinski et al. [1978], with permission of the publisher.
[b]Questionnaires completed by subjects or observers once daily for first 30 days of buprenorphine administration four times per day subcutaneously. Maximum response for any category with buprenorphine is 150.

Capacity to Produce Physical Dependence. The capacity of an opioid to produce morphine-like physical dependence is judged from (1) the subjective, behavioral, and physiological changes that occur following discontinuation after chronic administration and (2) the occurrence of an abstinence syndrome precipitated by the administration of an opioid antagonist such as naloxone during the period of chronic administration.

The effects of abrupt discontinuation and placebo substitution following 30–57 days of chronic administration of buprenorphine were evaluated by Jasinski et al. [1978]. The observations for the last 7 days of chronic buprenorphine administration and the first 10 days of placebo substitution are shown in Figure 1. In these studies, the primary measure of opiate abstinence was the Himmelsbach score. This

TABLE VII. Means and Standard Errors of Blood Laboratory Measures for Five Subjects and Their Comparison by Paired t Tests[a,b]

Measure	Normal ranges	Before drug	Drug	Before drug vs. drug	
				t	p
RBC, millions	4.6–6.2	5.23 ± 0.09	4.84 ± 0.06	5.16	.01
Hemoglobin, g/100 ml	14.6–16.6	16.2 ± 0.3	14.9 ± 0.4	3.99	.05
Hematocrit, %	43–49	47.2 ± 1.0	43.9 ± 1.1	3.23	.05
Sedimentation rate, mm/hr	0–10	13.1 ± 1.6	20.4 ± 3.2	4.35	.05
Total protein, g/100 ml	6.0–8.0	8.0 ± 0.2	7.4 ± 0.2	8.81	.01
Alkaline phosphatase, μU/ml	30–125	90.8 ± 8.3	72.2 ± 8.3	3.76	.05

[a]Reproduced from Jasinski et al. [1978], with permission of the publisher.
[b]Tests are for those laboratory measures in complete blood cell count and chemical analyses that showed significant differences during chronic buprenorphine administration. Normal ranges are those used in our laboratory.

score is based upon the autonomic dysfunction during opiate withdrawal and has been viewed as an objective measure of withdrawal [Jasinski, 1977]. During the first 10 days of withdrawal, Himmelsbach scores were minimal and less than the scores that followed the abrupt withdrawal of morphine and agonist–antagonist opioids with κ activity (Table VIII). The subjects reported little change in mood, feelings, perception, or thinking following placebo substitution, as shown in Figure 1 by responses on scales of chronic opioid symptoms, withdrawal symptoms, and dysphoric mood states (the "How do you feel?" question). Behaviorally, observers reported a minimal increase in withdrawal signs. The three subjects who completed the placebo substitution, however, did report some dysphoria and minimal withdrawal symptoms on the 15th day after placebo substitution, which were accompanied by slight increases in Himmelsbach scores. These signs and symptoms were relatively short-lived, lasting 1–2 days.

Naloxone, in doses ranging from 0.075 mg to 4 mg, was administered subcutaneously to subjects receiving 8 mg daily of buprenorphine [Jasinski et al., 1978]. The 4-mg dose of naloxone did not precipitate significant abstinence signs or symptoms compared with placebo. In contrast, a 4-mg dose of naloxone precipitated significant and discomforting withdrawal symptoms and signs in subjects receiving pentazocine (580 mg/day) or butorphanol (48 mg/day). This lack of response to naloxone in subjects treated chronically with buprenorphine is also different from the response in subjects who had received morphine (30 mg/day subcutaneously) and in whom a dose of 0.15 mg of naloxone precipitated a measurable abstinence response.

An additional study was conducted by Kosten et al. [1990], in which patients

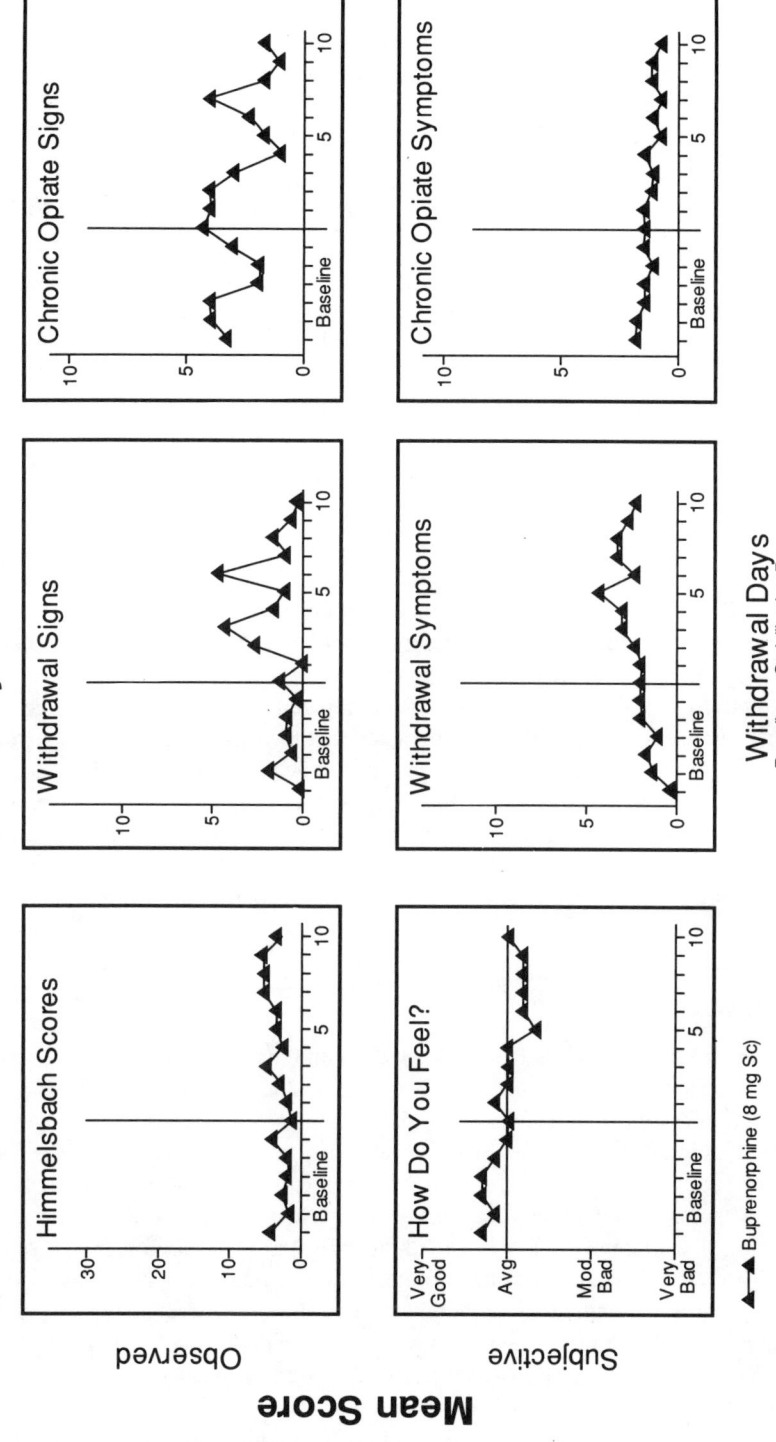

Fig. 1. Assessment of withdrawal measures for the last 7 days of chronic buprenorphine treatment (8 mg/day, s.c. baseline) and the first 10 days of abrupt placebo substitution. Values represent mean scale scores. From Jasinski et al. [1978] and unpublished data.

TABLE VIII. Comparison of Peak and Total Area Himmelsbach Scores for First Ten Days of Withdrawal of Buprenorphine, Morphine, Placebo, and Agonist–Antagonist Opioids[a]

Drug	Daily dose (mg)	Number subjects	Morphine equivalence (mg)	Total score area	Peak score
Morphine sulfate	240	8	240	198.1 ± 16.3	36.8 ± 2.7
Cyclazocine	13.2	6	260	103.6 ± 13.2	18.8 ± 2.9
Nalorphine	240	7	240	129.6 ± 10.6	18.2 ± 1.8
Nalbuphine	203	5	243	136.0 ± 6.4	24.4 ± 1.5
Pentazocine	580	6	145	106.0 ± 9.3	15.8 ± 1.8
Butorphanol	48	6	240	164.1 ± 15.2	26.3 ± 3.7
Buprenorphine	8	5	240	61.3 ± 4.2	11.0 ± 2.6
Placebo	0	8	—	34.5 ± 3.8	~6

[a]Reproduced from Jasinski et al. [1978], with permission of the publisher.

receiving either 2 mg or 3 mg/day of sublingual buprenorphine were given either 1 mg of oral naltrexone or 0.5 mg/kg of intravenous naloxone and placebo. Naltrexone challenges produced no increases in opioid withdrawal symptoms, plasma MHPG levels, or blood pressure. In contrast, naloxone produced significant signs and symptoms in blood pressure increases compared with placebo challenges. To put this response in context, it should be noted that the 0.5 mg/kg of naloxone dose in a 70-kg subject would represent 35 mg of intravenous naloxone. This dose of naloxone did produce a mild withdrawal syndrome reflected primarily in small increases in blood pressure and increases in MHPG levels. The investigators judged that this dose of naloxone produced a withdrawal response that was 35% less severe than the withdrawal response produced by a dose of naloxone 100 times lower in methadone patients.

Dependent Subjects

As part of the systematic study of the opiate withdrawal syndrome, it was recognized that opiates could alleviate the symptoms and signs of morphine abstinence [Jasinski, 1977, 1981]. When given to a subject who had been allowed to go into withdrawal, opiates suppressed withdrawal signs and symptoms; this has been called the suppression test. On the other hand, opiates administered to a morphine-dependent person prior to the emergence of morphine withdrawal prevented the occurrence of withdrawal; this was called the substitution test. Another type of test resulted from the discovery that the administration of opioid antagonists to opiate-dependent subjects resulted in a precipitated withdrawal syndrome. Subsequently, techniques were developed to assay this precipitated withdrawal. These precipitated abstinence techniques not only were then used to study dependence, but were later used to compare the antagonist potency and properties of various opioids.

In the process of screening large numbers of compounds with varying activities at opioid receptors, it was discovered that there were drugs that had both the ability to suppress or substitute for morphine on the one hand, and to act as an antagonist and precipitate abstinence on the other hand. Studies involving administration of buprenorphine to opioid-dependent subjects have been conducted to elucidate under which condition buprenorphine would either suppress or substitute for morphine or alternately precipitate abstinence.

The observation that buprenorphine, administered chronically to nondependent subjects, did not elicit a significant withdrawal syndrome suggested that substituted buprenorphine would also not be associated with a withdrawal syndrome. Jasinski et al. [1983, 1984b] conducted a series of substitution tests to determine if buprenorphine would substitute for and attenuate the expected abstinence when heroin- and methadone-dependent subjects were transferred to and maintained on buprenorphine (Table IX) and to determine if withdrawal would occur when the substituted buprenorphine was eventually terminated. In methadone-maintained patients transferred to buprenorphine (2 mg sublingual or subcutaneous), the buprenorphine attenuated and prevented the emergence of typical methadone abstinence. However, there was a mild transitory degree of withdrawal that lasted for about 3 days.

In another experiment, buprenorphine was administered daily to methadone-

TABLE IX. Substitution Studies With Buprenorphine in Methadone- and Heroin-Dependent Subjects[a,b]

Study	Maintenance drug	Number of subjects	Buprenorphine dose route and duration	Transition
1	Methadone 25–45 mg p.o.	8	2 mg s.l. q day for 28 days	Abrupt
2	Methadone 25–60 mg p.o.	4	2 mg s.l. q day for 28 days	Gradual with decreased doses of methadone
3	Methadone 51–60 mg p.o.	3	2 mg s.l. q day for 2 weeks	Abrupt
4	Methadone 25–60 mg p.o.	5	2 mg s.l. q day for 6 weeks	Abrupt
5	Methadone 45 mg p.o.	1	2 mg s.c. q day for 15 days and then for 105 days	Abrupt
6	Heroin followed by morphine 60 mg s.c. q day	9	2 mg s.l. q day for 2 weeks	Abrupt
7	Heroin followed by morphine 60 mg s.c. q day	6	2 mg s.c. q day for 2 weeks	Abrupt

[a]Data from Jasinski et al. [1983 and 1984b].
[b]Withdrawal in all studies was abrupt with placebo substitution.

dependent individuals who were receiving gradually decreasing doses of methadone [Jasinski et al., 1983, 1984b]. Buprenorphine administration did not reduce the withdrawal during the transition but may have intensified the withdrawal. Additional experiments were done in heroin-dependent persons who were transferred to and maintained on 60 mg/day of morphine given subcutaneously. Substitution of buprenorphine (2 mg either sublingually or subcutaneously) resulted in a mild transitory withdrawal syndrome that lasted approximately 3 days. After varying periods of time in each of these studies, the substituted buprenorphine was abruptly terminated with placebo substitution under double-blind conditions. A mild withdrawal syndrome resulted that peaked on the third day after termination of the substituted buprenorphine. This withdrawal syndrome was of greater intensity than the withdrawal syndrome observed following the termination of 8 mg of subcutaneous buprenorphine given chronically to nondependent subjects, which also suggested that the intensity of withdrawal decreased with lengthening periods of buprenorphine substitution. One interpretation of these results, especially in the methadone-dependent subjects, is that the withdrawal syndrome of substituted buprenorphine represented reemergence of the original withdrawal syndrome from methadone. In other words, the methadone withdrawal, which is usually of 4–6 weeks duration, had been suppressed by the buprenorphine and emerged after termination of the substituted buprenorphine to the level of untreated withdrawal.

Two studies have sought specifically to precipitate withdrawal with buprenorphine in subjects dependent upon methadone to study the antagonist properties of buprenorphine (Table X). In the first study, subjects receiving daily maintenance doses of 25–45 mg of methadone were given placebo condition, subcutaneous naloxone (0.5 mg) and were sublingually given buprenorphine (2 mg and 4 mg) [Jasinski et al., 1984b]. In these studies (Fig. 2), the subjects reported significant withdrawal sickness and showed typical signs of opiate withdrawal following administration of naloxone. On the other hand, when buprenorphine was given, there was no report of significant withdrawal sickness; however, their pupils dilated and blood pressure increased slightly. Interestingly, following buprenorphine administration the subjects reported symptoms resembling those of an opiate agonist, including feelings of stimulation, feelings of relaxation, sleepiness, some nervousness, upset stomach, and skin itching. More recently, Strain et al. [1992] found that

TABLE X. Precipitation Studies With Buprenorphine in Methadone Maintenance Patients

Study	Daily methadone dose p.o.	Maximum buprenorphine dose	Time after last dose of methadone	Reference
1	25–45	4 mg s.l.	3 hr	Jasinski et al. [1984b]
2	30	8 mg i.m.	20 hr	Strain et al. [1992]
3	60	8 mg s.l.	40 hr	Stitzer (personal communication)

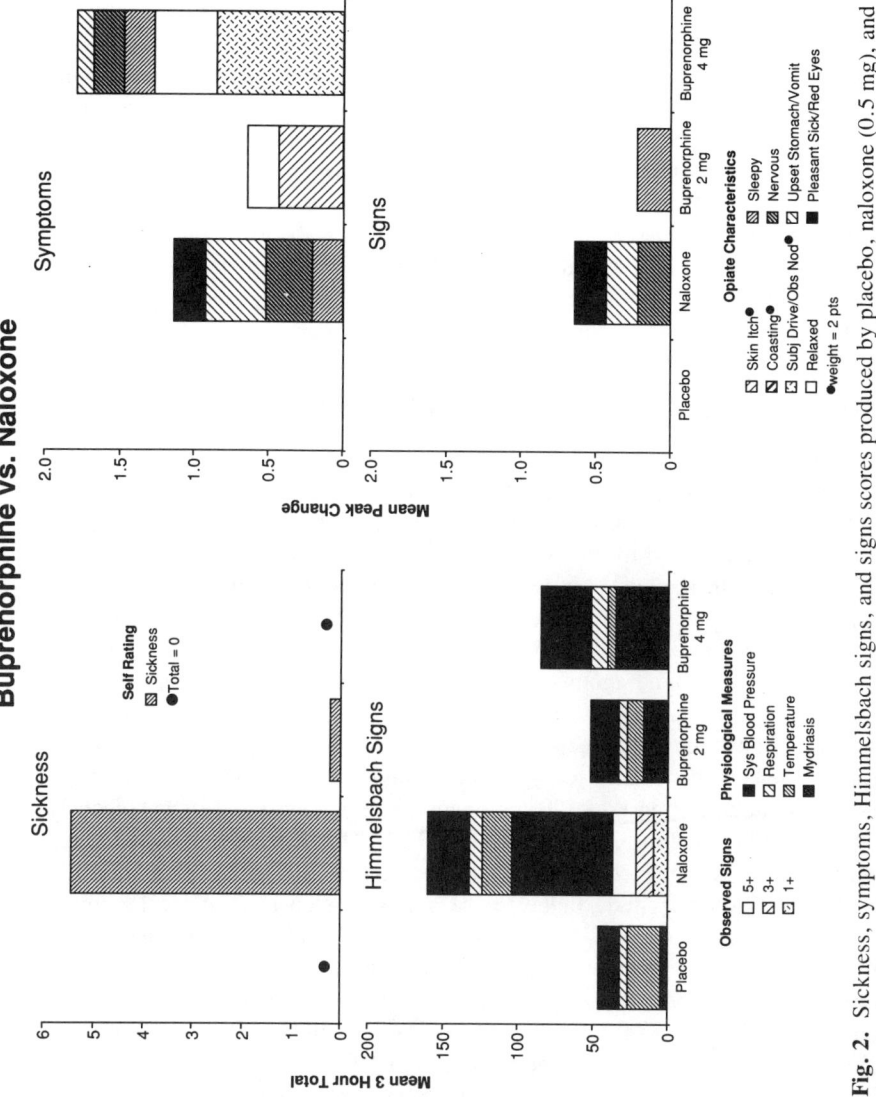

Fig. 2. Sickness, symptoms, Himmelsbach signs, and signs scores produced by placebo, naloxone (0.5 mg), and buprenorphine (2 and 4 mg) in methadone-dependent subjects from Jasinski et al. [1984b] and unpublished data. Individual item scores are shown for Himmelsbach signs (key below left panel) and for symptoms and signs (key below right panel).

in subjects dependent upon 30 mg/day of methadone, doses of buprenorphine to 8 mg administered 20 hr after the last methadone dose produced no signs of precipitated withdrawal, but did produce slight agonist effects similar to those of the opioid agonist hydromorphone.

The results of the two studies described above suggest that buprenorphine, administered to persons dependent upon methadone in doses less than 45 mg, lacks the ability to precipitate abstinence. This is consistent with the observation that buprenorphine has agonist activity equivalent to 30–60 mg of methadone. One other study has been conducted. Stitzer and her colleagues (personal communication) attempted to suppress abstinence from methadone (60 mg daily) by administering buprenorphine 40 hr after the last maintenance dose of methadone. In these subjects buprenorphine precipitated signs and symptoms of withdrawal, rather than suppressing abstinence. Thus, antagonist activity of buprenorphine has been demonstrated at higher methadone maintenance doses and at longer intervals between methadone and buprenorphine administrations.

INTERACTION STUDIES OF BUPRENORPHINE AND OPIOIDS

Interaction studies of buprenorphine and opioid agonists and opioid antagonists have been done for two purposes. First, opioid agonists have been administered to subjects receiving buprenorphine in order to test the blocking efficacy of buprenorphine and its ability to either attenuate or eliminate the response to such opioid agonists. The second set of experiments involved the simultaneous administration of buprenorphine and naloxone in order to determine if such a combination would have a lesser abuse potential. The studies involving the interaction of buprenorphine with opioid agonists and opioid antagonists are listed in Table XI.

Five experimental studies have assessed the ability of buprenorphine to block the subjective and physiological effects of opioid agonists (Table XI). In buprenorphine-dependent subjects receiving 8 mg of buprenorphine subcutaneously per day there was no or minimal response to 120 mg of morphine administered subcutaneously 1.5 hr after buprenorphine [Jasinski et al., 1978]. This blockade persisted when morphine (30 mg subcutaneously) was administered 25.5 hr after the last dose of buprenorphine. Another study involved the administration of 3, 6, or 18 mg of hydromorphone subcutaneously at 19–24 hr after the last maintenance dose of buprenorphine (2, 4, 8, and 16 mg/day sublingually) [Bickel et al., 1988]. There was a progressive blockade of hydromorphone effects with increasing doses of buprenorphine. Johnson et al. [1993] studied the effects of challenge doses of 3 mg and 6 mg of intravenous hydromorphone in groups of subjects who were receiving methadone (30 mg and 60 mg daily), and buprenorphine (2 mg, 8 mg, 12 mg, and 16 mg daily). Methadone and buprenorphine had been administered for 11 days before the hydromorphone challenge. The investigators concluded that doses greater than 8 mg sublingually per day of buprenorphine and 30 mg per day of methadone were required to block the subjective effects of intravenous hydromorphone. Walsh et al. [1993] studied the blocking ability of single sublingual doses of buprenorphine (0.5–32 mg) and single oral doses of methadone to 60 mg on the

TABLE XI. Studies of the Interaction of Buprenorphine With Other Opioids

Study	Purpose	Buprenorphine dose	Challenge drug and dose	Reference	Comments
1	Blockade of subjective withdrawal and behavioral effects	8 mg s.c. chronically	a. Morphine to 120 mg s.c. 1.5 hr after buprenorphine b. Morphine 30 mg s.c. 25.5 hr after buprenorphine	Jasinski et al. [1978]	Buprenorphine-dependent subjects
2	Blockade of subjective, behavioral, and physiological effects	2–16 mg s.l. chronically	Hydromorphone 6 and 18 mg s.c. at 19 hr	Bickel et al. [1988]	Buprenorphine-dependent subjects
3	Blockade of subjective and physiological effects of opioid agonists	2–16 mg s.l. for 11 days	Hydromorphone 3–6 mg i.v.	Johnson et al. [1993]	Buprenorphine-dependent subjects
4	Blockade of subjective and physiological effects of opioid agonist	0.5–32 mg s.l., single administration	Hydromorphone 1–4 mg i.m.	Walsh et al. [1993]	Nondependent subjects
5	Blockade of self-administration behavior	8 mg s.c. chronically	Heroin 21–40.5 mg i.v. per day	Mello et al. [1980, 1982]	Buprenorphine-dependent subjects
6	Alleviation of precipitated withdrawal	0.2 and 0.3 mg, single dose s.c.	Buprenorphine alone and in combination with naloxone 0.2 mg s.c.	Preston et al. [1988]	Methadone-dependent subjects
7	Blockade of acute effects of buprenorphine	0.4 and 0.8 mg	Buprenorphine alone and in combination with naloxone 0.4 and 0.8 mg	Weinhold et al. [1992]	Nondependent subjects

effects of hydromorphone (1 mg and 4 mg intramuscularly). The subjects were nondependent. The investigators found that 60 mg of methadone given orally and 8–32 mg of buprenorphine given sublingually attenuated the physiological and subjective effects of these doses of hydromorphone. Mello et al. [1980, 1982] demonstrated that buprenorphine (8 mg subcutaneously per day) would block the self-administration of heroin (21–40.5 mg intravenously per day), while placebo was ineffective in blocking heroin self-administration under the same experimental circumstances.

Two studies of combinations of buprenorphine and naloxone have been done (Table XI). The first involved the administration of combinations of buprenorphine (0.2 mg and 0.3 mg) and naloxone (0.2 mg) given subcutaneously to methadone maintenance patients to determine if simultaneous administration of buprenorphine would alter the ability of naloxone to precipitate an abstinence syndrome [Preston et al., 1988]. The investigators found a slight but nonsignificant alteration in the intensity of naloxone-precipitated abstinence with this combination and concluded that the naloxone added to buprenorphine would continue to act as an antagonist and lower the abuse potential of such a combination in opioid-dependent persons. A second set of experiments involved the administration of buprenorphine alone and in combination with naloxone to nondependent subjects [Weinhold et al., 1992]. The doses utilized were 0.4 and 0.8 mg of buprenorphine intramuscularly, 0.4 and 0.8 mg of buprenorphine in combination with naloxone (0.4 and 0.8 mg), and placebo condition. The combination of the low dose of buprenorphine (0.4 mg) and the high dose of naloxone (0.8 mg) resulted in complete blockade of the buprenorphine effects, which suggests a dose ratio of 2 to 1 for complete blockade in nondependent subjects. The combination of the large dose of buprenorphine (0.8 mg) and the large dose of naloxone (0.8 mg) produced a short-lived blockade of about 45 min duration. The investigators concluded that the combination of buprenorphine and naloxone may have lower abuse potential than buprenorphine alone.

SUMMARY AND CONCLUSIONS

Acute Effects

Administration of buprenorphine to subjects not physically dependent on opioids produces morphine-like subjective and stimulus effects. No evidence has been found of dysphoric effects similar to those produced by agonist-antagonists such as nalorphine, pentazocine, and butorphanol, that are believed to act primarily through the κ system. Unlike morphine and other morphine-like agonists, buprenorphine has been administered in extremely large doses to nondependent subjects without significant depression of the cardiovascular or respiratory systems. Other morphine-like adverse effects such as nausea, vomiting, urinary retention, and constipation, however, can be produced by buprenorphine. This lesser capacity to produce toxic and adverse effects is demonstrated by studies in which single doses of buprenorphine equivalent to 600 mg of parenteral morphine were administered safely. The major finding from metabolic studies in substance-abusing populations is that buprenorphine is excreted predominately in the feces with an enterohepatic circulation. Appropriate pharmacokinetic data are not available.

Chronic Effects

When administered repeatedly to volunteers who were initially not dependent on opioids, buprenorphine produced a profile of effects similar to that of morphine. The one exception was a decrease in blood pressure and pulse rate with repeated administration of buprenorphine, in contrast to the increases seen with repeated administration of morphine. Another exception was that buprenorphine had little, if any, capacity to produce physical dependence. Doses of naloxone that clearly precipitated abstinence in subjects who were dependent upon morphine-like agonists or nalorphine-like agonists were without effect in subjects receiving buprenorphine chronically. Only with massive doses of naloxone (approximately 200 times that necessary to produce withdrawal in morphine-dependent subjects) was there any suggestion of a precipitated withdrawal syndrome. In addition, when buprenorphine was abruptly withdrawn after chronic administration, slight, if any, withdrawal signs and symptoms occurred during the first 10 days of withdrawal. Modest withdrawal-like symptoms and signs occurred 14 or 15 days after abrupt withdrawal of buprenorphine in these same experiments.

Antagonist Activity

The opioid antagonist activity of buprenorphine was assessed under experimental circumstances only in subjects who were dependent upon methadone. The results of these studies were mixed. Buprenorphine precipitated withdrawal under some conditions but not others. When precipitated abstinence did occur, it was relatively mild and was more evident in autonomic signs than in subjective effects. The existing data suggest that the occurrence or intensity of precipitated withdrawal was dependent upon the dose of methadone and on the time interval between the administration of the maintenance dose of methadone and the subsequent administration of buprenorphine.

Substitution

Other experiments have involved the administration of buprenorphine to subjects who were physically dependent on morphine or methadone to determine the ability of buprenorphine to act as an agonist and substitute for morphine. In general, buprenorphine substituted for and prevented the withdrawal syndrome from either morphine or methadone when the subjects were transferred to buprenorphine. Under experimental circumstances a slight withdrawal syndrome, which lasted a few days, occurred during the period of transition to buprenorphine. When the substituted buprenorphine was withdrawn, there was a variable withdrawal response. The data suggested that the withdrawal syndrome from the substituted buprenorphine was less intense than the withdrawal syndrome that would have been expected from either morphine or methadone. In addition, available data suggest that the longer the period of substitution the less intense the withdrawal from the substituted buprenorphine.

One interpretation of the results of substitution studies would be that buprenorphine has a capacity to cause physical dependence. This interpretation contrasts with the virtual lack of a capacity to produce physical dependence that has been demonstrated in direct addiction studies.

Interaction Studies

The results of interaction studies of buprenorphine with other opioids have shown the ability of buprenorphine to attenuate the agonist effects of morphine-like opioids with no clear demonstration of the ability of buprenorphine to attenuate antagonist effects. When morphine, heroin, and hydromorphone were given to subjects receiving buprenorphine chronically, the effects of these morphine-like agonists have been attenuated or blocked by buprenorphine depending upon the dose of buprenorphine, the dose of the agonist, and the interval between the administration of the two drugs. Other experiments involved the administration of mixtures of naloxone and buprenorphine. In nondependent subjects, the naloxone attenuated the agonist effects of buprenorphine. In methadone-dependent subjects, buprenorphine possibly attenuated the naloxone-precipitated abstinence.

Mechanism of Action

The diverse effects of opioids in humans have been explained on the basis of actions at multiple opioid receptors. Conceptually, drugs have been viewed as being capable of acting at one or more of these receptors with differing degrees of intrinsic activity. The degree of intrinsic activity at various receptors was also used to explain the ability of a drug to act as either an agonist or an antagonist. The actions of buprenorphine observed in experimental studies in opioid abusers are generally consistent with the classification of buprenorphine as a morphine-like partial agonist. Buprenorphine appears to have a high degree of intrinsic activity because it has maximum agonist effects equivalent to 20–30 mg of morphine administered parenterally, and because it does not act as a strong antagonist.

Our own experiments and those of others have shown that 30 mg of morphine administered chronically will result in a definite withdrawal syndrome. The failure to see a dysphoric withdrawal syndrome in humans following abrupt cessation of repeated buprenorphine administration equivalent to 240 mg of morphine daily has been attributed to slow receptor dissociation and/or the enterohepatic circulation. Our own speculation is that this is not an adequate explanation for this phenomenon. Molecular studies and preclinical studies, however, have indicated that buprenorphine has the additional power to act as a very potent κ antagonist [e.g., Su, 1985; Gmerek et al., 1987; Leander, 1988; Negus and Dykstra, 1988]. We hypothesize that the lack of a withdrawal syndrome might be better explained by the potent κ antagonist activity.

Abuse Potential and Addiction Liability of Buprenorphine

Buprenorphine was discovered as a result of a systematic program to produce a selective analgesic. The original goal in developing selective substitutes for morphine was to retain the morphine-like analgesic effects while eliminating the capacity to cause physical dependence. This goal has been described as developing the "bee without the sting," as described by Himmelsbach [1978]. Buprenorphine's agonist effects as judged from analgesic studies and subjective effect studies are typically morphine-like. Buprenorphine is an effective substitute for morphine in most clinical situations. On the other hand, there is little if any capacity to produce physical dependence associated with its chronic use. In addition, there is clear evidence that buprenorphine does not have the capacity to produce respiratory depression to the point of apnea in normal subjects. Thus, buprenorphine has come closest to meeting the idea of the "bee without the sting." However, buprenorphine is not the "bee without the buzz."

The ability to produce morphine-like alterations in mood, feeling, thinking, and perception are the probable effects that lead to drug taking. Buprenorphine is not without a morphine-like reinforcing effect. Further, it is clear that buprenorphine can suppress and alleviate withdrawal from morphine and other opiates and maintain already established dependence, which could add to its reinforcing effects. Thus, from experimental pharmacology studies in addicts one would predict that the drug would have a liability to be abused but that the morbidity and mortality associated with its abuse should be lower than seen with other opioid agonists. Quite clearly the abuse of buprenorphine should not be associated with large numbers of deaths from respiratory depression, nor should it lead to large numbers of people who need detoxification. In practical terms, one would judge that the reinforcing efficacy of buprenorphine is equal to or slightly greater than that produced by such agonists as codeine and propoxyphene; however, buprenorphine does not have the capacity to induce physical dependence, respiratory depression, or toxicity equivalent to the effects of codeine and propoxyphene.

REFERENCES

Bickel WK, Stitzer ML, Bigelow GE, Liebson IA, Jasinski DR, Johnson RE (1988): Buprenorphine: Dose-related blockade of opioid challenge effects in opioid dependent humans. J Pharmacol Exp Ther 247:47–53.

Bigelow GE, Preston KL, Liebson IA (1992): Effects of pentazocine, nalbuphine and buprenorphine in humans trained to discriminate among saline, hydromorphone and butorphanol. NIDA Res Monogr 119:215.

Cami J (1991): Human drug abuse liability: Testing times ahead. Br J Addiction 86:1525–1526.

Cone EJ, Gorodetzky CW, Yousefnejad D, Buchwald WF, Johnson RE (1984): The metabolism and excretion of buprenorphine in humans. Drug Metab Dispos 12:577–581.

Eddy NB (1973): "The National Research Council Involvement in the Opiate Problem 1928–1971." Washington, DC: National Academy of Sciences.

Gmerek DE, Dykstra LA, Woods JH (1987): Kappa opioids in rhesus monkeys. III. Dependence associated with chronic administration. J Pharmacol Exp Ther 242:428–436.

Henningfield JE, Johnson RE, Jasinski DR (1987): Clinical procedures for the assessment of abuse potential. In Bozarth MA (ed): "Methods of Assessing the Reinforcing Properties of Drugs." Berlin: Springer-Verlag, pp 573–590.

Himmelsbach CK (1978): Summary of chemical, pharmacological, and clinical research. Drug addiction and the U.S. Public Health Service. In Martin WR, Isbell H (eds): "Proceedings of the Symposium Commemorating the 40th Anniversary of the Addiction Research Center at Lexington, Kentucky." DHEW Publication No. (ADM) 77-434, pp 13–24.

Jasinski DR (1977): Assessment of the abuse potential or morphine-like drugs (methods used in man). In Martin WR (ed): "Drug Addiction I." Vol 45-1. Heidelberg: Springer-Verlag, pp 197–258.

Jasinski DR (1979): Human pharmacology of narcotic antagonists. Br J Clin Pharmacol 7(Suppl 3):287S–290S.

Jasinski DR (1981): Opiate withdrawal syndrome: Acute and protracted aspects. Ann NY Acad Sci 362:183–186.

Jasinski DR (1984): Opioid receptors and classification. In Nimmo WS, Smith G (ed): "Opioid Agonist/Antagonist Drugs in Clinical Practice." Amsterdam: Excerpta Medica, pp 24–30.

Jasinski DR (1988): Chemical dependence. In Harvey AM, Johns RJ, McKusick VA, Owens AH, Ross RS (eds): "Principles and Practice of Medicine" (22nd ed). Norwalk, Connecticut: Appleton-Century-Crofts, pp 1144–1148.

Jasinski D (1990): Narcotic addiction: In Reier KA (ed): "Practical Clinical Management: Drug Abuse Education for Primary Care Physicians." Monograph of Conference Proceedings, Oct. 20–21, 1990, Baltimore. Baltimore: Medical and Chirurgical Faculty of Maryland, pp 19–25.

Jasinski DR (1991): History of abuse liability testing in humans. Br J Addiction 86:1559–1562.

Jasinski DR, Henningfield JE (1988): Conceptual basis of replacement therapies for chemical dependence. In Pomerleau OF, Pomerleau CS, Fagerstrom KO, Henningfield JE (eds): "Nicotine Replacement in Treatment of Smoking; a Critical Evaluation." New York: Alan R. Liss, pp 13–34.

Jasinski DR, Henningfield JE (1989): Human abuse liability assessment by measurement of subjective and physiological effects. NIDA Res Monogr 92:73–100.

Jasinski DR, Pevnick JS, Griffith JD (1978): Human pharmacology and abuse potential of the analgesic buprenorphine. Arch Gen Psychiatry 35:501–516.

Jasinski DR, Haertzen CA, Henningfield JE, Johnson RE, Makhzoumi HM, Miyasato K (1982): Progress report of the NIDA Addiction Research Center. NIDA Res Monogr 41:45–52.

Jasinski DR, Henningfield JE, Hickey JE, Johnson RE (1983). Progress report of the NIDA Addiction Research Center, Baltimore, Maryland, 1982. NIDA Res Monogr 43:92–98.

Jasinski DR, Johnson RE, Henningfield JE (1984a): Abuse liability assessment in human subjects. Trends Pharmacol, Sci 5:196–200.

Jasinski DR, Boren JJ, Henningfield JE, Johnson RE, Lange WR, Lukas SE (1984b): Progress report from the NIDA Addiction Research Center, Baltimore, Maryland. NIDA Res Monogr 49:69–76.

Jasinski DR, Fudala PJ, Johnson RE (1989): Sublingual versus subcutaneous buprenorphine in opiate abusers. Clin Pharmacol Ther 45:513–519.

Johnson RE, Risher-Flowers DL, Alim TN, Mastropaolo J, Vocci FJ, Deutsch SI (1993): Blockade of intravenous hydromorphone (H) by methadone (M) and Buprenorphine (B). Clin Pharmacol Ther 53:175.

Kosten TR, Krystal JH, Charney DS, Price LH, Morgan CH, Kleber HD (1990): Opioid antagonist challenges in buprenorphine maintained patients. Drug Alcohol Depend 25:73–78.

Leander JD (1988): Buprenorphine is a potent k-opioid receptor antagonist in pigeons and mice. Eur J Pharmacol 151:457–461.

Martin WR (1979): History and development of mixed opioid agonists, partial agonists and antagonists. Br J Clin Pharmacol 7(Suppl 3):273S–279S.

Mello NK, Mendelson JH (1980): Buprenorphine suppresses heroin use by heroin addicts. Science 207:657–659.

Mello NK, Mendelson JH, Kuehnle JC (1982): Buprenorphine effects on human heroin self-administration: An operant analysis. J Pharmacol Exp Ther 223:30–39.

Negus SS, Dykstra LA (1988): k Agonist properties of buprenorphine in the shock titration procedure. Eur J Pharmacol 156:77–86.

Preston KL, Bigelow GE, Liebson IA (1988): Buprenorphine and naloxone alone and in combination in opioid-dependent humans. Psychopharmacology 94:484–490.

Preston KL, Bigelow GE, Bickel WK, Liebson IA (1989): Drug discrimination in human post-addicts: Agonist–antagonist opioids. J Pharmacol Exp Ther 250:184–196.

Preston KL, Jasinski DR (1991): Abuse liability studies of opioid-antagonists in humans. Drug Alc Depend 28:49–82.

Preston KL, Liebson IA, Bigelow GE (1992): Discrimination of agonist–antagonist opioids in humans trained on a two-choice saline–hydromorphone discrimination. J Pharmacol Exp Ther 261:62–71.

Strain EC, Preston KL, Liebson IA, Bigelow GE (1992): Acute effects of buprenorphine, hydromorphone and naloxone in methadone-maintained volunteers. J Pharmacol Exp Ther 261:985–993.

Su T-P (1985): Further demonstration of kappa opioid binding sites in the brain: Evidence for heterogeneity. J Pharmacol Exp Ther 232:144–148.

Walsh SL, Preston KL, Stitzer ML, Bigelow GE, Liebson IA (1992): The acute effects of high dose buprenorphine in non-dependent humans. NIDA Res Monogr 119:245.

Walsh SL, Preston KL, Stitzer ML, Liebson IA, Bigelow GE (1993): Comparison of the acute effects of buprenorphine and methadone in non-dependent humans. NIDA Res Monogr 132:333.

Weinhold LL, Preston KL, Farre M, Liebson IA, Bigelow GE (1992): Buprenorphine alone and in combination with naloxone in non-dependent humans. Drug Alcohol Depend 30:263–274.

CLINICAL EFFICACY STUDIES OF BUPRENORPHINE FOR THE TREATMENT OF OPIATE DEPENDENCE

PAUL J. FUDALA
Department of Psychiatry, University of Pennsylvania School of Medicine, and Department of Veterans Affairs Medical Center, Philadelphia, Pennsylvania 19104

ROLLEY E. JOHNSON
Department of Psychiatry, The Johns Hopkins University School of Medicine, Baltimore, Maryland 21224

INTRODUCTION

Pharmacotherapies for treating dependence on narcotics have included such diverse agents as sodium bromide, cocaine, castor oil, and atropine. In 1929, Congress authorized the establishment of two hospitals, later located in Lexington, Kentucky, and Fort Worth, Texas, for the treatment of narcotics dependence. The use and gradual reduction of morphine or codeine to suppress the opiate withdrawal syndrome was reported by the Lexington hospital in 1939 [Maddux, 1978].

From the 1940s through the 1970s, the Addiction Research Center in Lexington assessed and evaluated numerous compounds for their abuse liability [Martin and Isbell, 1978]. These studies also provided information pertinent to the use of some of these compounds, such as methadone [Isbell et al., 1948], levo-alpha-acetylmethadol (LAAM [Fraser and Isbell, 1952], propoxyphene [Jasinski et al., 1977], naloxone [Jasinski et al., 1967], naltrexone [Martin et al., 1973b], and cyclazocine [Martin et al., 1966] as potential opiate-treatment medications.

Although many compounds have been evaluated, there are currently only three medications approved by the Food and Drug Administration for the treatment of opiate dependence: methadone, LAAM, and naltrexone. Methadone and LAAM are μ (morphine-like) opioid agonists, which are administered orally both to substitute

Buprenorphine: Combatting Drug Abuse With a Unique Opioid, pages 213–239
© 1995 Wiley-Liss, Inc.

for illicit opiates and to suppress the opiate-abstinence syndrome. Owing to pharmacokinetic differences between the two medications, methadone must be administered daily, while LAAM is administered three times weekly (e.g., Monday, Wednesday, Friday) or on an every-other-day basis. Naltrexone is an effective opiate-antagonist that has shown therapeutic utility in limited patient populations.

Buprenorphine is a μ opioid partial agonist that was first assessed in the 1970s for its abuse liability in humans and is currently under active investigation as a treatment for opiate dependence. The potential usefulness of buprenorphine as a treatment medication was first studied by Jasinski et al. [1978]. They reported subjective, behavioral, and physiological morphine-like effects resulting from acute, single doses of buprenorphine. These investigators went on to propose that buprenorphine be considered for the long-term treatment of opiate dependence because of its acceptability to the addict population, its long duration of action, and its ability to block certain effects of morphine, in addition to its limited physiological withdrawal syndrome following abrupt termination. Subsequently, clinical laboratory studies assessing the pharmacology, pharmacokinetics, and pharmacodynamics of buprenorphine provided a framework for later clinical evaluations.

This chapter reviews clinical efficacy studies, focusing primarily on induction, maintenance, and dosage reduction procedures as they relate to the use of buprenorphine as an opiate-dependence treatment. It discusses outpatient studies, completed or ongoing, that have been reported to date, and complements the review of human laboratory studies by Jasinski and Preston in this volume. Certain inpatient studies are reviewed to provide continuity and clarity to the present discussion. Issues related to buprenorphine's safety, treatment indications, and misuse are also considered.

DOSE INDUCTION PROCEDURES

An initially high dose of a partial agonist such as buprenorphine, administered to an opiate-dependent person, may appear to precipitate a withdrawal syndrome [Aceto, 1984]. If the initial dose of the partial agonist is too low, it can be expected that it might not provide morphine-like effects sufficient to prevent opiate withdrawal signs and symptoms.

A number of dose induction procedures have been used to transfer opiate-dependent individuals from either methadone or street opiates to buprenorphine [Jasinski et al., 1983; Reisinger, 1985; Seow et al., 1986; Bickel et al., 1988a; Kosten and Kleber, 1988; Johnson et al., 1989, 1992a; Kosten et al., 1991]. These procedures often were only a single component of studies designed to assess other treatment outcome parameters. Initial sublingual dosages ranged from 0.6 to 8 mg, with the induction period ranging from 1 day to 2 weeks. The studies that reported results for a dose induction phase are summarized in Table I.

Only one inpatient investigation has specifically reported on the transition from street opiates to buprenorphine [Johnson et al., 1989]. In that study data were reported from 19 subjects who were given buprenorphine in ascending daily doses of 2, 4, 8,

TABLE I. Dose Induction Procedures

Drug	No. of groups	Route	Daily dosage (mg)	Schedule	Induction duration (days)	Number of subjects	Type of subject	Design	Results	Reference
Buprenorphine Methadone	1	Sublingual Oral	2 25–45	Abrupt induction to buprenorphine	1	8	Methadone maintenance	Double-blind, double-dummy	Mild discomfort peaked by 3–4 days; gradually diminished	Jasinski et al. [1983]
Buprenorphine Morphine	1	Sublingual Subcutaneous	2 60	Abrupt induction to buprenorphine	1	9	Opiate-dependent	Double-blind, double-dummy	Mild withdrawal signs	Jasinski et al. [1984]
Buprenorphine Methadone	1	Sublingual Oral	2 25–60	Abrupt induction to buprenorphine Gradual reduction of methadone	1 9	4	Methadone maintenance	Double-blind double-dummy	Mild discomfort was not attenuated by gradual reduction of methadone	Jasinski et al. [1983]
Buprenorphine	3	Sublingual Sublingual Sublingual	2 4 8	Abrupt induction	1 1 1	10 4 2	8 methadone maintenance; 8 heroin-dependent	Open-label	Withdrawal symptoms 4 mg < 8 mg ≤ 2 mg	Kosten and Kleber [1988]
Buprenorphine	1	Sublingual	2 4 8	1st day 2nd day 3rd day	3	19	Opiate-dependent	Double-blind	Increased positive symptoms and decreased withdrawal symptoms over 3 days	Johnson et al. [1989]
Buprenorphine	4	Sublingual Sublingual Sublingual Sublingual Sublingual	2 2 3 4 6	1st and 2nd day 3rd day 3rd day 3rd day 3rd day	3	11 16 14 6 3	12 methadone maintenance; 27 heroin dependent	Open-label	Mild withdrawal symptoms	Kosten et al. [1991]

and 8 mg on study days 1, 2, 3, and 4, respectively. Outcome measures included a 20-item opiate-agonist-effects adjective checklist, a 15-item opiate-withdrawal symptom rating scale, a 9-item drug-effect question, three subscales from the Addiction Research Center Inventory, and various physiological parameters. The administration of buprenorphine was followed by changes in subject-reported states of feeling and physiological measures typical of those produced by μ opiate agonists.

Urinary opiate concentrations decreased and urinary buprenorphine levels increased over the induction period. The urinary excretion profile from seven of the subjects indicated that the majority of heroin metabolites were eliminated in the first 24 hr after the last reported administration of heroin. Overall, this investigation demonstrated that a daily sublingual dose as high as 8 mg could be achieved in only 3 days.

A previous outpatient investigation evaluating the transition from street opiates to buprenorphine [Kosten and Kleber, 1988] indicated that 4 mg could be effectively administered as an initial dose. It seems reasonable to assume, however, that the optimal dose of buprenorphine will depend on the dosage and elimination kinetics of the opiate on which the subject is maintained (e.g., methadone, LAAM) or is abusing (e.g., heroin).

The above studies have shown that the transition from street opiates to buprenorphine can be accomplished with minimal discomfort, as evidenced by the production of positive subjective effects, lack of withdrawal symptoms, and retention of subjects in treatment. The transition from street opiates to the partial agonist buprenorphine should be performed abruptly, with the maintenance dosage achieved as rapidly as possible. This recommended rapid achievement of the maintenance dosage is in contrast to the relatively slower dosage escalation suggested for the opioid agonists methadone and LAAM. Too rapid an induction with the latter medications could result in excessive sedation and respiratory depression.

The most effective buprenorphine induction procedure for persons maintained on methadone has not been well defined. The only studies to directly assess the transition from methadone to buprenorphine utilized a maximum buprenorphine dosage that was less than what is generally recognized to be effective for maintenance treatment today. It does appear, however, that a gradual reduction of methadone dosage along with stable buprenorphine administration does not result in a greater attenuation of withdrawal signs and symptoms [Jasinski et al., 1983]. From clinical experience gained from working with street-opiate addicts in conjunction with an understanding of methadone elimination kinetics, it seems reasonable that subjects maintained on 30–60 mg of methadone could be transferred to buprenorphine over five to seven days.

MAINTENANCE REGIMENS

A number of studies have examined the efficacy of buprenorphine given over varying periods of time and at different dosages [Jasinski et al., 1978; Mello and Mendelson, 1980; Reisinger, 1985; Seow et al., 1986; Bickel et al., 1988a; Kosten and Kleber, 1988; Kosten et al., 1991, 1993; Fudala et al., 1990; Johnson et al.,

TABLE II. Maintenance Regimens

Drug	No. of groups	Route	Dosage (mg)	Schedule	Maintenance duration (days)	Number of subjects	Type of subject	Design	Results	Reference
Buprenorphine	1	Subcutaneous	8	Daily	58	3	Nondependent	Double-blind	Morphine-like subjective effects equivalent to 40–60 mg methadone	Jasinski et al. [1978]
Buprenorphine	1	Sublingual	Not specified	Daily	60–510	65 entered; 34 followed	Opiate-dependent	Open-label	Not specified: "Improvement on psychological socio-professional and familial parameters"	Reisinger [1985]
Buprenorphine	2	Sublingual Sublingual	2 4	Daily Daily	Active dose weeks 1, 2, 4, 5; placebo dose week 3	32	Opiate-dependent	Randomized	Less withdrawal and opiate use in 4-mg group; observed mild withdrawal signs in week 3 with substitution of placebo	Seow et al. [1986]
Buprenorphine Methadone	2	Sublingual Oral	2 30	Daily Daily	21 21	22 23	Opiate-dependent	Randomized, double-blind, double-dummy	Buprenorphine equivalent to methadone	Bickel et al. [1988a]
Buprenorphine	3	Sublingual Sublingual Sublingual	2 4 8	Daily Daily Daily	30 30 30	10 4 2	Opiate-dependent	Open-label	78% of urine samples negative for opiates	Kosten and Kleber [1988]

(*continued*)

TABLE II. (*Continued*)

Drug	No. of groups	Route	Dosage (mg)	Schedule	Maintenance duration (days)	Number of subjects	Type of subject	Design	Results	Reference
Buprenorphine	2	Sublingual	8	Daily	36	9	Opiate-dependent	Randomized, parallel-group, double-blind	More subjective withdrawal symptomatology reported with alternate day dosing	Fudala et al. [1990]
Buprenorphine		Sublingual	8	Alternate days	36	10				
Buprenorphine	4	Sublingual	2	Daily	30	16	Methadone/heroin-dependent	Open-label	For completers, number of opiate-positive urine samples inversely related to number of days in treatment	Kosten et al. [1991]
		Sublingual	3	Daily	30	14	5/11			
		Sublingual	4	Daily	30	6	4/10			
		Sublingual	6	Daily	30	3	2/4			
							1/2			
Buprenorphine	3	Sublingual	8	Daily	119	53	Opiate-dependent	Randomized, parallel-group double-blind, double-dummy, stratified	Buprenorphine and methadone 60 mg were equally effective and both were more effective than methadone 20 mg	Johnson et al. [1992a]
Methadone		Oral	20	Daily	119	55				
Methadone		Oral	60	Daily	119	54				
Buprenorphine	4	Sublingual	2	Daily	168	28	Opiate-dependent	Randomized, parallel-group, double-blind, double-dummy	Buprenorphine 2 and 6 mg less effective than methadone 35 and 65 mg; both methadone dosages equally effective	Kosten et al. [1993]
Buprenorphine		Sublingual	6	Daily	168	28				
Methadone		Oral	35	Daily	168	34				
Methadone		Oral	65	Daily	168	35				

1992a] (Table II). Most of these investigations would generally be considered to be supportive for Food and Drug Administration approval of buprenorphine as an opiate-dependence treatment medication, and one [Johnson et al., 1992] is being evaluated as a pivotal trial. Two additional trials [Ling et al., 1993], currently ongoing, were designed as pivotal trials for the approval of buprenorphine.

In the first-reported clinical efficacy study, Jasinski et al. [1978] proposed that 8 mg daily of chronically administered, subcutaneous buprenorphine was equivalent to 40–60 mg of orally administered methadone. Reisinger [1985] noted that a program combining buprenorphine treatment (dosages in the range of 2–4 mg daily) and psychotherapy provided important advantages over standard detoxification programs, therapeutic communities, and methadone treatment. Seow et al. [1986] later proposed that buprenorphine dosages of 4 mg or greater (sublingual) would be necessary to afford sufficient blockade of subjective drug effects (e.g., euphoria) from illicitly administered opiates.

Bickel et al. [1988a] conducted the first randomized, double-blind trial comparing buprenorphine with methadone. Although the maintenance phase of this study lasted only 3 weeks, the study did provide important information regarding the relative effectiveness of buprenorphine compared with methadone. The data demonstrated that 2 mg of sublingual buprenorphine was less effective than 30 mg of oral methadone in its ability to block the physiological and subjective effects of a 6-mg intramuscular dose of hydromorphone. Buprenorphine and methadone were equivalent with respect to the results obtained for the following outcome measures: subject retention in treatment and reduction of illicit opiate use.

Data from the hydromorphone challenge portion of the above investigation formed the basis for a dose-response study assessing buprenorphine's opiate-blocking activity [Bickel et al., 1988b]. This crossover-design study compared the effects of 6- and 12-mg doses of intramuscular hydromorphone given to subjects maintained on 2, 4, 8, or 16 mg of buprenorphine. The authors found that an 8-mg sublingual dose was effective in blocking both subject-reported "high" and "drug effect" following intramuscular hydromorphone. Compared with lower buprenorphine doses, 8 mg was also associated with lower scores on an opiate-agonist adjective rating scale following administration of hydromorphone.

Two open-label studies [Kosten and Kleber, 1988; Kosten et al., 1991] also provided supportive evidence of buprenorphine's efficacy. The investigators reported good to excellent retention in treatment and reduced frequency of illicit opiate use.

Information from the above-mentioned studies, as well as from clinical laboratory studies described by Jasinski and Preston in this volume, was used to design an inpatient investigation [Fudala et al., 1990] to further assess the clinical pharmacology of buprenorphine for its use in the treatment of opiate dependence. The 8-mg dosage was chosen because it had been shown previously to effectively block the effects of intramuscular hydromorphone [Bickel et al., 1988b]. Higher doses were not considered because of a perceived increased risk of toxicity. This had been the only investigation reported to date that utilized both daily and alternate-day buprenorphine administration (8 mg sublingual), although subsequent studies have evaluated less-than-daily dosing regimens (Table III).

TABLE III. Ongoing and Recently Reported Studies

Drug	No. of groups	Route	Dosage (mg)	Schedule	Duration (days)	Number of subjects	Type of subject	Design	Results	Code	Reference
Buprenorphine	4	Sublingual	2	Daily	49	50	Dual opiate- and cocaine-dependent	Randomized, parallel-group, double-blind	80 patients enrolled as of August 1993	D	Study ongoing; PI: Montoya, ID
			8	Daily	49	50					
			16	Daily	49	50					
			16	Alternate day	49	50					
Buprenorphine	3	Sublingual	8	Daily	365	75	Opiate-dependent	Randomized, parallel-group, double-blind, double-dummy	On measure of craving for an opiate, buprenorphine 8 mg and methadone 30 mg were equivalent and methadone 80 mg was more effective; analysis ongoing.	B	Ling et al. [1993]
Methadone		Oral	30	Daily	365	75					
		Oral	80	Daily	365	75					
Buprenorphine	4	Sublingual	1	Daily	112	120	Opiate- or methadone-dependent	Randomized, parallel-group, double-blind	Multicenter trial; recruitment of subjects completed; data not analyzed.	B	Ling et al. [1993]; Study ongoing; PI: Ling, W
			4	Daily	112	120					
			8	Daily	112	120					
			16	Daily	112	120					
Buprenorphine	3	Sublingual	2	Daily days 1–31 double dose/placebo on alternate days 32–52	52	2	Opiate-dependent	Randomized, crossover, placebo control	When maintenance dose is doubled, buprenorphine can be administered effectively every 48 hr.	A	Amass et al. [1993]
			4		52	6					
			8		52	5					

Buprenorphine	6	Sublingual	2	Daily	27	∞	Opiate-dependent	Randomized, parallel-group, double-blind, double-dummy	Interim analysis showed doses of buprenorphine > 8 mg and methadone > 30 mg required to block the subjective effects of hydromorphone 6 mg. Data assessing the withdrawal syndrome associated with each medication are being analyzed.	F	Johnson et al. [1993]
			8	Daily		∞					
			12	Daily		∞					
			16	Daily		∞					
Methadone		Oral	30	Daily		∞					
			60	Daily		∞					
Buprenorphine	1	Sublingual	8	Daily maintenance	7	9	Opiate-dependent	Crossover, placebo control	Sublingual buprenorphine in combination with naloxone did not precipitate opiate withdrawal symptoms.	E	Jones and Mendelson [1993]
Buprenorphine + naloxone		Sublingual	8 + 0, 4, 8	Crossover							
Buprenorphine	1	Sublingual	4 or 8	Daily maintenance	30	20	Dual opiate- and cocaine-dependent	Randomized, crossover, single-blind	Daily doses of buprenorphine not associated with adverse side effects or toxic interactions with acute dose of intravenous cocaine or morphine.	C	Teoh et al. [1993]
Cocaine		Intravenous	30								
Saline		Intravenous	0								
Morphine		Intravenous	10	Crossover							

(*continued*)

TABLE III. (*Continued*)

Drug	No. of groups	Route	Dosage (mg)	Schedule	Duration (days)	Number of subjects	Type of subject	Design	Results	Code	Reference
Buprenorphine	4	Sublingual	4	Daily	182	Not specified	Dual opiate- and cocaine-dependent	Randomized, parallel-group, double-blind	Significant increase in days in treatment for buprenorphine 12 mg and methadone 65 mg groups; questionable decrease in cocaine use; opiate use not reported.	B	Schottenfeld et al. [1993]
Buprenorphine		Sublingual	12	Daily	182	Not specified					
Methadone		Oral	20	Daily	182	Not specified					
Methadone		Oral	65	Daily	182	Not specified					
Morphine	1	Intramuscular	15–120/day	Daily maintenance	Not specified	Not specified	Opiate-dependent	Crossover, double-blind	Agonist-like effects from buprenorphine at low levels and antagonist-like effects at high levels of morphine dependence.	G	Schuh et al. [1993]
Buprenorphine		Intramuscular	3	Crossover							
Naloxone		Intramuscular	0.3	Crossover							
Morphine		Intramuscular	30	Crossover							
Buprenorphine	2	Sublingual	8	Daily	140	52	Opiate-dependent	Randomized, parallel-group, double-blind	Recruitment of subjects completed; data not analyzed	A	Study ongoing; PI: Stitzer, ML
Buprenorphine		Sublingual	8	Alternate day	140	53					
Buprenorphine	2	Sublingual	8–16	Daily	112	24	Dual opiate- and cocaine dependent	Randomized, parallel-group, double-blind, double-dummy	Buprenorphine equally effective to methadone for opiate dependence; no difference observed between groups for cocaine use	D	Strain et al. [1993b]
Methadone		Oral	50–90	Daily	112	27					

Drug	N	Route	Dose	Schedule	Duration	N subjects	Population	Design	Results		Reference
Buprenorphine Clonidine Clonidine/Naltrexone	3	Sublingual Oral Oral	Not specified	Not specified	Not specified	33 31 33	Opiate-dependent	Not specified	Both buprenorphine and clonidine/naltrexone appeared to be more effective than clonidine alone	F	O'Connor et al. [1993]
Buprenorphine	2	Sublingual	Sufficient to suppress withdrawal symptoms	Rapid reduction Slow reduction	7–13 23–37	39 19	Opiate-dependent	Single-blind	Retention in treatment and heroin not different between groups	F	Pycha et al. [1993]
Buprenorphine	2	Sublingual	Range 4–10 doubled Tripled	Initial maint. dose alternate day Every third day	Not specified	23 10	Buprenorphine-dependent	Open-label	21 subjects continued heroin abstinence with a reduced dosing frequency of 3–4 days per week	A	Resnick et al. [1993]
Buprenorphine Cocaine Saline Morphine	1	Sublingual Intravenous Intravenous Intravenous	4 or 8/day 30 0 10	Daily maintenance Crossover Crossover Crossover	30	36	Methadone-dependent	Randomized, crossover, single-blind	Cocaine and opiate challenges during buprenorphine maintenance were not associated with adverse effects	C	Wapler et al. [1993]
Methadone Buprenorphine Naloxone Buprenorphine + naloxone	1	Oral Intravenous Intravenous Intravenous	40–60 0.2 0.1 0.2 0.1	Daily Crossover Crossover Crossover	Not specified	3	Methadone-dependent	Crossover, placebo control	Combination formulation produced more intense withdrawal effects than naloxone alone	E	Welm et al. [1993]

(*continued*)

TABLE III. (*Continued*)

Drug	No. of groups	Route	Dosage (mg)	Schedule	Duration (days)	Number of subjects	Type of subject	Design	Results	Code	Reference
Buprenorphine	2	Sublingual	8–16 × 0.9	Daily	98	84	Opiate-dependent	Randomized, parallel-group, double-blind, double-dummy	Buprenorphine 8.9 mg and methadone 54 mg equivalent as to retention in treatment and urine samples positive for opiates.	B	Strain et al. [1993, 1994a]
Methadone		Oral	50–80 × 54	Daily	98	90					
Buprenorphine	2	Sublingual	0.6–1.2	Daily	10	Not specified	Opiate-dependent	Randomized	Subjective and objective withdrawal signs and symptoms were less sustained and intense in individuals receiving buprenorphine.	F	Nigram et al. [1993]
Clonidine		Oral	0.3–0.9	Daily	10	Not specified					
Buprenorphine	1	Sublingual	0.15–0.6	Daily	1–2	15	Methadone-dependent	Open-label	Analgesic doses of buprenorphine were sufficient to treat withdrawal symptoms in 40% of subjects; possible ethnic differences noted.	F	Banys et al. [1994]

Codes refer to main focus or type of sudy: A = dosage regimen assessment; B = opiate treatment efficacy; C = safety evaluation; D = opiate/cocaine treatment efficacy; E = combination product assessment; F = detoxification; G = dose induction.

In the Fudala et al. [1990] study, data were reported from 19 subjects, randomly assigned to one of two groups upon admission. Following an 18-day induction/maintenance period, the subjects in one group continued to receive 8 mg of sublingual buprenorphine for an additional 18 days, while the subjects in the other group received the same buprenorphine dose on alternate days. No differences between the study groups for drug or withdrawal effects were reported, although the subjects who received buprenorphine on alternate days did report higher ratings on a scale measuring dysphoria on the days on which they received placebo. On a scale, administered on a daily basis, that assessed the urge for an opiate, the subjects who received buprenorphine every day consistently reported higher mean scores than subjects who received buprenorphine on alternate days. This finding was observed regardless of whether the alternate-day group was receiving buprenorphine or placebo. Within-group comparison of subjects in the alternate-day group found that a greater urge for an opiate was reported on placebo than on buprenorphine days. Additionally, for subjects in the alternate-day group, pupillary constriction (an objective measure of morphine-like effects) was observed on buprenorphine days but not on placebo days. This finding paralleled subjects' responses on an opiate-withdrawal adjective rating scale (Fig. 1). No differences in Himmelsbach scale scores (a traditional measure of opiate-withdrawal symptomatology) were observed between the two groups.

The results from the above inpatient study formed the basis for an outpatient efficacy trial that directly compared buprenorphine with methadone [Johnson et al., 1992a]. This 25-week, outpatient study assessed the efficacy of buprenorphine as a short-term maintenance/detoxification agent in a drug abuse treatment setting. The subjects were stratified by age, gender, and level of physical dependence (as assessed by a naloxone challenge procedure) in order to control more effectively for

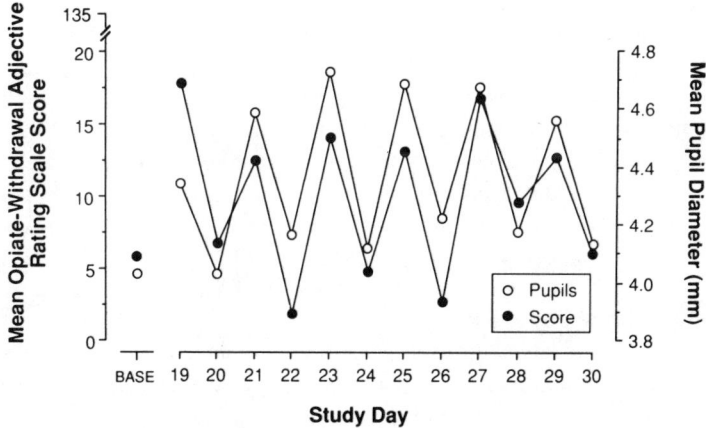

Fig. 1. Mean daily values from an opiate-withdrawal adjective rating scale and for pupillary diameter measured while subjects received buprenorphine on alternate days in the Fudala et al. [1990] study. Base (baseline) is the mean value of the data from study days 14 and 15. Each point represents the mean value of nine or ten subjects.

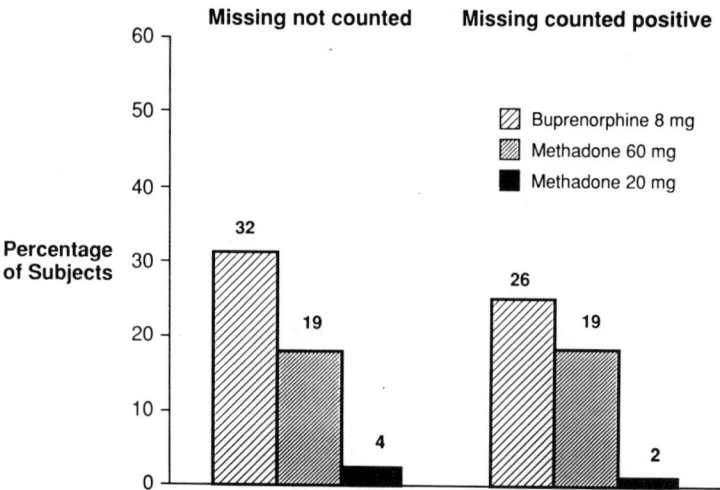

Fig. 2. Percentage of subjects in the Johnson et al. [1992a] study who submitted 12 consecutive urine samples that were negative for opiates. The results are shown as computed when missing samples are not counted and when they are counted as positive.

variables that might affect treatment outcome. A 20-mg dosage of methadone was chosen as an active control. A 60-mg methadone comparison dosage was selected, since it had been reported to be the median daily dosage used in maintenance therapy and one considered to produce effects distinguishable from the lower dosage of methadone. An 8-mg buprenorphine dosage was chosen, since it was hypothesized to be more effective than methadone 20 mg, and at least 80% as effective as methadone 60 mg [Jasinski et al., 1978]. This study was the first and largest clinical trial reported that demonstrated the effectiveness of buprenorphine as a treatment for opiate dependence.

In the Johnson et al. [1992a] study, the effectiveness of 8 mg of sublingual buprenorphine was generally equivalent to that seen for 60 mg of oral methadone and was significantly better than that observed with 20 mg of methadone in regard to treatment retention and decreased opiate use. Additionally, a post hoc analysis assessing subjects who provided at least 12 consecutive urine samples negative for opiates (over a minimum of 4 weeks) indicated that buprenorphine 8 mg and methadone 60 mg were superior to methadone 20 mg (Fig. 2; Johnson and colleagues, unpublished observations).

The subjects who received buprenorphine 8 mg or methadone 60 mg had approximately equal retention rates, 42% and 32% respectively, after the first 17 weeks of the study, while the retention rate of the subjects who received methadone 20 mg was 20% (Fig. 3). Subjects' enrollment in the study resulted in a rapid increase in opiate-negative urine samples during the second, compared with the first, treatment week, the extent of improvement being greater for buprenorphine than for either methadone group (Fig. 4). Nonetheless, a considerable amount of opiate-positive

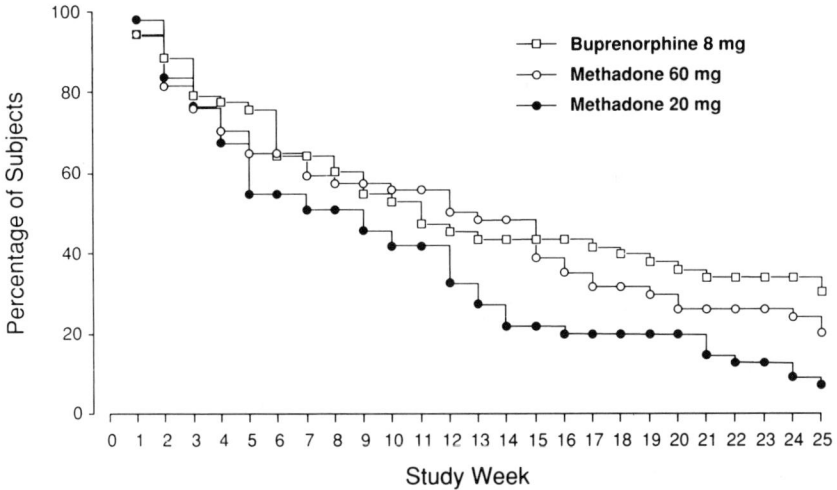

Fig. 3. Retention rates of subjects in the Johnson et al. [1992a] study. Each point represents the percentage of subjects in each group in the study who remained in treatment at the end of a given week.

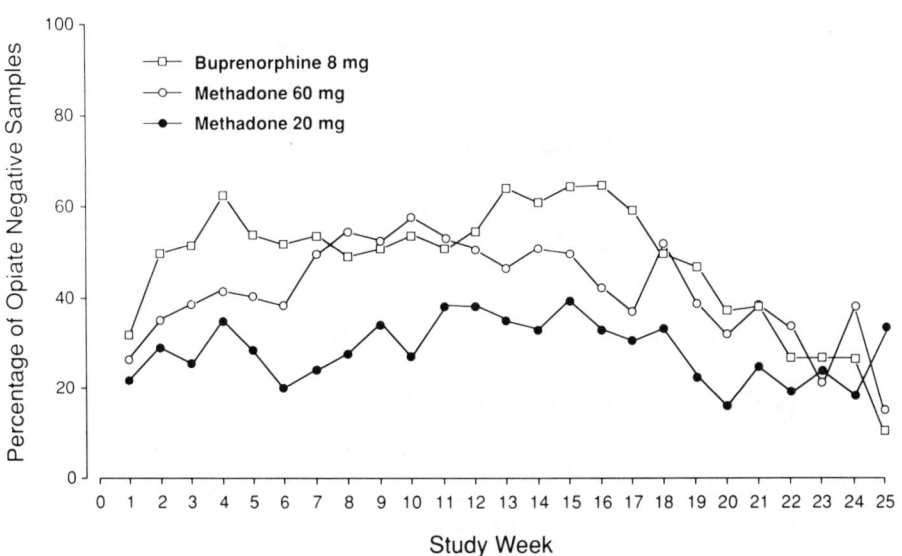

Fig. 4. Percentage of opiate-negative urine samples for each group in the Johnson et al. [1992a] study, by study week.

urine samples persisted during treatment with buprenorphine, suggesting that daily doses above 8 mg may be necessary to suppress illicit opiate use.

Another outpatient study that compared buprenorphine with methadone has also been reported [Kosten et al., 1993]. This trial compared the efficacies of buprenorphine (2 and 6 mg sublingual) and methadone (35 and 65 mg oral) by a double-blind/double-dummy procedure. Most analyses were conducted on information from 125 of the 140 subjects admitted to the study. Using data collapsed across each treatment medication, the authors concluded that methadone treatment was superior to buprenorphine. There were no statistical differences, however, between the two methadone groups or between the two buprenorphine groups with respect to subject retention or opiate-free urine samples and, surprisingly, the lower dosage level for each medication was associated with higher mean retention rates.

One recently completed investigation [Strain et al., 1993a, 1994], described in Table III, assessed flexible dosing schedules of buprenorphine (mean daily dose = 8.9 mg/day) and methadone (mean daily dose = 54.1 mg/day) in a randomized, double-blind/double-dummy evaluation that enrolled 164 subjects. Fifty-six percent of the subjects in each group completed the 16-week induction/maintenance phase of the study. No differences were observed between the two groups with respect to retention time in treatment or to urine samples found to be positive for opiates. These results parallel those reported by Johnson et al. [1992a] in their study using comparable fixed dosages of buprenorphine (8 mg) and methadone (60 mg).

The outpatient studies described above all utilized daily dosing schedules. An important consideration that requires further study is the determination of the optimum dose for safely maximizing the interdose interval for administration of buprenorphine. The only therapeutic alternative to methadone that has been shown to have a longer duration of action is LAAM. If buprenorphine could be dosed less often than once daily, its therapeutic efficacy might be enhanced.

DOSE REDUCTION PROCEDURES

Dose reduction procedures have ranged from those utilizing abrupt termination of buprenorphine to more gradual reductions. Data from three clinical laboratory studies (described by Jasinski and Preston in this volume and summarized below) provided initial information regarding both abrupt and gradual termination of buprenorphine in previously non-opiate-dependent persons, and also in heroin- and methadone-dependent subjects. Following the chronic administration of buprenorphine to non-opiate-dependent individuals, abrupt termination was associated with a mild-to-moderate withdrawal syndrome consisting primarily of subject-reported hot flashes and chills during the first 14 days of abstinence, followed by typical opiate-withdrawal signs and symptoms over the next 4 days [Jasinski et al., 1978]. In another study, Mello et al. [1982] found that a gradual dosage reduction (over 5 days) in subjects previously dependent on heroin was not associated with a withdrawal syndrome.

When data were compared across studies, buprenorphine was reported to be associated with a less severe withdrawal syndrome following its abrupt termination than the withdrawal syndrome associated with the abrupt termination of methadone [Martin et al., 1973a; Jasinski et al., 1984]. However, this comparison was based on a methadone dosage of 100 mg given daily for 8 weeks and buprenorphine 2 mg daily administered for 28 days.

Some subsequent clinical investigations of buprenorphine have incorporated a dose reduction component, although only a few [Fudala et al., 1990; Kosten et al., 1991; Johnson et al., 1992; Resnick et al., 1992] have directly evaluated potential dose reduction schedules (Table IV). Bickel et al. [1988a] reported the first outpatient study that compared gradual dosage reductions of buprenorphine with those of methadone. An analysis of outcome measures indicated that as the dosages of either medication were decreased, both treatments were associated with low retention rates and with increases in the proportions of opiate-positive urine samples. These results are typical of the retention rates and relapses to illicit opiate use reported for methadone detoxification [Resnick, 1983].

One study of 19 subjects [Fudala et al., 1990] considered the abrupt termination of buprenorphine treatment (8 mg daily or every other day) to reevaluate the differences regarding the onset, peak, and duration of withdrawal signs and symptoms reported by Jasinski and colleagues in 1978 and 1984. According to Fudala and colleagues, four subjects (two each from the daily and the alternate-day treatment groups) discontinued their participation following the abrupt termination of buprenorphine. Three subjects in each group received a total of two or fewer doses of temazepam (60 mg), propoxyphene napsylate (200 mg), or clonidine (0.2 mg). Two subjects in the daily and two in the alternate-day group, respectively, received 87% and 85% of all adjunct medication administered. Additionally, four subjects in each group received no pharmacotherapeutic intervention following the termination of buprenorphine.

When only data from subjects who received no additional pharmacotherapy were considered, no differences were found between groups for any physiological measure (e.g., pupillary diameter, blood pressure). Although there were no differences observed between groups for scores on the Himmelsbach scale (a measure assessing physiological withdrawal signs), there was a significant difference with respect to subject-reported responses on an opiate-withdrawal adjective rating scale (Fig. 1). Also, the subjects who had received buprenorphine daily reported less sleep following the termination of buprenorphine, and at discharge from the study the amount of sleep reported remained below baseline levels. Following the last dose of buprenorphine, the onset of withdrawal symptoms occurred within the first 3 days, peaked at 3–5 days, and gradually diminished over 8–10 days. Overall, for those subjects who received no therapeutic intervention, the withdrawal syndrome was described by the authors as mild to moderate. This study demonstrates the lack of sensitivity of scales that consider only autonomic and physiological signs of opiate withdrawal in comparison with subject self-reports.

In an open-label trial by Kosten et al. [1991], it was suggested that an abrupt discontinuation of buprenorphine might be more effective than current dose reduction techniques, such as methadone dosage tapering or clonidine substitution, for persons being transferred to oral, low-dose naltrexone therapy. This study reported

TABLE IV. Dose Reduction Procedures

Drug	No. of groups	Route	Dosage (mg)	Schedule	Reduction duration (days)	Number of subjects	Type of subject	Design	Results	Reference
Buprenorphine Methadone	2	Sublingual Oral	2 30	Reduction 33% week 1; 50% weeks 2, 3, 4	28 28	20 18	Opiate-dependent	Randomized, parallel-group, double-blind, double-dummy	Retention was similar in both groups; similar increase in opiate-positive urine samples was observed as dosages were reduced	Bickel et al. [1988a]
Buprenorphine	2	Sublingual	8 per day 8 every-other day	Abrupt termination	Followed for 20	Entered: 19; adjunct meds: 11; no adjunct meds: 8	Opiate-dependent	Randomized, parallel-group, double-blind	Mild to moderate withdrawal signs and symptoms reported for both groups, peak 3–5, duration 8–10 days	Fudala et al. [1990]
				Abrupt termination			Methadone/heroin-dependent 5/11	Open-label	Precipitated withdrawal effects from naltrexone were mild and di-	Kosten et al. [1991]

Drug		Dose (mg)	Route	Dosing schedule	Treatment duration (days)	Number of subjects	Subject type	Study design	Results	Reference
Buprenorphine	4	2 3 4 6	Sublingual Sublingual Sublingual Sublingual		30 30 30 30	16 14 6 3	4/10 2/4 1/2		rectly related to suprenorphine dose	
Buprenorphine Methadone	3	8 20 60	Sublingual Oral Oral	Equal percentage reduction across groups every 7 days	56 56 56	22 17 11	Opiate-dependent	Randomized, parallel-group, double-blind, double-dummy, stratified	No difference reported in withdrawal effects or illicit opiate use	Johnson et al. [1992a]
Buprenorphine	2	1.5–8 (mean = 5.5)	Sublingual	10% decrease twice weekly	35	29	Opiate-dependent	Randomized double-blind	Withdrawal symptoms appeared dose-related	Resnick et al. [1992]
Buprenorphine		1.5–8 (mean = 4.7)	Sublingual	Continued maintenance dose	35	22				
Buprenorphine	2	17 (total dose)	Sublingual	Three times daily for 3 days	3	10	Opiate-dependent	Randomized, parallel-group, double-dummy, double-blind	Buprenorphine more effective than clonidine	Johnson et al. [1992b]
Clonidine		2.7 (total dose)	Oral	Three times daily for 5 days	5	8				

that subjects could be given naltrexone as soon as 24 hr following the last buprenorphine dose while producing only minimal withdrawal symptoms.

Johnson et al. [1992a] described the first outpatient protocol in which individuals had been maintained on buprenorphine for as long as 17 weeks prior to dose reduction. This study directly compared buprenorphine with two dosages of methadone and incorporated an 8-week phase of gradual dosage reduction. Sixty percent of the subjects (30 of 50) completed the entire 8-week dose reduction phase (Fig. 3). Sixteen of 22 participants in the buprenorphine 8-mg group completed the dose reduction phase, compared with 11 of 17 and 3 of 11 subjects in the methadone 60-mg and 20-mg groups, respectively. As previously observed, increases in opiate-positive urine samples were seen during the dosage reduction phase. Sixty-seven percent of the samples from buprenorphine-treated subjects were positive for opiates, compared with 66% and 87% of the samples from the high- and low-dose methadone-treated subjects, respectively.

Resnick et al. [1992] reported the results from an outpatient study in which opiate-abstinent subjects were randomized to receive either a 5-week, 10% dosage reduction twice weekly of buprenorphine or to remain on their previous buprenorphine maintenance dosage. Relapse to heroin, increased heroin craving, or increased withdrawal symptoms were the basis for terminating individuals from the study. Three percent of the subjects who received dosage reductions did not relapse to heroin use and did not report increased withdrawal symptoms or craving for heroin. In contrast, 86% of the subjects who remained on their maintenance dosage did not relapse or report withdrawal symptoms or heroin craving.

In the report by Johnson et al. [1992b], described in Table III, results from an 18-day inpatient protocol comparing 3 days of sublingual buprenorphine treatment (total 17 mg) and 5 days of oral clonidine administration (total 2.7 mg) for rapid detoxification from street opiates were reported. Of the 25 subjects enrolled, 18 completed the protocol. A more rapid reduction in subject-reported opiate-withdrawal symptoms was seen during the first 3 days in buprenorphine-treated individuals. Overall, both subject-rated dysphoria and severity of withdrawal symptoms were significantly less in subjects given buprenorphine. The authors concluded that buprenorphine was more effective than clonidine for a rapid, opiate detoxification, which is congruent with results reported by others [O'Connor et al., 1993].

An ongoing inpatient study [Johnson et al., 1993] (Table III) is expected to be the first investigation conducted in a controlled environment that directly compares withdrawal syndromes following the abrupt termination of methadone and buprenorphine. Among other things, this study will assess the behavioral and physiological effects of terminating methadone (30 and 60 mg daily) and sublingual buprenorphine (2, 8, 12, and 16 mg daily).

SAFETY

The studies reviewed in this chapter indicate that more than 1,300 individuals have received buprenorphine. However, few of these investigations have systematically collected and reported data specific to the safety of buprenorphine when used as an opiate-dependence treatment.

Jasinski et al. [1978] related that in a study of five subjects, subcutaneous buprenorphine (8 mg/day), like morphine and methadone, produced a reversible anemia, an increase in sedimentation rate, and a decrease in total protein. However, these changes were not associated with values outside of the normal clinical range.

Mello et al. [1982] reported a profile of side effects typical of opioid agonists (e.g., constipation, decreased libido, sedation and drowsiness) and that two of seven subjects developed severe idiosyncratic reactions following administration of subcutaneous buprenorphine. These reactions were hypotension accompanied by feelings of panic in one person (following 2 and 4 mg/day); and nausea and vomiting, to which tolerance developed within 8 days, in another (following 0.5 mg/day).

In a trial comparing daily ($n = 9$) with alternate-day ($n = 10$) sublingual buprenorphine dosing (8 mg), no serious adverse reactions were reported and no reactions were considered to be definitely related to the study medication [Lange et al., 1990]. The median numbers of probable and possible adverse reactions were comparable between the two groups. Constipation and sedation/drowsiness (typical opiate effects) were the two symptoms most frequently reported by subjects. Although increases in liver function tests were noted for both groups over the duration of the study, the authors could not conclude that these results were directly attributable to buprenorphine.

In a study involving 162 subjects, 53 of whom received 8 mg of sublingual buprenorphine daily [Johnson et al., 1991], subjects were questioned every 2 weeks regarding whether they had experienced any of 14 adverse effects (based on Mello et al., 1982) during the past week. These effects were those that could be expected of the study medications, of illicitly administered opiates, or as part of an opiate withdrawal syndrome. They were decreased appetite, itching, difficulty urinating, decreased sexual urge or desire, anxiety, dizziness, dry mouth, ringing ears, headache, "couldn't sleep," sedation/drowsiness, constipation, vomiting, and nausea. A significant difference between groups for the frequencies of occurrence was reported for five effects: decreased appetite, difficulty urinating, anxiety, sedation/drowsiness, and constipation. There was no pattern of results that suggested any consistent effects, either between treatments or across time. Two deaths occurred in the study, one in a subject receiving buprenorphine (RE Johnson, personal communication; November 1989). Neither death was attributable to the study medication.

The profile of adverse effects associated with buprenorphine administration does not appear unlike that previously reported or expected from methadone, LAAM, or other μ opioid agonists. Although elevations in hepatic aminotransferase levels were observed in one inpatient study, these increases could not be directly related to buprenorphine administration. It is expected that currently ongoing or recently completed outpatient trials will provide additional data with respect to this finding.

Overall, the above results indicate that buprenorphine is not associated with an adverse effect profile that would limit its utility as an opiate-dependence treatment.

TREATMENT INDICATIONS

Based on the completed and ongoing studies reviewed in this chapter, buprenorphine appears to be an effective treatment for opiate dependence. In addition to its utility as a primary treatment medication, it may also be useful as a bridge

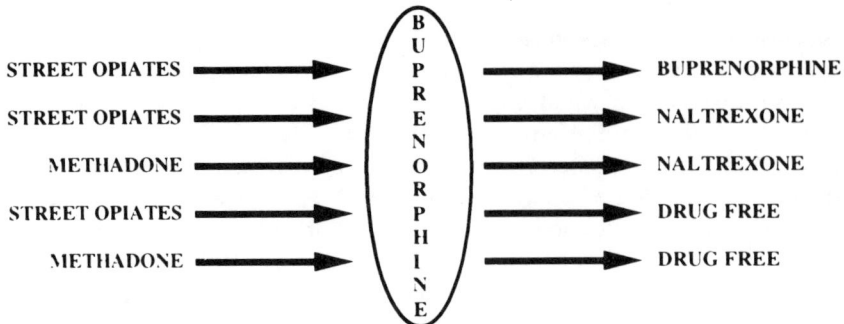

Fig. 5. Potential indications for buprenorphine as a treatment medication for opiate dependence.

to other pharmacotherapies and to a drug-free state. Its usefulness in facilitating the transfer of individuals from either street opiates or methadone to the opiate antagonist naltrexone is reviewed by Rosen and Kosten in this volume. Since buprenorphine treatment has been associated with fewer physiological withdrawal signs and symptoms than methadone [Jasinski et al., 1978; Mello and Mendelson, 1980; Mello et al., 1982; Fudala et al., 1990], a drug-free state should be more easy to achieve with buprenorphine. The potential indications for buprenorphine are depicted in Figure 5.

Buprenorphine treatment may not prove adequate for all individuals. Since it is a partial agonist, buprenorphine may not provide sufficient morphine-like subjective effects that are desired by some opiate addicts and are necessary to maintain them in treatment. For those persons, a full agonist (i.e., methadone or LAAM) may be a more appropriate therapy.

BUPRENORPHINE MISUSE

Because of its opiate agonist effects, buprenorphine could be expected to have a potential for misuse. While numerous anecdotal cases of abuse have been reported, few studies have directly assessed the abuse potential of buprenorphine.

One investigation [Jasinski et al., 1978] showed that both acute and chronic administration of subcutaneous buprenorphine produced euphoria and "drug liking." Maximal effects were observed in the dosage range of 0.8–1.2 mg s.c. (equivalent to 1.2–1.8 mg sublingual).

In a more recent study [Pickworth et al., 1993], buprenorphine administered intravenously in doses ranging from 0.3 mg to 1.2 mg produced increased scores on a euphoria scale extrapolated by the authors to be equivalent to those expected following 20 mg/70 kg of intravenous morphine. Drug "liking," and scores on a visual analog scale measuring "good effects," were reported to be equivalent to those anticipated after 30 mg of intravenous morphine.

Concerns about the potential misuse of buprenorphine led to the development of a buprenorphine/naloxone combination product currently marketed in New Zealand. These concerns may also influence the formulation that will eventually be developed for opiate-dependence treatment in the United States [Chiang and Hawks, 1993].

ONGOING AND RECENTLY REPORTED STUDIES

The results from 17 studies of buprenorphine, either ongoing or recently reported in the scientific literature, are summarized in Table III. The primary areas of study (and number of studies in each area) were the following: evaluation of dosage regimens (3), assessment of efficacy for opiate-dependence treatment (3), evaluation of safety (2), assessment of efficacy for dual opiate/cocaine dependence treatment (2), evaluation of a buprenorphine/naloxone combination formulation (2), opiate detoxification (4), and, assessment of dose induction procedures (1).

SUMMARY

Data from the studies reviewed in this chapter provide extensive evidence for the utility and effectiveness of buprenorphine as a medication for opiate-dependence treatment. The studies ranged in scope from those examining a single therapeutic issue (e.g., the effectiveness of buprenorphine compared with clonidine for rapid opiate detoxification) to those that considered multiple therapeutic applications (e.g., induction, maintenance, and dose reduction procedures). Also, some of these studies are expected to provide information that is pivotal to a New Drug Application for buprenorphine (for opiate-dependence treatment), while others are expected to provide efficacy or safety information.

Three pharmacotherapeutic components of buprenorphine administration are its induction, maintenance, and dosage reduction. Induction of individuals taking street opiates onto buprenorphine has been shown to be most effectively accomplished by a rapid, dose-escalating procedure. The rapid introduction of the medication appears optimal with respect to retaining individuals in treatment and providing for the greatest suppression of opiate-withdrawal symptoms. The differences between the elimination kinetics of methadone and most illicitly used opiates suggest that the induction of individuals from methadone to buprenorphine may potentially be more complex and protracted than the transition from other opiates.

The studies conducted to date have indicated that buprenorphine dosages of approximately 8 mg/day given sublingually will be required for the treatment of opiate dependence. This dosage has been shown by some investigators to be equivalent in efficacy to about 60 mg of orally administered methadone, although higher daily dosages may be required to adequately maintain certain individuals. It must be considered, however, that increased buprenorphine doses may not produce proportionally greater increases in agonist effects and may in fact produce no greater, or less of, an effect. Another factor affecting the therapeutic utility of buprenorphine is its dosing regimen. The usefulness of buprenorphine might be enhanced if it could be administered less often than once daily, an advantage that is currently under active investigation.

Data from studies assessing dosage reduction procedures suggest that the withdrawal syndrome following termination of buprenorphine is not insignificant. Furthermore, the similarity between reported subjective states of feeling following termination of buprenorphine and methadone should not be overlooked. The ongoing investigation by Johnson et al. [1993] is expected to further elucidate potential similitudes between the two medications. Of course, the magnitude and frequency of withdrawal signs and symptoms associated with buprenorphine, methadone, LAAM, or any opiate-substitution therapy are expected to be related to the dosage and rate of decrease in dosage.

There is a paucity of data regarding the use of buprenorphine as an agent for the rapid detoxification (3–10 days) of opiate addicts. If buprenorphine is shown to be useful in this regard, it can provide important advantages over current pharmacotherapies with respect to effectiveness and acceptance by patients.

This review has focused on the use of buprenorphine as an opiate-dependence treatment. There have been reports, though, that buprenorphine may also be effective in reducing cocaine use in opiate addicts. Two open-label evaluations by Kosten and colleagues [Kosten et al., 1989a,b] suggested that buprenorphine treatment was associated with less illicit cocaine use than was methadone treatment. However, two randomized, parallel-group, double-blind trials [Johnson et al., 1992; Strain et al., 1993b, 1994] found no evidence to substantiate the above hypothesis. The potential clinical utility of buprenorphine for reducing cocaine use is addressed by Mello and Mendelson in this volume.

CONCLUSIONS

The primary factors influencing the effectiveness of buprenorphine may be different for different individuals. For some, the maximal morphine-like effects obtainable from buprenorphine will be the most important consideration. For others, the blockade of effects from exogenously administered opiates or the number and degree of withdrawal symptoms accompanying buprenorphine dosage reduction may be paramount. Regardless of which factor is most important, the evidence reported thus far indicates that buprenorphine should be considered as a primary opiate-treatment medication, given its safety and pharmacological profile. This endorsement will be reevaluated to weigh the outcome of ongoing efficacy studies.

REFERENCES

Aceto MD (1984): Characterization of prototypical opioid antagonists, agonist–antagonists, and agonists in the morphine-dependent rhesus monkey. Neuropeptides 5:15–18.

Amass L, Bickel WK, Higgins ST, Hughes JR (1993): Administering twice the daily buprenorphine dose suppresses opioid withdrawal for 48 hours in opioid-dependent humans. Abstract, paper presented at the fifty-fifth annual scientific meeting of the College on Problems of Drug Dependence, Toronto, Canada, June 17, 1993.

Banys P, Clark HW, Tusel DJ, Sees K, Stewart P, Morgan L, Delucchi K, Callaway E (1994): An open trial of low-dose buprenorphine in treating methadone withdrawl. J Subst Abuse Treat 11:9–15.

Bickel WK, Stitzer ML, Bigelow GE, Liebson IA, Jasinski DR, Johnson RE (1988a): A clinical trial of buprenorphine: Comparison with methadone in the detoxification of heroin addicts. Clin Pharmacol Ther 43:72–78.

Bickel WK, Stitzer ML, Bigelow GE, Liebson IA, Jasinski DR, Johnson RE (1988b): Buprenorphine: Dose-related blockade of opioid challenge effects in opioid dependent humans. J Pharmacol Exp Ther 247:47–53.

Chiang CN, Hawks R (1993): Development of a buprenorphine-naloxone combination drug for the treatment of drug addiction. Abstract, paper presented at the fifty-fifth annual scientific meeting of the College on Problems of Drug Dependence, Toronto, Canada, June 16, 1993.

Fraser HF, Isbell H (1952): Actions and addiction liabilities of alpha-acetylmethadols in man. J Pharmacol Exp Ther 105:458–465.

Fudala PJ, Jaffe JH, Dax EM, Johnson RE (1990): Use of buprenorphine in the treatment of opioid addiction. II. Physiologic and behavioral effects of daily and alternate-day administration and abrupt withdrawal. Clin Pharmacol Ther 47:525–534.

Isbell H, Wikler A, Eisenman AJ, Daingerfield M, Frank K (1948): Liability of addiction to 6-dimethyl-amino-4-4-diphenyl-3-heptanone (methadon "amidone" or "10020") in man. Arch Intern Med 82:362–392.

Jasinski DR, Pevnick JS, Clark SC, Griffith JD (1977): Therapeutic usefullness of propoxyphene napsylate in narcotic addiction. Arch Gen Psychiatry 43:227–233.

Jasinski DR, Pevnick JS, Griffith JD (1978): Human pharmacology and abuse potential of the analgesic buprenorphine. Arch Gen Psychiatry 35:501–516.

Jasinski DR, Martin WR, Haertzen CA (1967): The human pharmacology and abuse potential of N-allylnoroxymorphone (naloxone). J Pharmacol Exp Ther 157:420–426.

Jasinski DR, Henningfield JE, Hickey JE, Johnson RE (1983): Progress report of the NIDA Addiction Research Center, Baltimore, Maryland, 1982. NIDA Res Monogr 43:92–98.

Jasinski DR, Boren JJ, Henningfield JE, Johnson RE, Lange WR, Lukas SE (1984): Progress report from the NIDA Addiction Research Center, Baltimore, Maryland, NIDA Res Mongr 49:69–76.

Johnson RE, Cone EJ, Henningfield JE, Fudala PJ (1989): Use of buprenorphine in the treatment of opiate addiction. I. Physiologic and behavioral effects during a rapid dose induction. Clin Pharmacol Ther 46:335–343.

Johnson RE, Fudala, PJ, Jaffe JH (1991): Outpatient comparison of buprenorphine and methadone maintenance. I. Effects on opiate use and self-reported adverse effects and withdrawal symptomatology. NIDA Res Monogr 105:585–586.

Johnson RE, Jaffe JH, Fudala PJ (1992a): A controlled trial of buprenorphine treatment for opioid dependence. J Am Med Assoc 267:2750–2755.

Johnson RE, Fudala PJ, Cheskin LR (1992b): A comparison of buprenorphine to clonidine for rapid inpatient opiate detoxification. Clin Pharmacol Ther 51:167.

Johnson RE, Risher-Flowers DL, Alim TN, Mastropaolo J, Vocci FJ, Deutsch SI (1993):

Blockade of intravenous hydromorphone (H) by methadone (M) and buprenorphine (B). Clin Pharmacol Ther 53:175.

Jones RT, Mendelson J (1993): Buprenorphine and naloxone interactions in opiate dependent volunteers. Abstract, paper presented at the fifty-fifth annual scientific meeting of the College on Problems of Drug Dependence, Toronto, Canada, June 17, 1993.

Kosten, TR, Kleber HD (1988): Buprenorphine detoxification from opioid dependence: A pilot study. Life Sci 635–641.

Kosten TR, Kleber HD, Morgan C (1989a): Role of opioid antagonists in treating intravenous cocaine abuse. Life Sci 44:887–892.

Kosten TR, Kleber HD, Morgan C (1989b): Treatment of cocaine abuse with buprenorphine. Biol Psychiatry 26:637–639.

Kosten TR, Morgan C, Kleber HD (1991): Treatment of heroin addicts using buprenorphine. Am J Drug Alcohol Abuse 17:119–128.

Kosten TR, Schottenfeld R, Ziedonis D, Falconi J (1993): Buprenorphine versus methadone maintenance for opioid dependence. J Nerv Ment Dis 181:358–364.

Lange WR, Fudala PJ, Dax EM, Johnson RE (1990): Safety and side-effects of buprenorphine in the clinical management of heroin addiction. Drug Alcohol Depend 26:19–28.

Ling W, Charuvastra C, Kintaudi K, Wesson DR (1993): Buprenorphine for opiate dependence: Two ongoing clinical trials. Abstract, paper presented at the fifty-fifth annual scientific meeting of the College on Problems of Drug Dependence, Toronto, Canada, June 16, 1993.

Maddux JF (1978): History of the hospital treatment programs, 1935–1974. In Martin WR, Isbell H (eds): "Drug Addiction and the U.S. Public Health Service." DHEW Publication No. (ADM)77-434. Washington, DC: U.S. Government Printing Office, pp 217–250.

Martin WR, Isbell H (eds) (1978): "Drug Addiction and the U.S. Public Health Service." DHEW Publication No. (ADM)77-434. Washington, DC: U.S. Government Printing Office.

Martin WR, Gorodetzky CW, McClane TK (1966): An experimental study in the treatment of narcotic addicts with cyclazocine. J Pharmacol Exp Ther 7:455–465.

Martin WR, Jasinski DR, Haertzen CA, Kay DC, Jones BE, Mansky PA, Carpenter RW (1973a): Methadone—A reevaluation. Arch Gen Psychiatry 28:286–295.

Martin WR, Jasinski DR, Mansky PA (1973b): Naltrexone, an antagonist for the treatment of heroin dependence. Arch Gen Psychiatry 28:784–791.

Mello NK, Mendelson JH (1980): Buprenorphine suppresses heroin use by heroin addicts. Science 207:657–659.

Mello NK, Mendelson JH, Kueknle JC (1982): Buprenorphine effects on human heroin self-administration: An operant analysis. J Pharmacol Exp Ther 223:30–39.

Nigram AK, Ray R, Tripathi BM (1993): Buprenorphine in opiate withdrawal: A comparison with clonidine. J Subst Abuse Treat 10:391–394.

O'Connor PG, Waugh ME, Weiss EC, Carroll KM, Rounsaville BJ, Kosten TR, Schottenfeld RS (1993): A comparison of clonidine, clonidine/naltrexone, and buprenorphine in opioid detoxification. Abstract, paper presented at the fifty-fifth annual scientific meeting of the College on Problems of Drug Dependence, Toronto, Canada, June 16, 1993.

Pickworth WR, Johnson RE, Holicky BA, Cone EJ (1993): Subjective and physiologic effects of intravenous buprenorphine in humans. Clin Pharmacol Ther 53:570–576.

Pycha C, Resnick RB, Galanter M (1993): Buprenorphine: Rapid and slow dose-reductions for heroin detoxification. Abstract, paper presented at the fifty-fifth annual scientific meeting of the College on Problems of Drug Dependence, Toronto, Canada, June 16, 1993.

Reisinger M (1985): Buprenorphine as new treatment for heroin dependence. Drug Alcohol Depend 16:257–262.

Resnick R (1983): Methadone detoxification from illicit opiates and methadone maintenance. In Cooper JR, Altman F, Brown BS, Czechowicz D (eds): "Research on the Treatment of Narcotic Addiction. State of the Art." DHHS Publication No. (ADM)87-1281. Rockville, MD: U.S. Department of Health and Human Services, pp 160–167.

Resnick RB, Galanter M, Pycha C, Cohen A, Grandison P, Flood N (1992): Buprenorphine: An alternative to methadone for heroin dependence treatment. Psychopharmacol Bull 28:109–113.

Resnick RB, Pycha C, Galanter M (1993): Buprenorphine maintenance: Reduced dosing frequency. Abstract, paper presented at the fifty-fifth annual scientific meeting of the College on Problems of Drug Dependence, Toronto, Canada, June 16, 1993.

Schottenfeld RS, Kosten TR, Pakes J, Ziedonis D, Oliveto A (1993): Buprenorphine vs. methadone maintenance for combined cocaine and opioid dependence. Abstract, paper presented at the fifty-fifth annual scientific meeting of the College on Problems of Drug Dependence, Toronto, Canada, June 17, 1993.

Schuh KJ, Walsh SL, Stitzer ML, Preston KL, Bigelow GE (1993): Effects of buprenorphine in morphine-dependent subjects. Abstract, paper presented at the fifty-fifth annual scientific meeting of the College on Problems of Drug Dependence, Toronto, Canada, June 16, 1993.

Seow SSW, Quigley AJ, Ilett KF, Dusci LJ, Swensen G, Harrison-Stewart A, Rappeport L (1986): Buprenorphine: A new maintenance opiate? Med J Australia 144:407–411.

Strain EC, Stitzer ML, Liebson IA, Bigelow GE (1993a): An outpatient trial of methadone versus buprenorphine in the treatment of opioid dependence. NIDA Res Mongr 132:99.

Strain EC, Stitzer ML, Liebson IA, Bigelow GE (1993b): An outpatient trial of methadone versus buprenorphine in the treatment of combined opioid and cocaine dependence. Abstract, paper presented at the fifty-fifth annual scientific meeting of the College on Problems of Drug Dependence, Toronto, Canada, June 17, 1993.

Strain EC, Stitzer ML, Liebson IA, Bigelow GE (1994): Comparison of buprenorphine and methadone in the treatment of opioid dependence. Am J Psychiatry 151:1025–1030.

Teoh SK, Mendelson JH, Mello NK, Kuehnle J, Sintavanarong P, Rhoades EM (1993): Acute interactions of buprenorphine with intravenous cocaine and morphine: An investigational new drug phase I safety evaluation. J Clin Psychopharmacol 13:87–99.

Wapler M, Mendelson JH, Weiss RD, Teoh SK, Mello NK, Kuehnle J (1993): Interaction of buprenorphine with cocaine and morphine challenge on craving self-reports. Abstract, paper presented at the fifty-fifth annual scientific meeting of the College on Problems of Drug Dependence, Toronto, Canada, June 16, 1993.

Welm S, Jones RT, Meldelson J, Batki S, Upton R (1993): Buprenorphine and naloxone interactions in methadone maintained patients. Abstract, paper presented at the fifth-fifth annual scientific meeting of the College on Problems of Drug Dependence, Toronto, Canada, June 16, 1993.

BUPRENORPHINE TREATMENT OF COCAINE AND HEROIN ABUSE

NANCY K. MELLO and JACK H. MENDELSON
Alcohol and Drug Abuse Research Center, McLean Hospital—Harvard Medical School, Belmont, MA 02178

INTRODUCTION

In 1972, the structure and pharmacology of buprenorphine were described by Lewis [1974] and Cowan [1974] at a conference on narcotic antagonists. In the ensuing 20 years, buprenorphine has been the focus of a broad spectrum of clinical and basic research. Buprenorphine's unique combination of μ opioid agonist effects and antagonist properties suggested that it might be a potent analgesic with minimal capacity to induce physical dependence [Cowan, 1974; Lewis, 1974]. Buprenorphine's promise as a safe and effective analgesic has been fulfilled [Houde, 1979; Jaffe and Martin, 1990] but, in addition, there is an evolving recognition of the potential usefulness of buprenorphine for the treatment of opiate dependence as well as dual dependence on cocaine and opiates [see Mello and Mendelson, 1985, 1993, for review]. In 1992, our understanding of the ways in which this opioid mixed agonist-antagonist analgesic affects both opiate and cocaine abuse remains very limited. However, the questions raised by these empirical findings may suggest new approaches to the pharmacotherapeutic treatment of substance abuse.

In this review, we will summarize some clinical and preclinical studies from our laboratory that examined the effectiveness of buprenorphine for the treatment of heroin abuse and of dual dependence on cocaine and opiates. We began studies of the effects of buprenorphine on heroin self-administration by heroin-dependent men in 1978 after Jasinski and co-workers described its long-acting antagonism of the physiological and subjective effects of morphine [Jasinski et al., 1978; Mello and Mendelson, 1980; Mello et al., 1982]. The rationale for using buprenorphine was

---

Buprenorphine: Combatting Drug Abuse With a Unique Opioid, pages 241–287
© 1995 Wiley-Liss, Inc.

similar to that for the long-acting opioid antagonist naltrexone, which also antagonized the subjective and physiological effects of opiates for 24–48 hr [Martin et al., 1973; Julius and Renault, 1976; Verebey et al., 1976]. Naltrexone effectively reduced heroin self-administration by heroin-dependent men in inpatient clinical studies [Meyer and Mirin, 1979; Mello et al., 1981], but unfortunately it has not been widely effective in the treatment of heroin dependence because it has been difficult to retain patients in naltrexone treatment programs [Meyer and Mirin, 1979; Schecter, 1980]. In contrast, treatment with metadone, an opioid agonist, has been very effective. Methadone and other opioid agonists such as levo-α-acetylmethadol produce subjective effects that are similar to heroin and induce cross-tolerance, which attenuates the euphorogenic response to heroin and other opiates [Dole and Nyswander, 1965; Blaine et al., 1978, 1981; Jaffe, 1990].

Buprenorphine combines the characteristics of opioid agonist and antagonist pharmacotherapies for opiate addiction and offers some advantages over using either agonists or antagonists alone. Unlike opioid agonists, abrupt discontinuation of buprenorphine treatment does not produce severe and protracted withdrawal signs and symptoms in humans [Jasinski et al., 1978; Fudala et al., 1989]. Buprenorphine is also safe, since its antagonist component appears to prevent lethal overdose even at approximately 10 times the analgesic therapeutic dose [Banks, 1979]. This reduces the risk for overdose deaths, which are often associated with illicit methadone abuse [Kreek, 1978]. Finally, the opioid agonist component of buprenorphine appears to be important for acceptance by patients and is therefore the primary advantage of buprenorphine over naltrexone treatment.

Although buprenorphine significantly reduced heroin self-administration in inpatient studies [Mello and Mendelson, 1980; Mello et al., 1982], we were unable to proceed with outpatient clinical trials because buprenorphine was not approved for outpatient treatment. We found that buprenorphine also reduced opiate self-administration by rhesus monkeys and the concordance of our clinical and preclinical data stimulated our interest in using the primate drug self-administration model to evaluate the effectiveness of new pharmacotherapies for drug abuse treatment [Mello et al., 1983]. We reasoned that if an evaluation conducted in a primate drug self-administration model showed good concordance with clinical studies, then this model could be used to predict the potential effectiveness of those pharmacotherapies. One illustration of the potential value of a primate model for pharmacotherapy evaluation is our discovery that buprenorphine significantly reduced cocaine self-administration by rhesus monkeys [Mello et al., 1989, 1990a]. This finding has now led to confirmatory clinical trials in men with dual dependence on cocaine and heroin [Gastfriend et al., 1992, 1993; Mello and Mendelson, 1993]. Our rationale for these studies and our major findings are summarized in the remainder of this review. Some possible neurobehavioral, neuroendocrine, and pharmacological mechanisms related to buprenorphine's reduction of cocaine will also be discussed. Our studies of the behavioral pharmacology of buprenorphine in rhesus monkeys are not described because these have been reviewed elsewhere [Mello and Mendelson, 1985, 1992].

BUPRENORPHINE'S EFFECTS ON OPIATE SELF-ADMINISTRATION

Clinical Studies of Buprenorphine in Opioid-Dependent Men

In 1978 Jasinski and coworkers reported the first inpatient clinical studies of the effects of buprenorphine on the subjective and physiological effects of opiates. They reported that buprenorphine antagonized the physiological and subjective effects of high doses of morphine (60–120 mg/day) for up to 29.5 hr [Jasinski et al., 1978]. These data suggested that buprenorphine might be useful for the treatment of opioid dependence.

In 1979 we conducted the first inpatient trials to evaluate buprenorphine's effects on heroin self-administration by heroin-dependent men [Mello and Mendelson, 1980; Mello et al., 1982]. Ten men who had abused heroin for an average of 10.4 years volunteered for participation in this study. All volunteers had failed in conventional treatment programs and none were under any legal constraints. All men were in good health as determined by clinical and laboratory examinations. Seven were unemployed and three had worked recently at semiskilled jobs before selection for the study. These men had an average of 12.4 years of formal education. Subjects were recruited in groups of four and lived on the Clinical Research Ward of the Alcohol and Drug Abuse Research Center at McLean Hospital–Harvard Medical School throughout a 40-day study.

The effects of buprenorphine on heroin self-administration were compared with placebo-buprenorphine under double-blind conditions. The buprenorphine induction, maintenance, and withdrawal schedules were identical to those used by Jasinski et al. [1978]. After a 5-day drug-free baseline, the subjects were given ascending doses of buprenorphine (or its placebo) for 14 days. An ascending dose schedule was used to assess the physiological and behavioral effects of various doses of buprenorphine, as well as to ensure the safety of the subjects. An initial buprenorphine dose of 0.5 mg/day administered subcutaneously was gradually increased in 0.5-mg increments over 12 days and 1-mg increments for 2 days to a final maintenance dose of 8 mg/day. The subjects were maintained on 8 mg/day of buprenorphine (or placebo) for 10 days during which they could work at a simple operant task for intravenous heroin. Then the subjects were gradually withdrawn from buprenorphine over 5 days in a dose-reduction sequence of 7, 6, 5, 3, and 1 mg/day. The placebo-buprenorphine subjects who used heroin were detoxified with methadone in progressively decreasing doses (25 to 5 mg/day) over 5 days. An aftercare program was developed for each subject at an outpatient treatment program near his home. The subjects were given an opportunity to select outpatient maintenance on the long-acting opioid antagonist naltrexone, since buprenorphine was not yet approved for outpatient treatment.

Heroin Self-Administration Procedures. Medical and ethical considerations precluded studies of spontaneous patterns of unrestricted heroin self-administration as in our previous studies of the operant acquisition of marijuana [Mendelson et al.,

1974], alcohol [Mello and Mendelson, 1965, 1972], and alcohol plus marijuana [Mello et al., 1978]. The subjects could acquire a maximum of three heroin injections each day, once every 8 hr (at 9 AM, 5 PM, and 1 AM). They could refuse to take any heroin dose earned, but they could not take fractional doses. A total of 21 mg/day of heroin was available for the first 5 days and 40.5 mg/day for the second 5 days. The subjects were not told that the heroin dose would be increased on Day 6.

Operant behavioral procedures were used to evaluate the effects of buprenorphine on heroin self-administration. The subjects could work for points for heroin on the last day of the buprenorphine induction period (Day 14) and on each day of heroin availability. Points earned for heroin could be accumulated and spent throughout the 10-day period of heroin availability. The subjects could work for money throughout the 40-day study. Operant points earned for money and for heroin were not interchangeable. The subjects chose whether to work for money or for heroin each time they turned on their operant instruments by pressing the appropriate button on an instrument panel in the dayroom. Any pause in responding of 5 min or more required reactivation of the operant instrument. The subjects could work at the operant task any time for as long as they wished. This design permitted us to examine the effects of buprenorphine (or placebo) and heroin on operant performance for money and for heroin.

The subjects earned points for money or for heroin by pressing the button on a portable operant manipulandum on a second-order schedule of reinforcement, an FR 300 (FI 1-sec: S). Only the first response after 1 sec had elapsed was recorded as an effective response by the programming circuitry. Responses emitted at a rate faster than one per second had no programmed consequence. Each effective response was followed by a brief stimulus light flash (S) on the operant panel. Three hundred effective responses on the FI 1-sec schedule of reinforcement earned one purchase point in about 5 min. One heroin injection cost 18 purchase points, which required approximately 90 min of sustained operant performance. The subjects had to work about 4.5 hr to earn the 54 purchase points necessary to buy all three heroin injections available each day. A record of points earned for heroin and for money was continuously available on counters located on the operant instrument panel in the dayroom. Whenever a subject accumulated 18 purchase points, he could press a drug request button to inform staff that he wished to take the next available dose of heroin. The requisite number of purchase points was then subtracted from his total heroin point accumulation.

The subjects also could acquire $1.50 in cash for 90 min of operant performance, to be paid upon completion of the study. Points earned for heroin that were not spent could not be exchanged for money. However, to avoid penalizing subjects who were assigned to buprenorphine maintenance and elected not to use heroin, the points earned for heroin on the last day of the buprenorphine induction period could be exchanged for money at the end of the study. This simple schedule of compensation was effective in maintaining cooperation and retaining subjects in this study as well as in our previous studies designed to evaluate naltrexone's effects on heroin self-administration [Mello et al., 1981].

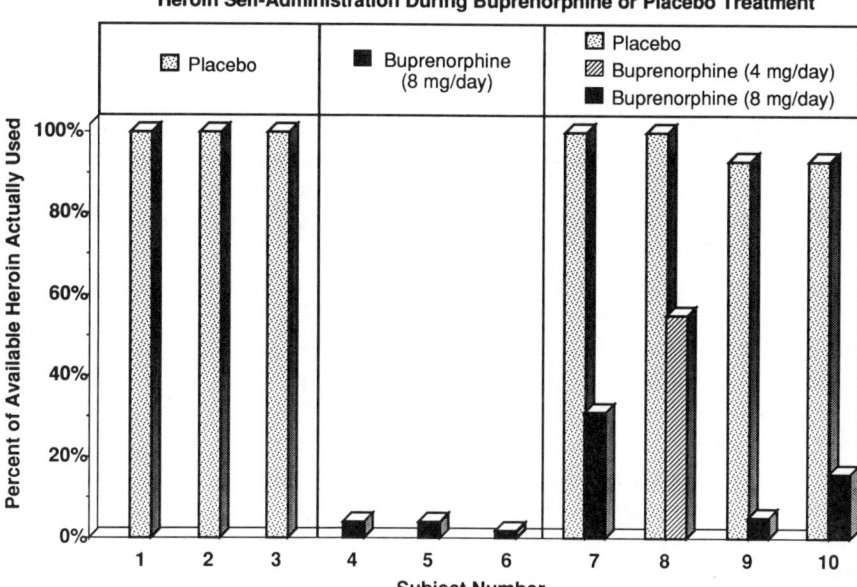

Fig. 1. Percentage of available heroin used during 10 days of treatment with buprenorphine or placebo buprenorphine. Three subjects were maintained on placebo-buprenorphine (left columns); three subjects were maintained on buprenorphine (8 mg/day, s.c.) (middle columns), and four subjects were studied under both buprenorphine and placebo-buprenorphine conditions (right columns). Adapted from Mello et al. [1982].

Buprenorphine's Effects on Heroin Self-Administration. The effects of buprenorphine on heroin self-administration by individual subjects are shown in Figure 1. The subjects who were maintained on placebo-buprenorphine took all of the heroin available each day over 10 days (Fig. 1, left). Buprenorphine significantly reduced heroin self-administration in all subjects. Three subjects maintained on 8 mg/day of buprenorphine self-administered between 2% and 4% of the available heroin (Fig. 1, middle). Four subjects participated in two separate studies and received both buprenorphine and placebo-buprenorphine in a counterbalanced order. These subjects self-administered 93%–100% of all the heroin available during placebo-buprenorphine maintenance. However, during buprenorphine maintenance (8 mg/day), heroin self-administration was reduced by 69%–98% of the heroin self-administered during placebo-buprenorphine maintenance (Fig. 1, right). One subject became hypotensive and was maintained on 4 mg rather than 8 mg of buprenorphine per day. This subject took 45% less heroin during low-dose buprenorphine maintenance than during placebo treatment [Mello and Mendelson, 1980; Mello et al., 1982].

Buprenorphine maintenance did not impair operant performance for money. The placebo-buprenorphine subjects also showed no impairment of operant performance for heroin or for money during heroin intoxication. The subjects titrated operant work to acquire the desired amount of heroin and then resumed working for money. There were no significant differences between the buprenorphine- and the placebo-maintained group in total operant points earned or total hours worked during any phase of the study [Mello and Mendelson, 1980; Mello et al., 1982].

Subjective and Physiological Effects of Buprenorphine in Opioid-Dependent Men. Subcutaneous injection of buprenorphine did not produce a rapid high or rush similar to an intravenous heroin injection. All subjects reported that buprenorphine had opiate-like effects on mood characterized by a generalized feeling of contentment. These data are consistent with previous observations that 8 mg/day s.c. of buprenorphine produced subjective effects and euphoria equivalent to that produced by 120 mg/day of morphine (30 mg q.i.d.) or 40–60 mg of methadone [Jasinski et al., 1978]. Subsequent reports that buprenorphine also significantly decreased depressive symptoms in opiate abusers as well as in depressed patients attests to its mood-elevating effects [Emrich et al., 1982; Kosten et al., 1990; Nutt et al., this volume]. Although our study was conducted under double-blind conditions, the buprenorphine-maintenance group probably knew that they were receiving an active drug. On the first day that subjects were allowed to work for heroin (Day 14 of the buprenorphine induction period) the buprenorphine group earned significantly fewer heroin points than the placebo-buprenorphine group. Buprenorphinemaintained subjects who took heroin on the first day of heroin availability reported no acute subjective effects of heroin [Mello and Mendelson, 1980; Mello et al., 1982].

Buprenorphine produced somatic and sedative effects similar to the effects of other opiate agonists [Mansky, 1978]. Constipation, urinary hesitancy, and drowsiness were reported by our subjects, as well as by five subjects previously studied by Jasinski et al. [1978]. Our subjects developed tolerance to urinary hesitancy within 9 days and to the constipating effects of buprenorphine within 21 days. Only three subjects reported feeling drowsy during buprenorphine induction, and tolerance developed rapidly to this sedative effect. Other somatic effects, such as itching, headache, dizziness, tinnitus, and dry mouth, were reported only occasionally during the early phase of buprenorphine induction. Reports of changes in appetite and sleep patterns were infrequent and transient. Several subjects reported decreased libido during buprenorphine induction and maintenance, and this persisted during the first 3 days of buprenorphine detoxification [Mello et al., 1982].

Although most subjects tolerated buprenorphine well and had no persistent debilitating side effects, two men had idiosyncratic reactions to buprenorphine that were severe enough to require modification of the protocol. One subject developed persistent hypotension (90–100 mm Hg systolic/50–70 mm Hg diastolic) and was maintained on 4.0 mg/day. A second subject had severe nausea and vomiting after 0.5 mg/kg of buprenorphine, and buprenorphine doses were increased more slowly over the first 7 days of induction. However, this subject reported that he frequently had severe nausea and vomiting after using opiates. Nausea and vomiting were also reported by subjects in both acute and chronic buprenorphine administration studies conducted by Jasinski et al. [1978]. No other clinically significant side effects were reported and no subject left the study because of adverse reactions to buprenorphine [Mello et al., 1982].

When the buprenorphine maintenance dose was gradually reduced from 7 to 1 mg/day over 5 days, no opioid withdrawal signs were observed and the subjects did not complain of withdrawal symptoms. We maintained contact with most subjects after discharge and none reported withdrawal signs or symptoms over a 30-day period after termination of buprenorphine maintenance. Jasinski et al. [1978] reported that abrupt termination of 8 mg/day of buprenorphine produced minimal withdrawal signs and symptoms within the first 2 or 3 days, but somewhat more marked withdrawal symptoms were observed 14–15 days after cessation of buprenorphine maintenance [Jasinski et al., 1978]. However, after 36 days of sublingual buprenorphine treatment (8 mg/day), mild withdrawal symptoms were reported within 3–5 days after the last dose of buprenorphine [Fudala et al., 1989]. The different rates of onset of withdrawal signs and symptoms in these two studies may be due to differences in the effective dose of buprenorphine and duration of treatment as well as the procedures used to measure withdrawal signs and symptoms. The delayed appearance of mild withdrawal signs and symptoms after buprenorphine maintenance probably reflects the slow dissociation of buprenorphine from the opiate receptor [Lewis et al., 1983].

Summary and Conclusions. Buprenorphine significantly reduced heroin self-administration by heroin-dependent men who had abused heroin for over 10 years. Because this evaluation of buprenorphine was based on a direct behavioral measure of heroin self-administration, rather than on retrospective recall or an anticipatory self-report, we have more confidence in the conclusion that daily buprenorphine administration effectively suppressed heroin use by heroin-dependent men. Buprenorphine treatment was well accepted by these subjects, and tolerance to its opioid agonist side effects developed gradually. Moreover, buprenorphine did not compromise operant performance for money in comparison with placebo-buprenorphine maintenance conditions [Mello and Mendelson, 1980; Mello et al., 1982].

Buprenorphine maintenance may offer two advantages over the opioid agonist methadone for treatment of opiate addiction: (1) Buprenorphine does not induce significant physical dependence in humans and (2) the possibility of lethal overdose is remote owing to the opiate antagonist properties of buprenorphine [see Lewis et al., 1983, for review]. However, the potential abuse liability of buprenorphine relative to methadone remains to be determined. Buprenorphine produces opiate agonist-like subjective effects that are comparable to those of morphine and methadone [Jasinski et al., 1978], but in contrast to these opiate agonists subcutaneous injection of buprenorphine did not produce a rapid high or rush similar to the effect of an i.v. heroin injection [Mello et al., 1982]. Buprenorphine's agonist properties make it more acceptable to patients than the long-acting antagonist naltrexone, and patients' acceptance is a crucial component of the effectiveness of any pharmacotherapy. Buprenorphine is currently assigned to Schedule V by the FDA because there is little evidence of illicit diversion in the United States. Yet, in countries where buprenorphine is available with minimal restrictions, there is recent evidence of increased abuse by heroin abusers (see San et al., 1992, for review). For example, in Spain abuse of relatively low doses of buprenorphine (averaging 2.0 mg/day) was reported in 29 former heroin abusers. Buprenorphine use was verified by drug urine screens [San et al., 1992]. Similarly, in Ireland buprenorphine was used to prevent heroin withdrawal symptoms when heroin was unavailable [O'Connor et al., 1988], but buprenorphine was rarely the preferred drug. Although buprenorphine consistently maintains behavior that leads to its administration in rhesus monkeys and baboons [Mello and Mendelson, 1992], progressive ratio studies indicate that it is less reinforcing than heroin or pentazocine [Yanagita et al., 1982; Mello et al., 1988]. In 1982 we concluded that the safety and potential therapeutic benefits of buprenorphine probably outweigh the possible risks associated with its abuse potential [Mello et al., 1982]. In the absence of compelling information to the contrary, we would arrive at a similar conclusion a decade later.

Primate Studies of Buprenorphine's Effects on Opiate Self-Administration

In preclinical studies we compared the effectiveness of buprenorphine and methadone in reducing the self-administration of heroin and hydromorphone [Mello et al., 1983]. The effects of methadone and buprenorphine treatment were studied over an ascending and a descending dose series. Treatment doses were extrapolated from the clinically effective dose range. However, it was necessary to exceed the clinical dose range for buprenorphine and methadone to significantly reduce opiate self-administration by rhesus monkeys. Buprenorphine was studied over a dose range of 0.014–0.789 mg/kg/day, which is equivalent to 1–56 mg/day in humans. Methadone was studied over a dose range of 0.179–11.86 mg/kg/day, which is equivalent to 12.5–800 mg/day in humans. Each dose was studied for 20 sessions, but the severity of adverse reactions to methadone limited the maximum dose studied in individual monkeys.

Heroin (0.01 and 0.02 mg/kg/inj), hydromorphone (0.02 mg/kg/inj), and food (1 g banana pellet) self-administration were maintained on a second-order schedule of reinforcement [FR 4 (VR 16:S)]. An average of 64 responses was required for delivery of each food pellet or drug injection. Buprenorphine, methadone, and saline control solutions were administered at 9:30 AM, 2.5 hr before the first morning opiate self-administration session at 12 noon. The four daily opiate and food sessions were 1.5–16 hr after methadone or buprenorphine administration. Food sessions began at 11 AM, 3 PM, 7 PM, and 11 PM each day and drug sessions began 1 hr later, at 12 noon, 4 PM, 8PM, and 12 midnight. Each drug or food session lasted for 1 hr or until 20 drug injections or 65 food pellets had been delivered.

Buprenorphine significantly suppressed opiate self-administration at doses of 0.282–0.675 mg/kg/day. These doses were equivalent to 20–48 mg/day in humans or 2.5–7 times the dose that reduced heroin self-administration by humans [Mello et al., 1982]. In contrast, methadone doses of 1.43–11.86 mg/kg/day failed to suppress opiate self-administration in four of five monkeys. This methadone dose range is equivalent to 100–800 mg/day in humans. The distribution of opiate self-administration across drug availability sessions did not account for the absence of methadone suppression, because monkeys took 43% of their total daily opiate injections during the first daily drug sessions, 2.5 hr after methadone administration. These preclinical data were consistent with our clinical evaluations of buprenorphine [Mello and Mendelson, 1980; Mello et al., 1982] as well as previous inpatient studies of the effects of methadone on opiate self-administration by opiate-dependent men [Martin et al., 1973; Jones and Prada, 1977]. During maintenance on 50–100 mg/day of methadone, some heroin abusers continued to work for hydromorphone (4 mg, i.v.) for approximately 2.5 months, by riding an exercycle for 10 miles within 1 hr [Martin et al., 1973; Jones and Prada, 1975]. In rhesus monkeys, methadone attenuated heroin self-administration, but only at doses associated with severe debilitation, depression, and death [Harrigan and Downs, 1981; see Mello and Mendelson, 1992, for review].

Food self-administration was also affected differently by buprenorphine and methadone. Maintenance on buprenorphine (0.014–0.789 mg/kg/day) for over 6.5 months was not associated with reduced food intake, and two monkeys showed significant increases in food intake at higher buprenorphine doses [Mello et al., 1983]. Monkeys maintained on buprenorphine remained generally healthy and alert. Methadone suppressed food intake significantly below baseline in all monkeys. Moreover, monkeys appeared very debilitated by the prolonged course of high-dose methadone exposure. Comparable methadone toxicity has been reported by several other investigators [Crowley, et al., 1975; Snyder et al., 1977; Harrigan and Downs, 1981].

BUPRENORPHINE'S EFFECTS ON COCAINE SELF-ADMINISTRATION: PRECLINICAL STUDIES

Preclinical Evaluation of New Pharmacotherapies for Drug Abuse Treatment

Primates will self-administer most drugs that are abused by humans; consequently this model has been valuable for the prediction of drug abuse liability [Thompson and Unna, 1977; Griffiths et al., 1979; Brady and Lukas, 1984]. We believe that this model should also be useful for the preclinical evaluation of new treatment medications, since drug self-administration is the target behavior that is reduced by an effective pharmacotherapy [Mello, 1992]. It is reasonable to assume that if a new pharmacotherapy reduces drug self-administration by monkeys, it is more likely to be effective in humans than a pharmacotherapy that has no effect or that increases drug self-administration [Mello, 1992]. If medications that significantly reduce drug self-administration in monkeys also prove to be effective in human drug abusers, this could facilitate the rapid identification of promising new pharmacotherapies for the treatment of drug abuse. An emerging database shows good concordance between clinical and preclinical studies of pharmacotherapy evaluation [Mello et al., 1983; Mello, 1991]. Recently, there has been increasing interest in developing preclinical models to evaluate the potential clinical effectiveness of new pharmacotherapies for treatment of drug abuse and drug withdrawal [Balster et al., 1992; Mello, 1992; Spealman, 1992; Woolverton and Kleven, 1992].

The primate model offers several advantages over clinical trials for evaluation of new pharmacotherapies: (1) Compliance with the drug treatment regimen is ensured; (2) the treatment drug effects cannot be influenced by unreported polydrug abuse; (3) the treatment drug effects cannot be modulated by expectancy (i.e., placebo responding); (4) accurate baseline measures of the daily dose and pattern of drug self-administration are available for comparison before, during, and after treatment; (5) targeted preclinical trials are more cost-effective than extensive clinical trials [Mello, 1992]. There is one important caveat in using the drug self-administration model to evaluate potential pharmacotherapies. It is necessary to concurrently examine operant responding maintained by a nondrug reinforcer such as food. If a pharmacotherapy reduces drug self-administration with only transient effects on behavior maintained by another reinforcer, this would suggest that the treatment drug has *selective* effects on drug self-administration. However, if the pharmacotherapy reduces *both* drug and food self-administration in a parallel and dose-dependent manner, this might indicate that the treatment drug suppressed behavior generally or that the subject was sedated or sick [Mello, 1992].

Buprenorphine's Effects on Cocaine and Food Self-Administration by Rhesus Monkeys

We reported that daily buprenorphine treatment significantly reduced cocaine self-administration with minimal effects on food self-administration [Mello et al., 1989]. The effectiveness of this opioid mixed agonist–antagonist in reducing cocaine self-administration was somewhat surprising, since the reinforcing properties of cocaine appear to be modulated primarily by dopaminergic rather than endogenous opioid systems in brain [Fischman, 1987; Ritz et al., 1987; Kuhar et al., 1988; Gawin and Ellinwood, 1988; Johanson and Fischman, 1989]. We were prompted to examine buprenorphine's effects on cocaine self-administration, in part, because our studies of cocaine's effects on the neuroendocrine system suggested the possible relevance of dopaminergic–endogenous opioid interactions. Luteinizing hormone (LH) is a gonadotropin that is regulated in part by endogenous opioid inhibitory control of hypothalamic LHRH (luteinizing hormone-releasing hormone). LH is stimulated by opioid antagonists such as naloxone or naltrexone, which antagonize endogenous opioid inhibition of LHRH [Yen et al., 1985; Mendelson et al., 1986, 1991b]. We found that acute administration of cocaine (0.4 and 0.8 mg/kg, i.v.) also resulted in a rapid and significant increase in LH in female rhesus monkeys [Mello et al., 1990b]. Cocaine (30 mg, i.v.) also stimulated a rapid and persistent increase in LH in men with a history of concurrent cocaine and opiate abuse [Mendelson et al., 1992b]. Cocaine's stimulation of LH is consistent with evidence that the neuroendocrine regulation of LH is comodulated by both endogenous opioid and dopaminergic systems in brain [Yen, 1986; Kuljis and Advis, 1989]. Therefore, exploration of the effects of buprenorphine on cocaine-maintained behavior seemed warranted. However, as discussed more fully in a later section (Mechanisms of Buprenorphine's Actions), the ways in which endogenous opioid and dopaminergic systems may converge to affect cocaine self-administration are unclear (see Mello, 1992, for review).

Buprenorphine Maintenance Procedures. We compared the effects of 15 days of treatment with buprenorphine (0.237–0.70 mg/kg/day) and saline on cocaine and food self-administration by rhesus monkeys [Mello et al., 1989, 1990a]. These monkeys had self-administered cocaine for an average of 8 months. The monkeys worked for intravenous cocaine (0.05 or 0.10 mg/kg/inj) and for food (1-g banana pellets) on a second-order fixed-ratio 4, variable-ratio 16:S schedule of reinforcement. An average of 64 responses was required for delivery of each food pellet or cocaine injection. Food and cocaine each were available during four 1-hr sessions each day. Food sessions began at 11 AM, 3 PM, 7 PM, and 7 AM Cocaine sessions began at 12 noon, 4 PM, 8 PM, and 8 AM. Each food or drug session lasted for 1 hr or until 20 drug injections or 65 food pellets had been delivered. The total number of cocaine injections was limited to 80 per day to minimize the possibility of any adverse drug effects [Johanson et al., 1976]. A nutritionally fortified banana pellet diet was supplemented with fresh fruit, vegetables, biscuits, and multiple vitamins each day.

Buprenorphine (or an equal volume of saline control solution) was administered each day beginning at 9:30 AM. Buprenorphine or saline was gradually infused at a rate of 1 ml every 12 min and flushed through the catheter with sterile saline in a volume that exceeded the catheter dead space. Buprenorphine was administered at two doses (0.40 and 0.70 mg/kg/day) that effectively suppressed opiate self-administration in our previous studies in the primate model [Mello et al., 1983]. Subsequently, a lower dose of buprenorphine (0.237 mg/kg/day) was also studied. Each dose of buprenorphine and saline was studied for 15 consecutive days. Then buprenorphine treatment was abruptly discontinued and daily saline treatment was resumed.

Figure 2 shows cocaine and food self-administration during 15 days of baseline saline treatment and six successive 5-day periods of buprenorphine treatment [Mello et al., 1989]. Monkeys self-administered relatively high doses of cocaine during baseline saline treatment (2.1–4 mg/kg/day; group average of 3.07 ± 0.17 mg/kg/day). This dose of cocaine is comparable to that often reported by human cocaine abusers; 1–2 g of cocaine per week is equivalent to 2.04–4.08 mg/kg/day in a 70-kg man [Mendelson et al., 1988]. All monkeys reduced cocaine self-administration significantly during buprenorphine treatment. Average cocaine self-administration decreased by 49% during the first 5 days of buprenorphine treatment to an average dose of 1.60 ± 0.25 mg/kg/day. Average cocaine self-administration decreased to 77% and 83% below baseline during Days 6–10 and 11–15 of buprenorphine treatment. During 15 days of buprenorphine treatment at 0.40 mg/kg/day, cocaine self-administration averaged 0.98 ± 0.11 mg/kg/day [Mello et

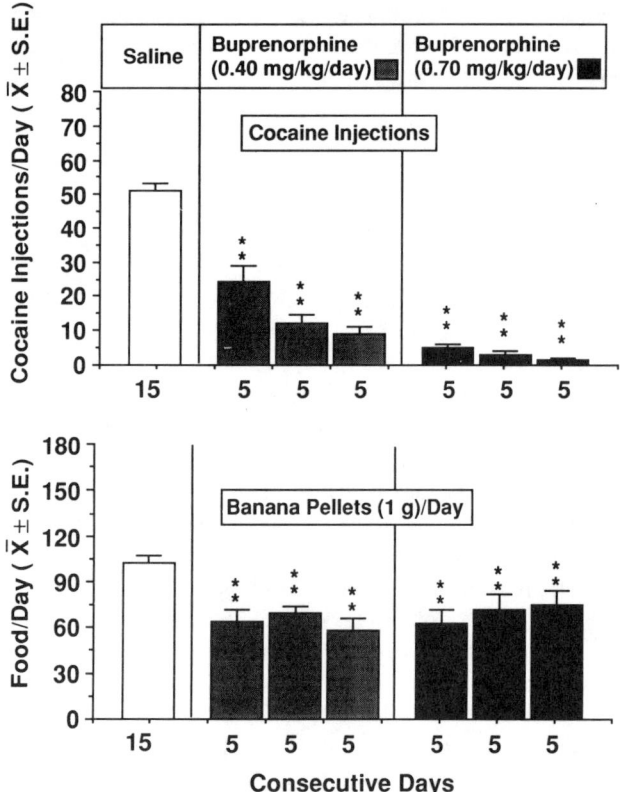

Fig. 2. Buprenorphine reduction of cocaine self-administration by rhesus monkeys. The effects of single daily infusions of buprenorphine or a saline control solution on cocaine and food self-administration. Saline treatment is shown as an open bar in the left panel and buprenorphine treatment 0.40 mg/kg/day in the middle panel and 0.70 mg/kg/day in the right panel. The average number of cocaine injections self-administered is shown in the top row. The average number of food pellets self-administered is shown in the second row. The number of days that each treatment condition was in effect is shown on the abscissa. Each data point is the mean ±SEM for five subjects. The statistical significance of each change from the saline treatment as determined by analysis of variance for repeated measures and Dunnett's tests for multiple comparisons is shown by asterisks (**$p < 0.01$). Reproduced from Mello et al. [1989], with permission of the publishers. Copyright 1989 by the AAAS.

al., 1989]. During the second 15 days of buprenorphine treatment at a higher dose (0.70 mg/kg/day), cocaine self-administration decreased to between 91% and 97% below baseline levels. Monkeys self-administered an average of 0.19 ± 0.03 mg/kg/day of cocaine.

After abrupt cessation of 30 days of buprenorphine treatment, cocaine-maintained responding was suppressed for at least 15 days in all animals. Individual monkeys returned to baseline levels of cocaine self-administration at different rates that ranged from 15 to 58 days (mean 30.5 ± 10 days). This prolonged reduction of cocaine self-administration after termination of buprenorphine treatment is similar to clinical reports of a delayed onset of buprenorphine withdrawal signs and symptoms [Jasinski et al., 1978; Fudala et al., 1989].

Food-maintained responding was also reduced by 31% during the first 15 days of buprenorphine treatment (Fig. 2). During the second 15 days of treatment with a higher dose of buprenorphine, food self-administration gradually recovered to average 20% below baseline. Although these changes in food-maintained responding were statistically significant, it is unlikely that they were biologically significant. There were no correlated changes in body weight and the animals continued to eat food—daily fruit and vegetable supplements. Moreover, there were no consistent buprenorphine effects on patterns of food self-administration. Buprenorphine treatment, in comparison with saline, did not change the overall daily distribution of food self-administration across sessions. Food intake during the 11 AM session immediately after buprenorphine treatment (9:30 to 10:30 AM) was not suppressed in comparison with saline treatment. The animals did not appear sedated during buprenorphine treatment and activity levels were normal. We concluded that buprenorphine treatment selectively reduced cocaine-maintained responding but did not produce a generalized suppression of behavior [Mello et al., 1989, 1990a].

Onset and Duration of Buprenorphine's Effects on Cocaine Self-Administration. The first dose of buprenorphine usually reduced cocaine self-administration significantly below saline treatment baseline levels. Figure 3 shows the effects of buprenorphine (0.237 and 0.40 mg/kg/day) on daily cocaine self-administration by six individual monkeys [Mello et al., 1990a]. The lower dose of buprenorphine suppressed cocaine self-administration by 49%–96% in five of the six monkeys on the first day of treatment. By the second day of buprenorphine administration, cocaine self-administration fell to 50% or more below baseline in all monkeys. The higher dose of buprenorphine also reduced cocaine self-administration by 50%–95% in five monkeys on the first day of treatment. On the second buprenorphine treatment day, cocaine self-administration was 64%–95% below the saline treatment baseline in five monkeys. One monkey (679 C) was initially resistant to the effects of 0.40 mg/kg/day of buprenorphine and 6 days of treatment were required to reduce cocaine self-administration by 50%. Cocaine self-administration remained 40%–100% below baseline except on three occasions. On one day, two monkeys increased cocaine self-administration to average 12% and 22% below baseline (CH 61 and 199 C) and one monkey exceeded saline treatment baseline levels by 46% on Day 4 of treatment [Mello et al., 1990a].

Since the eventual clinical application of buprenorphine for drug abuse treatment will involve chronic maintenance, we subsequently asked if tolerance developed to buprenorphine effects on cocaine-maintained responding. Tolerance to many of buprenorphine's effects has been reported in humans, monkeys, pigeons, and rodents (see Lewis et al., 1983, for review). In clinical studies, tolerance developed to the opioid agonist side effects of daily buprenorphine treatment (8 mg, s.c.) within 10–21 days. In rhesus monkeys, tolerance developed to buprenorphine's suppres-

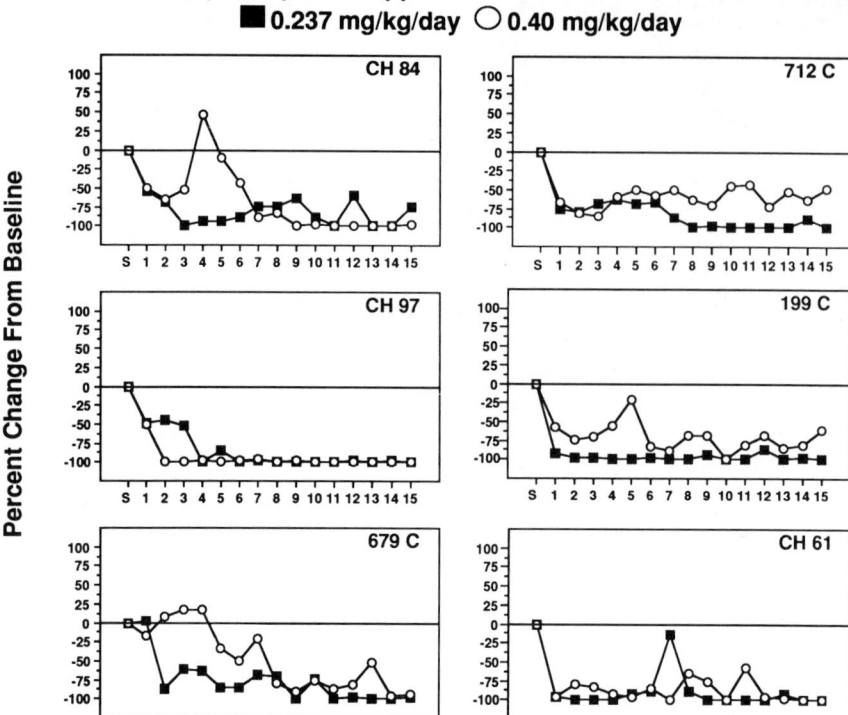

Fig. 3. Rate of buprenorphine's suppression of cocaine self-administration by individual monkeys. Data are shown as the percentage of change in cocaine injections from the saline treatment baseline. Saline control treatment (S) and the first 15 days of buprenorphine treatment (0.237 and 0.40 mg/kg/day) are shown on the abscissa. Percentage of change from the average number of cocaine injections self-administered during saline treatment (S) is shown on the ordinate. Buprenorphine doses of 0.237 mg/kg/day are shown as squares and 0.40 mg/kg/day as circles. Reproduced from Mello et al. [1990a] with permission of the publisher.

sive effects on food self-administration during chronic buprenorphine treatment [Mello et al., 1983; Mello and Mendelson, 1985; Lukas et al., 1988; see Mello and Mendelson, 1992, for review].

We examined the effects of 30–120 days of buprenorphine treatment (0.32 mg/kg/day) on cocaine and food self-administration in six rhesus monkeys [Mello et al., 1992a]. These monkeys had a long history of i.v. cocaine self-administration that averaged 415 days (range 96–781 days) at the beginning of the study. Saline control treatment was in effect for 15 days before and 15 days after the conclusion of buprenorphine treatment. Cocaine (0.05 or 0.10 mg/kg/inj) and food (1-g banana pellet) self-administration were maintained on an FR 4 (VR 16:S) schedule of reinforcement.

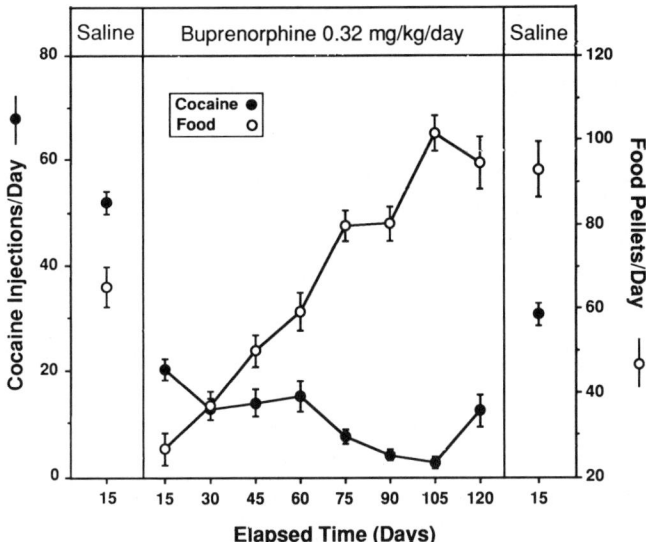

Fig. 4. Effects of 3–4 months of daily buprenorphine treatment (0.32 mg/kg/day) on cocaine and food self-administration. Each of the data points for cocaine injections (filled circles) and food pellets (open circles) during the prebuprenorphine saline control period is the average ±SE of four monkeys over 15 days. The first 100 days of buprenorphine treatment are an average of data from four monkeys and Days 101–120 are an average of data from three monkeys. Adapted from Mello et al. [1992].

During the prebuprenorphine saline treatment baseline, the six monkeys self-administered an average of 52.3 (±2.05) cocaine injections per day and 65.4 (±4.7) food pellets per day. Buprenorphine treatment was followed by a prompt and sustained decrease in cocaine self-administration ($p < 0.0001$). During the first 15 days of buprenorphine treatment cocaine self-administration averaged 60% below the saline treatment baseline levels ($p < 0.01$). Cocaine self-administration remained suppressed by 70%–94% over the period of observation ($p < 0.01$). Although food-maintained responding was initially suppressed by an average of 58% ($p < 0.01$), it gradually returned to and significantly exceeded saline treatment baseline levels by 22%–55% ($p < 0.05-< 0.01$). After saline was substituted for buprenorphine, cocaine self-administration resumed and averaged between 31 (±2.15) and 43 (±7.8) injections per day. Figure 4 shows average cocaine and food self-administration by four of the six monkeys that completed 100–120 days of daily buprenorphine treatment [Mello et al., 1992].

We interpreted these data as indicating that tolerance did not develop to buprenorphine's effects on cocaine self-administration by rhesus monkeys during 100–120 days of daily treatment. The resumption of cocaine self-administration after termination of buprenorphine treatment strengthened the conclusion that the significant decrease in cocaine self-administration observed was due to buprenorphine treatment and not to other uncontrolled variables. These data confirm and extend our previous observations that 15 days of buprenorphine treatment

(0.237–0.70 mg/kg/day) reduced cocaine self-administration by 72%–93% in rhesus monkeys [Mello et al., 1989, 1990a].

Subsequently we examined the effects of intermittent buprenorphine treatment on cocaine and food self-administration by rhesus monkeys over 60 days [Mello et al., 1993b]. In contrast to our findings during 4 months of *daily* buprenorphine administration (Fig. 4), when buprenorphine (0.40 mg/kg/day) was administered once every 48 hr, there was a gradual increase in cocaine self-administration over time ($p < 0.0001$). Food self-administration did not increase or decrease over the period of observation. We interpreted these data as suggesting that intermittent buprenorphine administration might be less effective than daily buprenorphine administration for treatment of cocaine-dependent polydrug abusers. These preclinical data are consistent with clinical observations that the effect of buprenorphine treatment (8 mg, s.l.) on alternate days was less effective than daily treatment in reducing interest in drugs [Fudala et al., 1990]. On days when they were given placebo instead of buprenorphine, heroin-dependent men reported a "greater urge for an opiate" and more dysphoria than on days when they received buprenorphine [Fudala et al., 1990].

Buprenorphine Plasma Levels During Daily Buprenorphine Administration.
Buprenorphine plasma levels averaged 18 (±2.84) ng/ml during daily buprenorphine (0.32 mg/kg) treatment. Buprenorphine plasma levels were quite consistent within each monkey but varied between monkeys and averaged 10.9–30 ng/ml. These average buprenorphine plasma levels are comparable with the mean peak level (11.0 ng/ml) measured in six men during treatment with 1.2 mg of buprenorphine [Cone et al., 1992]. Higher buprenorphine plasma levels were not associated with lower levels of cocaine self-administration; rather, the lowest buprenorphine plasma levels (10.9 ng/ml) were associated with the greatest average decrease in cocaine self-administration (97%). The highest average buprenorphine plasma levels (30 ng/ml) were associated with the smallest reduction in cocaine self-administration (72.3%). Buprenorphine plasma levels usually decreased by 50% or more within 27 hr after the last buprenorphine dose. However, low levels of buprenorphine in plasma (0.10–0.19 ng/ml) were measured for 30–60 days after abrupt termination of daily buprenorphine treatment. But these low buprenorphine plasma levels did not reduce cocaine self-administration [Mello et al., 1992]. The slow dissociation of buprenorphine from the opiate receptor and its lipophilic characteristics may contribute to its persistence in plasma [Hambrook and Rance, 1976; Lewis et al., 1983].

BUPRENORPHINE'S EFFECTS ON CONCURRENT OPIATE AND COCAINE ABUSE

Buprenorphine's selective and persistent reduction of cocaine self-administration by rhesus monkeys suggested that it might be a useful pharmacotherapy for the treatment of dual dependence on opiates and cocaine [Mello et al., 1989, 1990a, 1992].

We previously had found that buprenorphine reduces heroin abuse by heroin-dependent men [Mello and Mendelson, 1980; Mello et al., 1982], and further clinical studies of buprenorphine treatment of polydrug abusers seemed warranted. In 1989 Kosten and co-workers reported that opioid-dependent patients treated with daily sublingual doses of buprenorphine for 1 month (average 3.2 mg/day; range 2–8 mg) had significantly fewer cocaine-positive urines than patients treated with methadone [Kosten et al., 1989a,b]. Accordingly, we began an open trial of the effects of buprenorphine treatment on men who were dually dependent on cocaine and heroin. The trial consisted of a 30-day inpatient phase and a 26-week outpatient phase. The details of our studies and recent findings concerning buprenorphine's safety and efficacy are summarized below.

Inpatient Studies of Buprenorphine's Safety and Effectiveness

Men who met DSM-III-R diagnostic criteria for concurrent cocaine and opiate dependence were admitted to a NIDA-supported Treatment Research Unit (TRU) at the Alcohol and Drug Abuse Research Center, McLean Hospital and Harvard Medical School. Volunteer subjects responded to newspaper advertisements and provided informed consent for participation in the study. All subjects had physical evidence of intravenous drug use (i.e., needle tracks) and opiate- and cocaine-positive urines. All subjects were in good physical health as determined by medical, psychiatric, and laboratory screening examinations and all were negative for the HIV antibody. Men with another DSM-III-R Axis I diagnosis or concurrent medical disorders were not accepted for these studies.

The subjects lived in groups of four on the TRU clinical research ward for 30 days. After methadone detoxification, they were drug-free for 9 days. Then they were given ascending doses of sublingual buprenorphine (1, 2, 4, 6, 8 mg/day) over 5 days. After buprenorphine induction, the subjects were randomly assigned to a chronic maintenance dose of 4 or 8 mg/day of buprenorphine. The maintenance dose was continued for 16 days until discharge. The patients could elect to transfer to the outpatient program of the TRU and remain on their maintenance dose of buprenorphine or transfer to a naltrexone or methadone maintenance clinic or a drug-free treatment program.

During the 30-day inpatient phase, studies were undertaken to evaluate the safety and effectiveness of buprenorphine for the treatment of persons with dual dependence on heroin and cocaine [Mendelson et al., 1991a; Teoh et al., 1993]. Since many drug abusers continue to use some drugs during maintenance treatment [Kreek, 1987, 1991], it was important to determine if buprenorphine had any potentially adverse interactions with cocaine or morphine. Buprenorphine alone (1, 2 and 4 mg sublingual) did not change blood pressure, pulse, respiratory rate, or body temperature in men with a history of opiate abuse who were studied on a clinical research ward [Jasinski et al., 1989]. However, it is well established that cocaine can induce significant changes in cardiovascular function [Fischman et al., 1985; Kumor et al., 1988; Foltin and Fischman, 1991]. These cardiovascular changes may have adverse medical consequences in susceptible persons [Cregler and Mark, 1986; Isner et al., 1986]. Cocaine abuse has been associated with myocardial infarction, ventricular arrhythmias, and stroke [Isner et al., 1986; Ascher et al., 1988]. In contrast, morphine has minimal effects on cardiovascular function in reclining subjects, but orthostatic hypotension may occur in ambulatory subjects [Mansky, 1978; Jaffe and Martin, 1990]. It is well-established that morphine has dose-related suppressive effects on respiration and also lowers body temperature [Mansky, 1978; Jaffe and Martin, 1990].

The physiological effects of a single-blind challenge dose of cocaine (30 mg, i.v.), morphine (10 mg, i.v.), and i.v. saline placebo were measured before and during buprenorphine maintenance [Teoh et al., 1993]. Buprenorphine maintenance (4 or 8 mg/day) did not have adverse effects on cardiovascular function during or following acute intravenous administration of cocaine (30 mg) or morphine (10 mg). Figure 5 shows that the cocaine-induced increases in heart rate and systolic and diastolic blood pressure were identical during buprenorphine-free and buprenorphine-maintenance conditions. The time course and magnitude of cocaine-induced changes in measures of cardiovascular function were similar to those reported by others [Foltin and Fischman, 1991]. Peak plasma cocaine levels averaged 159 ± 46 and 258 ± 33 ng/ml during the drug-free baseline, and equivalent plasma cocaine levels were measured during buprenorphine treatment [Teoh et al., 1993].

Respiration and temperature changes in response to cocaine also were equivalent before and during buprenorphine maintenance. Respiratory rates were slightly lower following morphine administration during maintenance on 8 mg of buprenorphine, but this difference was not statistically significant. No abnormalities in EKG waveform were observed during the study before or during buprenorphine therapy. Normal sinus rhythm was observed before and during buprenorphine maintenance.

Buprenorphine's Effects on EEG Sleep Patterns. Sleep disorders are often associated with substance abuse problems and there is an extensive literature that shows that many abused drugs have direct disruptive effects on sleep [Oswald et al., 1969]. The eventual clinical utility of buprenorphine will depend in part on its effects on sleep. Consequently, we examined the effects of buprenorphine on EEG sleep patterns during the inpatient phase of the open trial [Lukas et al., 1992]. Each subject served as his own control during the prebuprenorphine drug-free period (study Days 4 and 5) and after 8 and 9 days of buprenorphine maintenance. Seven subjects were maintained on 4 mg/day, s.c. of buprenorphine and seven were maintained on 8 mg/day s.c. of buprenorphine. All subjects were withdrawn from methadone before the first day of the drug-free period.

Each subject slept in his own bed on the TRU clinical research ward during the sleep-recording nights. The subjects were prepared with standard scalp and muscle electrodes for polysomnography and all night sleep recordings were taken. EEG, EMG, and EOG activity were amplified and digitized at the bedside, and then encoded as a video signal and transmitted via coaxial cable to the McLean Hospital Sleep Disorders Center. Sleep records were scored blind by a computer-assisted sleep-scoring program, and the results were verified by visual inspection of raw data by established criteria.

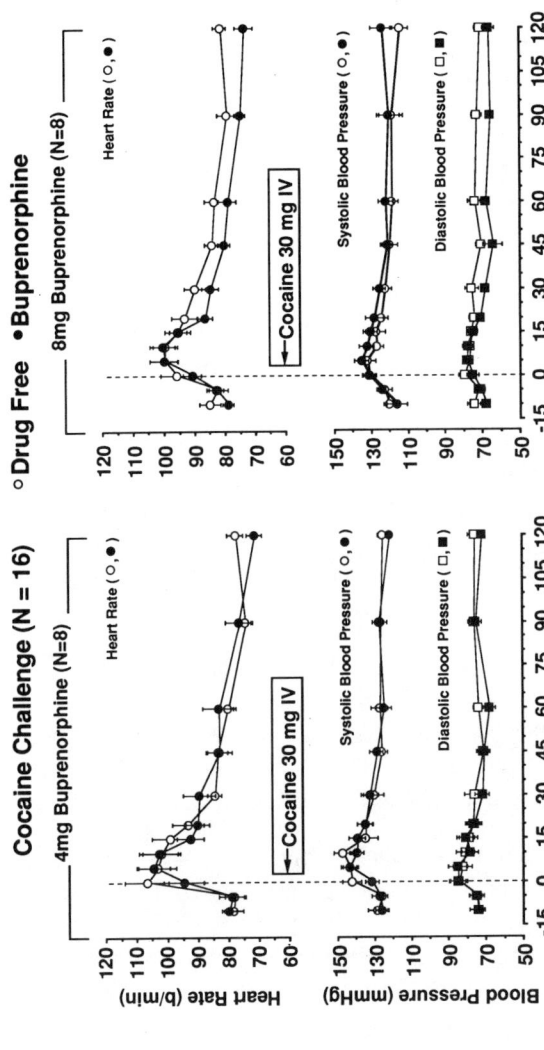

Fig. 5. Cardiovascular effects of cocaine before and during maintenance on 4 or 8 mg/day of buprenorphine. Heart rate and blood pressure changes in response to intravenous cocaine administration (30 mg) during drug-free (open symbols) and buprenorphine maintenance conditions (closed symbols). Cardiovascular responses during 4 mg/day of buprenorphine are shown on the left; cardiovascular responses during 8 mg/day of buprenorphine are shown on the right. Each of the data points is the average (±SE) for eight subjects. Heart rate (bpm) before and during buprenorphine maintenance is shown in the top row of each panel (open and filled circles). Systolic blood pressure (mm Hg) (open and filled circles) and diastolic blood pressure (mm Hg) (open and filled squares) are shown in the bottom panel. Intravenous cocaine was administered between −1 min and time 0. The vertical line denotes the end of the cocaine injection. Measures were acquired every 5 min for the first 20 min after administration of cocaine, then at 15 min intervals for the remainder of the first hour after administration of cocaine. Additional measures were collected at 90 and 120 min after cocaine administration.

Before administration of buprenorphine all 14 subjects had disordered sleep patterns. Sleep onset and latency to REM sleep were delayed and REM sleep time and slow-wave sleep were low in comparison with normal patterns. These abnormal sleep patterns were similar to those observed in persons during opiate or cocaine withdrawal [Kay et al., 1969; Weddington et al., 1990]. Objective measures of sleep disturbances were well correlated with subjective complaints.

Sleep patterns were markedly improved after 8 and 9 days of buprenorphine maintenance. Latency to sleep onset, total sleep time, and non-REM sleep were improved during maintenance on both doses of buprenorphine. However, latency to REM sleep and duration of slow-wave sleep were improved only during maintenance on 4 mg/day of buprenorphine. In contrast, both duration of slow-wave sleep and total REM sleep decreased during 8-mg/day buprenorphine maintenance in comparison with prebuprenorphine, drug-free measures. These data suggest that 4 mg/day of buprenorphine is more effective in normalizing sleep patterns in abstinent polydrug users than 8 mg/day of buprenorphine. Moreover, these data suggest that high-dose buprenorphine maintenance may disrupt sleep after a relatively brief period of exposure [Lukas et al., 1992]. Further studies will be necessary to clarify buprenorphine's long-term effects on sleep.

Buprenorphine's Side Effects. Induction on gradually increasing doses of buprenorphine was associated with relatively few side effects in men with a history of polydrug abuse [Teoh et al., 1993]. These findings are consistent with previous reports that buprenorphine has relatively mild side effects in opiate abusers [Lange et al., 1990], and tolerance develops to most of these within 21 days [Mello et al., 1982]. Some subjects occasionally reported headaches, sedation, and drowsiness during buprenorphine induction. The 8-mg/day dose of buprenorphine was associated with reports of dry mouth, cough, and nasal discharge, but these symptoms were infrequent during buprenorphine maintenance. Abdominal discomfort was greater in the 4-mg/day buprenorphine group. Constipation was rarely reported in the first 20 subjects studied, although constipation was a frequent complaint in previous studies [Mello et al., 1982; Lange et al., 1990]. Other somatic complaints, including anxiety, were also reported but were absent by Day 14 of buprenorphine maintenance. Excessive sweating, lacrimation, and pruritis were not reported. Hypotension was observed in one of 20 men studied to date and in one of 10 men in our previous series [Mello et al., 1982; Teoh et al., 1993]. However, doses of 2 and 4 mg/day of buprenorphine were well tolerated by these patients.

Safety of Buprenorphine Maintenance. The safety of buprenorphine has been repeatedly demonstrated [Jasinski et al., 1978; Mello and Mendelson, 1980, 1985; Mello et al., 1982; Bickel et al., 1988; Johnson et al., 1991; Teoh et al., 1993]. The antagonist component of this mixed agonist–antagonist appears to exert a protective effect such that overdose was not possible even at 10 times the therapeutic analgesic dose [Banks, 1979]. The relative safety of buprenorphine in comparison with an opioid agonist such as methadone is one argument that has been advanced for its potential clinical utility for drug abuse treatment [Mello and Mendelson, 1985, 1993]. Since buprenorphine in combination with an acute high dose of cocaine or of morphine did not accentuate the physiological changes induced by either drug alone, we concluded that buprenorphine was potentially safe for the treatment of persons with concurrent opioid and cocaine dependence [Teoh et al., 1993]. It is of interest that buprenorphine has been shown to reduce lethality associated with cocaine-induced convulsions in mice [Witkin et al., 1991]. These data were interpreted as suggesting that toxic interactions between buprenorphine and cocaine were unlikely in clinical studies [Witkin et al., 1991].

Outpatient Studies of Buprenorphine Treatment of Concurrent Opiate and Cocaine Abuse

After completion of the 30-day inpatient study, the subjects could elect to participate in a 26-week outpatient open trial of buprenorphine maintenance treatment. They remained on their final maintenance dose of buprenorphine (2, 4, or 8 mg/day) and reported to the Alcohol and Drug Abuse Research Center TRU Outpatient Clinic at the Massachusetts General Hospital. The TRU Outpatient Clinic is open 7 days each week. These studies are now in progress and preliminary findings on the first cohort of subjects are summarized below.

Sixteen subjects were randomly assigned to a dose of 4 or 8 mg/day of buprenorphine during the inpatient phase of the study [Teoh et al., 1993]. All subjects had a long history of concurrent intravenous use of heroin and cocaine and had failed in previous drug abuse treatment programs. In the 2- to 4-mg/day buprenorphine maintenance group, the average dose was 0.05 ± 0.004 mg/kg/day. One man was maintained on 2 mg/day (0.027 mg/kg/day) because of hypotensive side effects. In the 8-mg/day buprenorphine maintenance group, the average dose was 0.10 ± 0.006 mg/kg/day.

Figure 6 summarizes group data for eight men maintained on 2 or 4 mg/day of buprenorphine for 10 weeks and eight men maintained on 8 mg/day of buprenorphine for 10 weeks. These data are expressed as the percentage of total possible days of attendance at the clinic and as the percentage of the total number of urines collected that were positive for opiates, cocaine, and other drugs. Urine collections for drug screening were scheduled to be unpredictable for the subjects, and urines were collected at least once a week but not more than three times a week.

Clinic attendance was high (88% and 85%) in both groups. The lower dose of buprenorphine was somewhat more effective than the higher dose in reducing opiate and cocaine abuse. In the 2- or 4-mg/day buprenorphine maintenance group, 36% of the urines collected were positive for opiates and 52% were positive for cocaine. Since these men had a long history of *daily* intravenous use of both drugs, this means that opiate use was reduced by 64% and cocaine use was reduced by 48%. In the 8-mg/day buprenorphine maintenance group, 42% of urines were positive for opiates and 68% were positive for cocaine. This means that opiate use was reduced by 58% and cocaine use was reduced by 32%. Other drug use (benzodiazepine abuse and occasional marijuana use) was equivalent in both buprenorphine dose

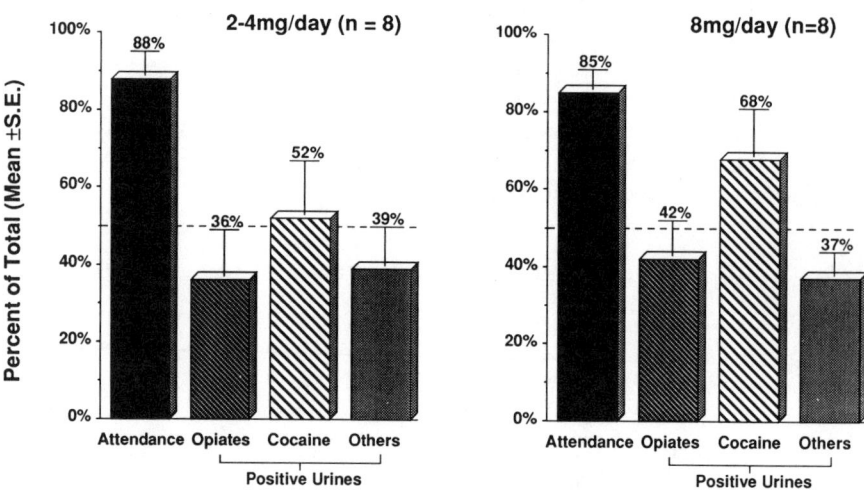

Fig. 6. Effects of 10 weeks of buprenorphine treatment on drug abuse. Data from the first 10 weeks of an outpatient open trial of buprenorphine treatment (2–4 mg/day and 8 mg/day, s.l.) are shown for 16 patients. Each of the data points is the average for eight men who reported daily intravenous use of cocaine and heroin before treatment. The average percentage of total possible days attended over 10 weeks is shown at the left of each panel. The average percentage of the total urine drug screens collected that were positive for opiates, cocaine, and other drugs is shown in columns 2 through 4 of each panel.

groups and averaged 39% and 37%, respectively. This pattern of polydrug use was consistent with each man's reported drug abuse history.

The effects of buprenorphine on needle use and needle sharing in these 16 subjects during the first 10 weeks of treatment are summarized in Figure 7. In the 4-mg/day buprenorphine treatment group, all subjects reported previous needle use and six subjects reported previous needle sharing. Needle use decreased appreciably during the first 10 weeks of buprenorphine treatment and only one subject reported needle sharing. In the 8-mg/day buprenorphine treatment group, all subjects reported previous needle use and needle sharing before treatment. Needle use continued at a high rate in this group but needle sharing decreased.

Clinical attendance and opiate, cocaine, and other drug use by two individual subjects during 26 weeks of buprenorphine treatment (4 mg/day) are shown in Figure 8. Clinic attendance by subject S-00390 averaged over 90%. Fewer than 10% of his urine screens were positive for opiates and no opiate use was detected during Weeks 7–12 and 19–24. Fewer than 10% of urine screens were positive for cocaine during the first 6 weeks of treatment and no cocaine use was detected during Weeks 7–12. However, during Weeks 13–18, this subject resumed cocaine use and 80% of the urines collected were positive for cocaine. Cocaine use subsequently decreased

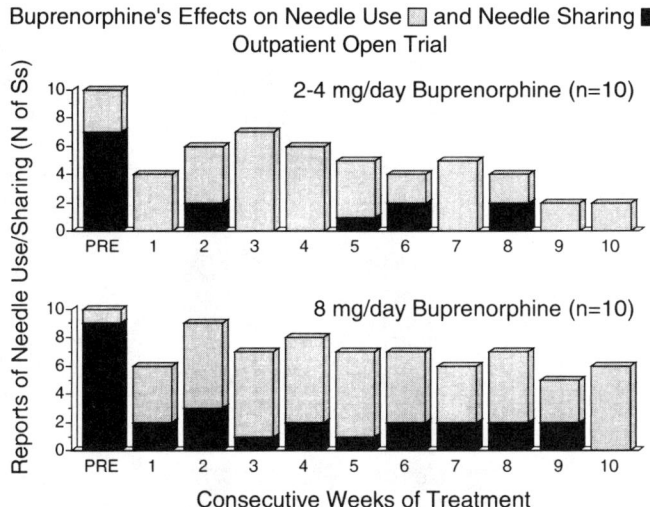

Fig. 7. Reports of needle use and needle-sharing before and during the first 10 weeks of outpatient buprenorphine treatment. Data are based on self-reports by eight subjects maintained on 2–4 mg/day, s.l., of buprenorphine and eight subjects maintained on 8 mg/day, s.l., of buprenorphine in an outpatient open trial.

to fewer than 20% positive urines during Weeks 19–24. Other drug use, primarily benzodiazepines and marijuana, continued at high levels throughout buprenorphine treatment.

A more consistent effect of buprenorphine on opiate and cocaine use was observed in subject S-00190. Clinic attendance averaged 60% over 26 weeks of treatment. Opiate-positive urines averaged 30% over this time period but increased to almost 50% during Weeks 19–24. Cocaine-positive urines averaged 15% over 26 weeks of treatment. During the first 6 weeks of treatment cocaine-positive urines averaged 20 percent of the total collected. During weeks 19–24 no urines were positive for cocaine or for other drugs, but this coincided with an increase in heroin use (Fig. 8).

Conclusions and Implications

These preliminary results of an open trial of buprenorphine treatment for dual cocaine and heroin dependence are very encouraging. On the basis of the apparent safety of buprenorphine for the treatment of polydrug abuse, the Food and Drug Administration (FDA) granted a compassionate extension of the approved period of outpatient buprenorphine treatment from 26 to 52 weeks. This extension was requested by patients who had a positive response to buprenorphine during the first 26 weeks of treatment and who did not want to transfer to a naltrexone, methadone, or drug-free treatment program. Of the first 22 men enrolled in this outpatient study, the retention rate has been 91% [Gastfriend et al., 1992, 1993]. Psychometric and

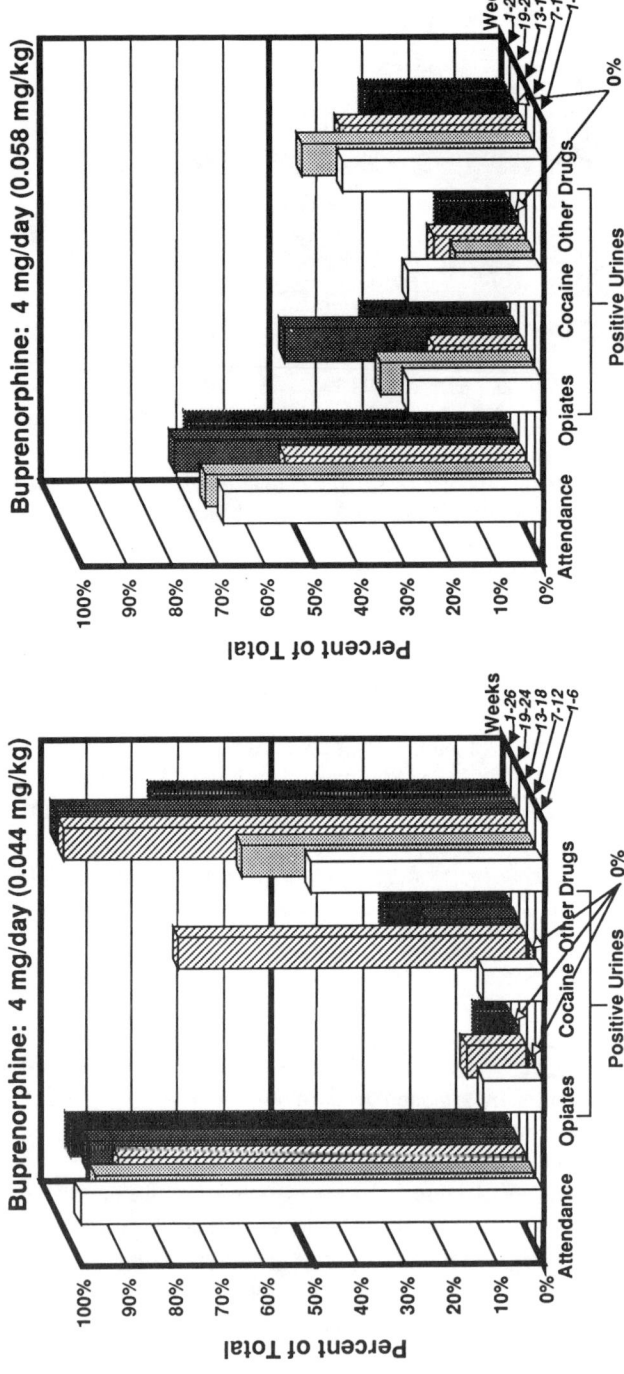

Fig. 8. Clinical attendance and drug use by individual subjects during 26 weeks of buprenorphine treatment (4 mg/day, s.l.). The average percentages of total possible days of attendance at the clinic over Weeks 1–6, 7–12, 13–18, and 19–24, and for all 26 weeks, are shown at the left of each panel. The average percentages of total urine drug screens collected that were positive for opiates, cocaine, and other drugs over the same time periods are shown at the right of each panel. Data are from an outpatient open trial of buprenorphine treatment for dual dependence on opiates and cocaine.

self-report measures of drug use were concordant with the objective data based on urine screens. Craving for both opiates and cocaine decreased by 76% and 69%, respectively [Gastfriend et al., 1992, 1993]. The Addiction Severity Index drug severity subscale scores also improved by 43% from average pretreatment levels [Gastfriend et al., 1992, 1993]. This open trial of buprenorphine is still ongoing and final data should be available in 1994. Subsequently, these studies will be extended to a double-blind comparison of buprenorphine and methadone treatment for dual dependence on cocaine and opiate abuse.

These preliminary clinical data confirm our previous findings that buprenorphine decreases opiate self-administration by opiate-dependent men [Mello and Mendelson, 1980; Mello et al., 1982]. These data are also concordant with our preclinical evaluations of buprenorphine's effects on opiate and cocaine self-administration by rhesus monkeys [Mello and Mendelson, 1993; Mello et al., 1983, 1989, 1990a]. These findings are also consistent with those of a 30-day outpatient open trial of buprenorphine treatment conducted by Kosten and co-workers [1989a,b]. Opiate-dependent patients treated with buprenorphine (average 3.2 mg/day; range 2–8 mg) had significantly fewer cocaine-positive urines than patients treated with methadone [Kosten et al., 1989a,b]. However, it is important to recognize that these clinical studies are open trials of buprenorphine [Gastfriend et al., 1992, 1993; Kosten et al., 1989a,b], and controlled double-blind trials have not yet been done in patients who meet DSM-III-R criteria for dual dependence on cocaine and opiates. Preliminary data from a double-blind comparison of buprenorphine (8 mg/day) and methadone (60 mg/day) treatment of opiate-dependent patients found no differences in occasional cocaine use [Fudala et al., 1991].

Occasional cocaine use also continued during buprenorphine treatment in our patients [Gastfriend et al., 1992, 1993]. Although buprenorphine significantly reduced opiate-positive urines by 58% ($p < 0.001$) and cocaine-positive urines by 48% ($p < 0.06$) in the first 22 patients studied [Gastfriend et al., 1992, 1993], it did not completely eliminate cocaine and heroin use. However, it is important to recognize that no pharmacotherapy is likely to induce complete drug abstinence. It is unfortunate that the effectiveness of pharmacotherapies for drug abuse treatment are often evaluated by more stringent criteria than are used to evaluate medications for the treatment of other disorders. For example, if a medication produced a 30%–50% reduction in anginal pain, or in degree of hypertension, or in malignant tumor growth, it would be considered very effective. This perspective should also be used to evaluate the effectiveness of pharmacotherapies that reduce drug abuse. The *occasional* cocaine and heroin use observed during buprenorphine treatment is a significant change from *daily* cocaine and heroin abuse. Since any reduction in intravenous drug use concomitantly reduces the risk for HIV infection associated with needle sharing, we conclude that buprenorphine treatment has had a significant impact on these patients. Continued use of multiple drugs is common among patients in treatment for drug abuse problems [Kreek, 1978, 1987, 1991], but during the relatively drug-free interval associated with pharmacotherapeutic treatment, the drug abuser is likely to be more responsive to counseling, social service interventions, job skills training, and other forms of assistance.

MECHANISMS OF BUPRENORPHINE'S ACTIONS

Buprenorphine consistently reduces opiate self-administration by opioid-dependent men and polydrug abusers [Mello and Mendelson, 1980; Mello et al., 1982; Kosten et al., 1989a,b; Fudala et al., 1990; Johnson et al., 1991, 1992; Gastfriend et al., 1992a,b] as well as in a primate model of drug self-administration [Mello et al., 1983]. Buprenorphine also blocks the acute subjective and physiological effects of opiates under many experimental conditions [Jasinski et al., 1978; Lewis et al., 1983; Bickel et al., 1988]. These effects were predicted from buprenorphine's long-acting μ opioid antagonist properties [Cowan, 1974; Lewis, 1974] and are consistent with our current understanding of its basic pharmacology and opiate receptor activity [Lewis et al., 1983; Jaffe and Martin, 1990].

In contrast, there is no simple pharmacological explanation for buprenorphine's reduction of cocaine self-administration. Cocaine is a stimulant that acts like a dopamine agonist in most systems. There is considerable evidence that the reinforcing and discriminative stimulus effects of cocaine are modulated by dopaminergic neural systems [Fischman, 1987; Ritz et al., 1987; Gawin and Ellinwood, 1988; Kuhar et al., 1988; Johanson and Fischman, 1989]. Since buprenorphine is an opioid mixed agonist–antagonist analgesic with affinity for μ opioid receptors, its effects on cocaine self-administration may reflect, in part, its influence on dopaminergic neural systems as well as endogenous opioid systems. As we discussed in an earlier section (Buprenorphine's Effects on Cocaine Self-Administration), neuroendocrine evidence of dopamine and endogenous opioid interactions encouraged us to examine buprenorphine's effects on cocaine self-administration in rhesus monkeys [Mendelson et al., 1986; Kuljis and Advis, 1989; Mello et al., 1990b,c].

Although there are several different hypotheses about how buprenorphine reduces cocaine self-administration, the basic findings appear to be replicable and robust. As described earlier, 15 days of treatment with buprenorphine (0.237–0.70 mg/kg/day) selectively reduced cocaine self-administration by rhesus monkeys, and tolerance to buprenorphine's effects on cocaine self-administration did not develop over 30–120 days of treatment [Mello et al., 1989, 1990a, 1992]. In our ongoing outpatient clinical studies, buprenorphine doses of 0.05–0.10 mg/kg/day were associated with significant reductions in both cocaine and opiate self-administration by men with a long history of dual dependence on cocaine and opiates [Gastfriend et al., 1992, 1993] (see Figs. 6 and 8 above). Kosten and coworkers [1989a,b] also reported that buprenorphine was significantly more effective than methadone in reducing cocaine abuse by heroin abusers. The concordance between preclinical data and open clinical trials of buprenorphine effects on cocaine self-administration illustrates the potential value of the primate drug self-administration model for predicting the clinical effectiveness of pharmacotherapies for drug abuse treatment [Mello, 1992]. However, evaluation of the clinical utility of buprenorphine will require more extensive controlled double-blind trials with cocaine- and heroin-dependent polydrug abusers. In the remainder of this section, we will describe data relevant to several hypotheses about buprenorphine's actions that have been derived from clinical and basic studies.

Behavioral Studies of Buprenorphine and Cocaine Interactions

Does Buprenorphine Increase Cocaine's Aversive Effects? Buprenorphine's selective reduction of cocaine self-administration by rhesus monkeys led us to conclude that buprenorphine attenuated cocaine's reinforcing efficacy. This interpretation was consistent with reports by opioid-dependent men that cocaine use was followed by dysphoria during buprenorphine treatment, whereas during methadone treatment cocaine use was followed by pleasant feelings and somnolence [Kosten et al., 1989b]. These dysphoric effects of cocaine might be related to buprenorphine's antagonism of heroin's effects in speedball users because some patients reported that the dysphoric effects of cocaine were diminished when it was used simultaneously with heroin [Kosten et al., 1989b].

In an effort to clarify the behavioral interactions between cocaine and buprenorphine, we studied the subjective effects of cocaine and morphine before and during buprenorphine maintenance on a clinical research ward [Teoh et al., 1994]. Twenty-six men who met DSM-III-R criteria for concurrent opioid and cocaine dependence were given an acute intravenous dose of cocaine (30 mg), morphine (10 mg), and saline before and during daily treatment with 4 or 8 mg of sublingual buprenorphine [Teoh et al., 1994]. These men had average histories of 10 years of cocaine abuse and over 11 years of opiate abuse. The latencies to first detection and to certainty of a drug effect after the intravenous injection were measured and the subjects were asked to rate drug intensity and quality at 5- to 15-minute intervals over 2 hr. Thirty minutes after drug injection, the subjects also were asked to provide a global impression of drug quality and intensity. All subjects discriminated saline from cocaine and morphine before buprenorphine treatment and correctly identified each drug condition.

Buprenorphine maintenance (4 and 8 mg) completely eliminated detection of morphine in 18 of the 26 subjects [Teoh et al., 1994]. These data confirm previous findings that buprenorphine blocks the acute effects of an opioid challenge [Bickel et al., 1988; Jasinski et al., 1978]. However, six subjects correctly identified the morphine challenge dose as morphine and the remaining two subjects identified morphine as cocaine. Only one of the subjects who correctly identified morphine during buprenorphine maintenance reported that drug quality and intensity were unchanged. The others felt that morphine quality and intensity were slightly decreased.

In contrast, all 26 subjects identified cocaine correctly before and during buprenorphine maintenance. The time required to detect a cocaine effect also was equivalent before and during buprenorphine maintenance. The latency from detection to certainty of a drug effect was consistently shorter during buprenorphine maintenance. Cocaine quality and intensity were rated as unchanged or slightly deceased by 15 subjects (nine subjects on 4 mg/day of buprenorphine and six subjects on 8 mg/day of buprenorphine). Six of the 4-mg buprenorphine group and five of the 8-mg buprenorphine group reported a slight enhancement of cocaine quality and intensity. However, none of these differences were statistically significant. Buprenorphine also did not significantly alter the dysphoric effects of cocaine. Dysphoric effects of cocaine were consistently reported before and during buprenorphine maintenance. Dysphoria ratings began immediately following cocaine detection and were maximal between 2 and 40 min following administration of cocaine.

It was surprising that buprenorphine had so little effect on the accuracy of cocaine identification or the perceived quality and intensity of the drug experience [Teoh et al., 1994]. These data appeared to be at variance with outpatient data that indicated that buprenorphine maintenance reduces cocaine abuse, as shown earlier in Figures 6 and 8. Consequently, we evaluated the extent to which reports of subjective responses to an acute dose of cocaine or morphine during *inpatient* buprenorphine maintenance predicted drug use during an *outpatient* open trial [Teoh et al., 1994]. The percentage of drug-positive urines during the first 4 weeks of buprenorphine maintenance was examined in a subset of subjects who agreed to participate in an outpatient open trial. Nine patients whose cocaine-positive urines averaged 21% of total (range 0%–50%) were compared with seven patients whose cocaine-positive urines averaged 94% of total (range 71%–100%). The actual dose of buprenorphine (corrected for body weight) was 0.07 ± 0.01 mg/kg for the relative abstainers and 0.10 ± 0.01 mg/kg for the persistent cocaine abusers.

There were no significant differences between these two groups in latency to detection of cocaine and certainty during inpatient buprenorphine maintenance. These groups also did not differ significantly in terms of reported drug intensity and drug quality (euphoria and dysphoria). A similar dissociation between the inpatient and outpatient response to buprenorphine was observed in men who significantly reduced opiate use (17% opiate-positive urines) or continued opiate use (69% opiate-positive urines) during the first 4 weeks of treatment. Thirteen of the sixteen subjects were unable to detect 10 mg of morphine during inpatient buprenorphine maintenance. The three men who detected morphine rarely used opiates during the first 4 weeks of outpatient treatment; their opiate-positive urines were 0%, 14%, and 43%, respectively. In contrast, seven men who were unable to detect morphine during inpatient buprenorphine maintenance continued to use opiates during outpatient maintenance and their opiate-positive urines averaged 50%–88% of the total urines collected. These data suggest that response to an acute dose of cocaine or morphine during inpatient buprenorphine maintenance does not reliably predict drug use behavior in an outpatient treatment trial [Teoh et al., 1994].

The apparent discrepancy between verbal behavior and actual drug use behavior illustrates the limitations of self-report data in predicting outpatient drug self-administration. Fischman and coworkers [1990] noted a similar dissociation between cocaine self-administration behavior and reports of cocaine's subjective effects in their controlled studies of the effects of desipramine maintenance. Desipramine (at a blood level of 125 ng/ml over 3–4 weeks) had no effect on cocaine self-administration (8, 16, or 32 mg) response rates or latency to the first response in comparison with the pre-desipramine baseline cocaine self-administration measures. However, subjects' reports of "I want cocaine" as well as Profile of Mood States (POMS) measures of arousal and vigor were significantly reduced during desipramine maintenance. Fischman et al. [1990] and Fischman and coworkers [1990] concluded that desipramine did not affect the reinforcing properties of cocaine but may have interfered with its other stimulus properties, since there was an increase in POMS confusion, anger, and anxiety ratings.

Does Buprenorphine Increase Cocaine's Reinforcing Properties? An alternative to the dysphoria hypothesis is the notion that buprenorphine may reduce cocaine self-administration because it enhances cocaine's reinforcing properties and therefore less cocaine is required to produce a salient positive effect. A slight but nonsignificant enhancement of cocaine's effects was reported by 11 of our 26 subjects during inpatient buprenorphine maintenance in comparison with pre-buprenorphine conditions [Teoh et al., 1994]. The subjective effects of intranasal cocaine (2 mg/kg) were initially enhanced in five patients after 3 days of buprenorphine treatment (2 mg/day), but this effect diminished after 5 days of treatment [Rosen et al., 1992]. These data were interpreted as suggesting a dissociation between the acute and chronic effects of buprenorphine [Rosen et al., 1992].

Preclinical evidence in support of the reinforcement enhancement hypothesis comes from recent studies in squirrel monkeys and in rats [Brown et al., 1991; Kamien and Spealman, 1991; Spealman and Bergman, 1992]. Pretreatment with relatively low *acute* doses of buprenorphine (0.001–0.01 mg/kg) shifted the cocaine discrimination dose-response curve for three squirrel monkeys threefold or more to the left [Kamien and Spealman, 1991]. Subsequent studies in squirrel monkeys confirmed that buprenorphine (3 and 5.6 µg/kg), as well as a series of µ agonists, shifted the cocaine discrimination dose-response curve to the left [Spealman and Bergman, 1992]. Buprenorphine also potentiated the rate-increasing effects of cumulative doses of cocaine (0.03–0.30 mg/kg) on a fixed-interval (FI 3) schedule of shock termination [Kamien and Spealman, 1991].

We recently examined the effects of daily saline or buprenorphine (0.1, 0.3, and 1.0 mg/kg/day) treatment on cocaine's reinforcing properties over a wide dose range (0.001–0.3 mg/kg/inj) [Lukas et al., 1992; Drieze et al., 1993]. Cocaine and saline self-administration were maintained on an FR 4 (VR 16:S) schedule of reinforcement. During treatment with 0.1 mg/kg of buprenorphine, the cocaine dose-response curve remained parallel to the placebo curve and was shifted one-half log unit to the right. The effects of the higher dose of buprenorphine (0.3 mg/kg/day) were bimodal: Intake of lower cocaine doses increased while intake of higher unit doses decreased, resulting in a marked change in the slope. These data are summarized in Figure 9 and suggest that buprenorphine decreases the reinforcing potency of intermediate and high doses of cocaine but may increase the reinforcing efficacy of low cocaine doses. Acute administration of buprenorphine (1.0–3.2 mg/kg) also produced a downward shift in the cocaine dose-response curve in rhesus monkeys [Winger et al., 1992]. Whether these conflicting findings in squirrel and rhesus monkeys are most parsimoniously attributed to a species difference or to procedural differences (drug discrimination vs. drug self-discrimination) or to the effect of *acute* versus *chronic* buprenorphine treatment is unclear. Species differences in the distribution of opioid receptor types might influence buprenorphine's effects in rhesus versus squirrel monkey and rats [Mansour et al., 1986].

In rats, the combination of an *acute* dose of cocaine (1.5 mg/kg) and a low dose of buprenorphine (0.01 mg/kg) elicited conditioned place preference (CPP), whereas these doses of each drug alone did not elicit CPP [Brown et al., 1991]. Brown and coworkers [1991] concluded that cocaine and buprenorphine interacted syner-

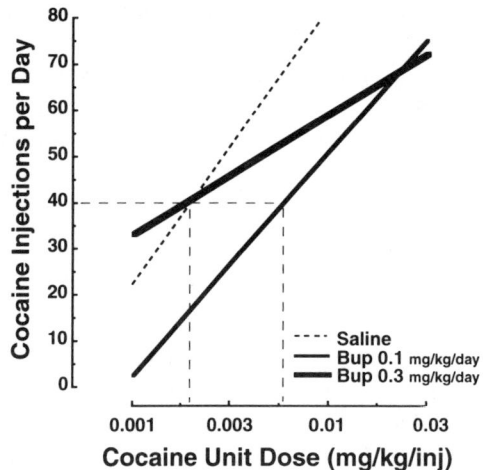

Fig. 9. Effects of daily treatment with saline and buprenorphine (0.10 and 0.30 mg/kg/day) on cocaine's reinforcing potency over a dose range of 0.001–0.30 mg/kg/inj. Cocaine doses and saline were studied in an irregular order for at least 5 days or until cocaine or saline self-administration was stable for 4 days (±20%). The results of a regression analysis of the resulting ascending limb of the cocaine dose-response curves were calculated for five monkeys by means of the formula described by Tallarida and Murray [1981]. Correlation coefficients for cocaine dose-response curves during daily treatment with saline and 0.1 and 0.3 mg/kg of buprenorphine were 0.87, 0.80, and 0.78, respectively. The slope of the regression line for the higher dose of buprenorphine was significantly ($p < 0.01$) different from the regression line for saline or for the lower dose of buprenorphine. Adapted from Lukas et al. [1992] and Drieze et al. [1993].

---

gistically to elicit CPP, and that therefore buprenorphine may enhance rather than attenuate cocaine's reinforcing properties. However, subsequent CPP studies in rats by Kosten et al. [1991] have not replicated the observations of Brown et al. [1991]. The effects of *chronic* buprenorphine treatment (0.5 mg/kg, s.c. twice a day) on CPP elicited by cocaine (15 mg/kg, i.p.) were compared with vehicle control treatment. Cocaine CPP training began after 1 week of chronic buprenorphine or vehicle control treatment, and treatment continued during 5 days of training. Buprenorphine treatment significantly reduced cocaine's ability to elicit CPP in 8 rats, compared with 12 vehicle-control-treated animals. Moreover, buprenorphine alone did not elicit CPP [Kosten et al., 1991].

Several procedural differences may account for these conflicting findings reported by Brown and coworkers [1991] and Kosten and coworkers [1991]. The buprenorphine dose (0.5 mg/kg) studied by Kosten et al. [1991] was within the range that effectively reduced cocaine self-administration by rhesus monkeys (0.237–0.70 mg/kg/day) [Mello et al., 1989, 1990a, 1992], whereas Brown et al. [1991] used considerably lower doses of buprenorphine (0.01–0.075 mg/kg). The training dose of cocaine (15 mg/kg, i.p.) used by Kosten and coworkers [1991] was higher than the training doses (1.5 and 5 mg/kg) used by Brown and coworkers [1991]. However, the most critical procedural difference is that Kosten et al. did not pair the buprenorphine injection with cocaine CPP training, whereas Brown et al. trained rats on CPP with an acute dose of cocaine (5.0 mg/kg) and an acute dose of buprenorphine (0.075 mg/kg). Kosten and coworkers carried out cocaine CPP training approximately 7 hr after the daily buprenorphine injection. This long interval between buprenorphine treatment and cocaine training minimized any possible confounding influence of the sedating effects of buprenorphine on CPP. The effects of buprenorphine and cocaine combinations on general activity were not discussed by Brown and coworkers [1991].

We have argued previously that it seemed unlikely that a cocaine reinforcement enhancement hypothesis could account for the effects of chronic buprenorphine treatment on cocaine self-administration by rhesus monkeys, because monkeys often took no cocaine injections during treatment [Mello et al., 1990a, 1992; Mello and Mendelson, 1993]. In an effort to evaluate the cocaine reinforcement enhancement hypothesis in the context of buprenorphine's effects on cocaine-maintained responding by rhesus monkeys, we reexamined individual data from our previous studies. Daily patterns of responding for cocaine by individual monkeys during 15 days of treatment with buprenorphine at doses of 0.237, 0.40, and 0.70 mg/kg/day and during 100–120 days of treatment at 0.40 mg/kg/day were examined, and the number of days on which monkeys self-administered 0, 1, 2, 3, 4, and 5 cocaine injections was tabulated. The results of this analysis are shown in Figure 10. It is

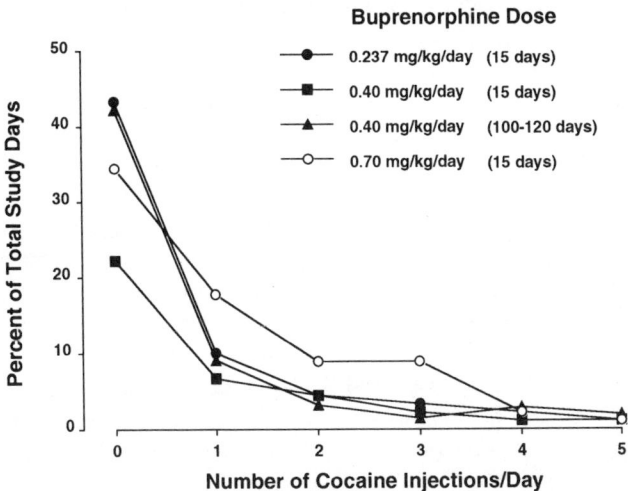

Fig. 10. Effects of daily buprenorphine treatment on cocaine self-administration by rhesus monkeys. Data are shown as the percentage of total days of buprenorphine treatment (ordinate) on which five or fewer cocaine injections were self-administered (abscissa). Each of the data points for 15 days of buprenorphine treatment is the average for six monkeys. Each of the data points for 100–120 days of buprenorphine treatment is the average for five monkeys. This graph is based on a reanalysis of data from Mello et al. [1990a, 1992].

apparent that subjects took zero cocaine injections on 22%–43% of the total number of buprenorphine treatment days. Moreover, monkeys took fewer than six injections per day on 64%, 38%, and 73% of the buprenorphine treatment days at doses of 0.237, 0.40, and 0.70 mg/kg/day. We conclude that it is unlikely that a cocaine reinforcement enhancement hypothesis can account for the absence of cocaine self-administration by rhesus monkeys during *chronic* buprenorphine treatment.

Pharmacological Interactions Between Buprenorphine and Cocaine

As we discussed in the section Buprenorphine's Effects on Cocaine Self-Administration, neuroendocrine evidence of dopamine and endogenous opioid interactions encouraged us to examine buprenorphine's effects on cocaine self-administration in monkey [Kuljis and Advis, 1989; Mello et al., 1990b,c; Mendelson et al., 1986]. Further evidence of interactions between dopaminergic and endogenous opioid systems in brain comes from reports that exposure to cocaine increases opiate receptor density and changes [^{3}H]naloxone binding in brain [Hammer, 1989; Ishizuka et al., 1988]. Moreover, cocaine increases opiate receptor density in brain areas associated with drug reinforcement [Hammer, 1989]. Evidence of dopaminergic and endogenous opioid interactions also comes from several behavioral studies [Blumberg and Ikeda, 1978; Bozarth and Wise, 1981; Shippenberg and Herz, 1987; Kiritsky-Roy et al., 1989]. For example, cocaine and opioids stimulate increased release of dopamine and activate D_1 receptors in brain [Herz and Shippenberg, 1989].

There is emerging evidence that both cocaine and opioids stimulate increased release of dopamine and activate D_1 receptors in brain [Herz and Shippenberg, 1989]. *In vivo* microdialysis studies show that both buprenorphine (0.01 mg/kg, i.p.) and cocaine (5.0 mg/kg, i.p.) increase extracellular dopamine in the nucleus accumbens by 100% and 82%, respectively, in rats [Brown et al., 1991]. When the same doses of cocaine and buprenorphine were given in combination, dopamine levels increased by 163% in a time course similar to that measured after administration of cocaine alone. Buprenorphine's enhancement of cocaine-stimulated increases in extracellular dopamine were interpreted as suggesting that buprenorphine may enhance, rather than decrease, cocaine's reinforcing properties and thereby reduce the number of self-administered cocaine injections [Brown et al., 1991].

Buprenorphine's μ Opioid Agonist-Antagonist Effects. We have suggested elsewhere that buprenorphine's *unique* combination of opioid agonist and antagonist effects is probably critical for its selective reduction of cocaine self-administration [Mello and Mendelson, 1993, Mello et al., 1989, 1990a, 1992]. This conclusion was based on observations that neither opioid agonists nor opioid antagonists alone have been as effective as buprenorphine in reducing cocaine self-administration. However, the results from clinical and primate studies of opioid agonist and antagonist effects on cocaine self-administration are often inconsistent [Mello and Mendelson, 1993; Mendelson and Mello, 1992].

The opioid antagonist naltrexone has a long duration of action (like buprenorphine), and it also significantly reduced heroin self-administration by heroin-dependent men [Meyer and Mirin, 1979; Mello et al., 1981]. Naltrexone treatment (100–150 mg, three times per week) of opiate-dependent polydrug abusers also significantly reduced cocaine-positive urines in comparison with treatment with methadone [Kosten et al., 1989a]. Naltrexone (0.32–3.20 mg/kg/day) also significantly reduced cocaine self-administration over 15 days by an average of 25%–28% in rhesus monkeys [Mello et al., 1990a]. These effects of naltrexone were selective for cocaine, since food self-administration decreased significantly below baseline (24%) only during Days 6–10 of low-dose naltrexone treatment and then returned to, and exceeded, baseline levels during high-dose naltrexone treatment [Mello et al., 1990a]. We interpreted these data as suggesting the importance of the antagonist component of buprenorphine in decreasing cocaine use. Naltrexone and diprenorphine (the antagonist constituent of buprenorphine) have a similar duration of action and both have antagonist effects at the μ receptor [Jaffe and Martin, 1990].

Our conclusion that the μ opioid *antagonist* component of buprenorphine may be important in decreasing cocaine self-administration [Mello et al., 1990a] has been challenged by preliminary data from our ongoing studies of concurrent buprenorphine and naltrexone administration [Mello et al., 1993c,d]. We compared the effects of buprenorphine alone (0.40 mg/kg/day) and in combination with ascending doses of naltrexone (0.05, 0.10, 0.20, and 0.40 mg/kg/day) on cocaine and food self-administration under conditions identical to those of our earlier studies of buprenorphine treatment. Each treatment condition was in effect for 10 days. Buprenorphine alone significantly ($p < 0.01$) reduced cocaine self-administration by an average of 53% in comparison with the saline treatment baseline. When saline was again substituted for buprenorphine, each monkey returned to pre-buprenorphine levels of cocaine self-administration. When buprenorphine and naltrexone were administered simultaneously, naltrexone attenuated buprenorphine's suppressive effects on cocaine self-administration. Ascending doses of naltrexone in combination with buprenorphine reduced cocaine self-administration by an average of 30%, 30%, 23%, and 23% ($p < 0.05 - < 0.01$) [Mello et al., 1993c,d]. When naltrexone was administered 20 min before buprenorphine, there was a significant ($p < 0.05 - < 0.01$) naltrexone dose-dependent decrease in buprenorphine's reduction of cocaine self-administration. Illustrative data from one monkey are shown in Figure 11. Food self-administration remained equivalent to or significantly higher than levels during the saline treatment baseline in all conditions ($p < 0.05$) [Mello et al., 1993c,d].

These data suggest that naltrexone, a μ opioid antagonist, antagonized the μ agonist component of buprenorphine and that μ opioid receptor activity may be an important factor in buprenorphine–cocaine interactions [Mello et al., 1993c]. These findings were surprising because of clinical reports that treatment with the opioid agonist methadone did not reduce cocaine-positive urines in heroin-dependent patients as effectively as buprenorphine or naltrexone [Kosten et al., 1987, 1989a]. But in contrast to these clinical reports, morphine pretreatment suppressed cocaine self-administration in a dose-dependent manner in squirrel monkeys [Stretch, 1977]. These inconsistent findings illustrate the complexity of cocaine–opioid inter-

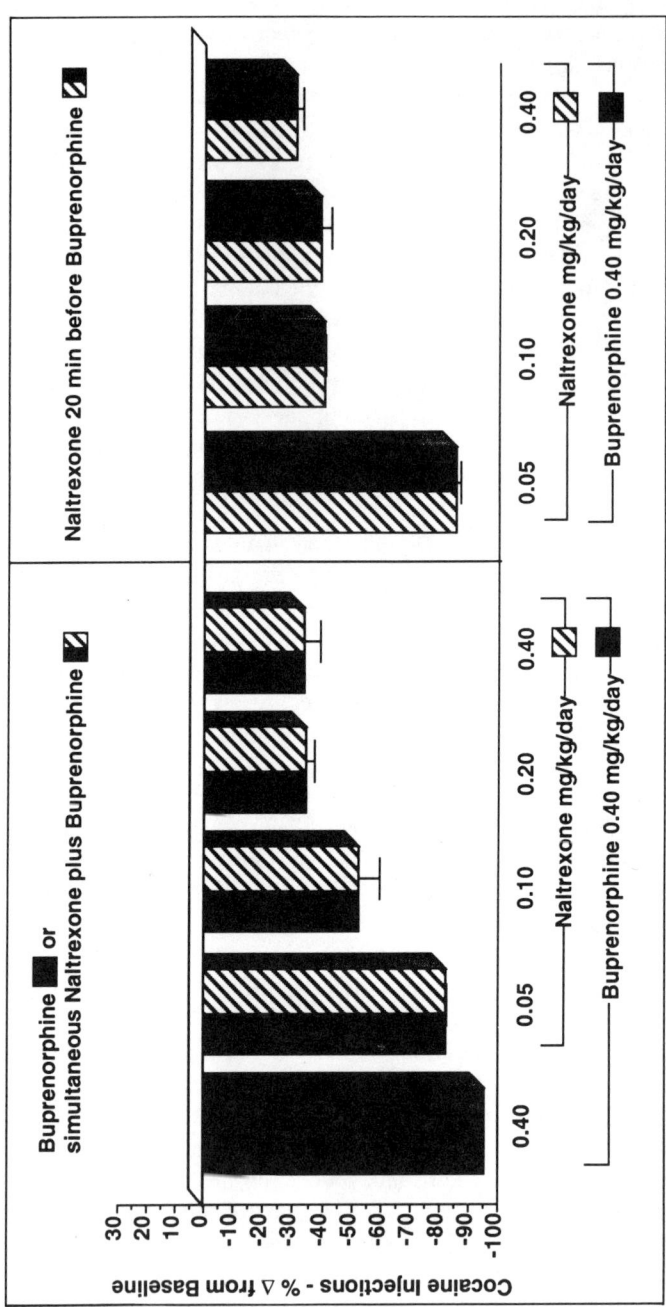

Fig. 11. Comparison of the effects of buprenorphine only with buprenorphine–naltrexone combinations on cocaine self-administration: The average number of cocaine injections (mean ±SEM) self-administered by a single representative monkey are shown as percentage change from the saline treatment baseline before buprenorphine administration. Successive 10-day periods of buprenorphine treatment (black bar) and of buprenorphine and simultaneous administration of ascending doses of naltrexone (0.05–0.40 mg/kg/day) (black and striped bars) are shown in the left panel. Successive 10-day periods of ascending doses of naltrexone (0.05–0.40 mg/kg/day) administered 20 min before buprenorphine (striped and black bars) are shown in the right panel.

actions. However, our findings in rhesus monkey are consistent with a recent report that naltrexone antagonized buprenorphine's reduction of cocaine-induced lethality in mice [Witkin et al., 1991]. The protective effect of buprenorphine (0.3–3.0 mg/kg) against the lethal effect of cocaine-induced convulsions was antagonized by low doses of naltrexone (0.3–1.0 mg/kg). The opioid agonists, methadone and morphine, also reduced cocaine-induced lethality, and the dose-effect functions for those drugs were similar to that of buprenorphine [Witkin et al., 1991].

Kappa Antagonist Effects of Buprenorphine. There is compelling physiological and behavioral evidence that buprenorphine has κ antagonist effects as well as partial μ agonist and μ antagonist activity. These data raise the intriguing possibility that buprenorphine's κ antagonist effects may contribute to its suppressive effects on cocaine self-administration [see Mello and Mendelson, 1993, for review]. Recent data concerning buprenorphine's κ antagonist properties are reviewed by Dykstra and Negus (this volume). To briefly summarize, buprenorphine antagonized the diuretic effects of a κ agonist (bremazocine) [Richards and Sadee, 1985] and precipitated withdrawl in κ agonist-dependent monkeys [Gmerek et al., 1987]. Buprenorphine also antagonized the effects of a selective κ agonist (U50,488) in both a drug discrimination and a shock-titration procedure in squirrel monkeys [Negus and Dykstra, 1988; Negus et al., 1991]. However, buprenorphine appears not to have κ agonist-like properties, since it did not substitute for U50,488 in a drug discrimination paradigm [Negus et al., 1991]. Buprenorphine pretreatment also produced dose-dependent antagonism of the discriminative stimulus properties of U50,488 in rats [Negus et al., 1990].

Inferential support for the notion that buprenorphine's κ antagonist properties may be an important factor in its selective effects on cocaine self-administration comes from our recent studies of other opioid mixed agonist–antagonist analgesics with different opioid receptor affinities [Mello et al., 1993a]. Nalbuphine and butorphanol are also opioid mixed agonist–antagonist analgesics, but each has κ *agonist* activity [Schmidt et al., 1985; Jaffe and Martin, 1990], whereas buprenorphine has κ *antagonist* activity. The profile of butorphanol is complex, since it can antagonize the effects of both μ and κ agonists on shock-maintained behavior [Dykstra, 1990], and it was also more potent than nalbuphine as a κ antagonist in a shock titration behavioral paradigm [Dykstra, 1990]. We found that both nalbuphine (0.1–3.0 mg/kg/day) and butorphanol (0.01–0.3 mg/kg/day) significantly decreased cocaine self-administration by rhesus monkeys [Mello et al., 1993a]. However, in contrast to buprenorphine, the effects of nalbuphine and butorphanol were not selective for cocaine, since food self-administration also decreased significantly in a dose-dependent manner. The concurrent and sustained reductions in food-maintained behavior as well as the short duration of action of nalbuphine and butorphanol compared with buprenorphine led us to conclude that these opioid mixed agonist–antagonist analgesics may have limited utility for cocaine abuse treatment. Presumably, differences in the effects of nalbuphine, butorphanol, and buprenorphine on cocaine and food self-administration reflect differences in their respective opioid receptor affinities [Mello et al., 1993a].

Buprenorphine's Effects on Cocaine-Induced Adrenocorticotropin (ACTH) Secretion.
It is also possible that buprenorphine's effects on cocaine self-administration are related to its suppressive effects on cocaine-induced stimulation of ACTH secretion. We found that intravenous administration of cocaine (30 mg) significantly increased plasma ACTH to 105% above baseline within 5 min in men who were concurrently dependent on cocaine and heroin [Mendelson et al., 1992a]. The time course of ACTH stimulation coincided with peak levels of cocaine in plasma (200 ng/ml), and all subjects reported a significant degree of euphoria within 2–3 min after injection of cocaine. Subsequently, these subjects were maintained on 4 mg/day of buprenorphine for at least 10 days; then the effects of cocaine on ACTH levels and subjective responses were reexamined. As shown in Figure 12, buprenorphine maintenance was associated with a significant suppression of plasma ACTH levels after intravenous administration of cocaine. Cocaine-induced euphoria ratings were also significantly lower during buprenorphine maintenance. In contrast, buprenorphine had no effect on plasma cocaine levels or cardiovascular responses to cocaine administration in comparison with levels prior to administration of buprenorphine. Consequently, the decrease in cocaine-induced changes in plasma ACTH levels and mood states cannot be accounted for by these physiological or pharmacological variables. Since placebo-cocaine administration did not change plasma ACTH levels or mood states, it is unlikely that expectancy could account for these findings.

A possible relationship between cocaine's effects on mood and ACTH levels is an intriguing prospect. It is well established that cocaine stimulates ACTH in experimental animals [Moldow and Fischman, 1987; Kiritsky-Roy et al., 1988] and that

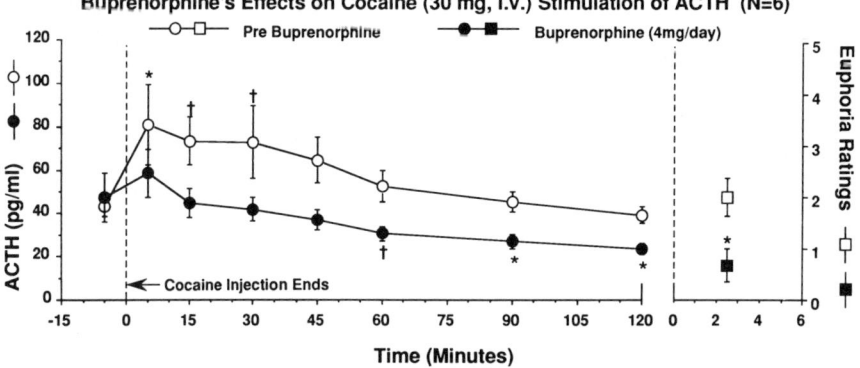

Fig. 12. Buprenorphine attenuation of cocaine stimulation of ACTH and euphoria. Time courses of changes in plasma ACTH levels following intravenous cocaine administration (30 mg) before and during buprenorphine treatment (4 mg/day, s.l.) are shown at the left (circles). Ratings of euphoria at the time that the subjects were certain of a cocaine drug effect are shown at the right (squares). Each of the data points before (open symbols) and during (filled symbols) buprenorphine maintenance is the average ($\pm$SE) for six men (*$p < 0.01$, †$p < 0.05$). Adapted from Mendelson et al. [1992a].

pretreatment with CRF antiserum completely blocked cocaine-induced stimulation of ACTH in rats [Rivier and Vale, 1987]. The latter investigation provided compelling evidence that cocaine stimulates ACTH through CRF activation. Although the exact mechanisms that underlie cocaine-induced stimulation of CRF and ACTH are unclear, recent studies suggest that cocaine stimulation of the dopaminergic neural systems that regulate ACTH secretion may be one component of this process [Borowsky and Kuhn, 1991]. Since the reinforcing properties of cocaine appear to be mediated by its effects on dopaminergic neural systems [Ritz et al., 1987; Kuhar et al., 1988; Johanson and Fischman, 1989], cocaine's stimulation of CRF secretion in brain may contribute to its reinforcing effects. CRF is present in many brain regions and not only activates ACTH secretion, but also has multiple neuromodulatory actions in the central nervous system [Vale et al., 1983; Rivier and Vale, 1987]. Alcoholic intoxication was also associated with increases in ACTH that paralleled reports of euphoria and increased EEG alpha power during the ascending limb of the blood alcohol curve [Lukas and Mendelson, 1988]. However, much remains to be learned about cocaine and other drug-induced stimulation of the hypothalamic–pituitary–gonadal axis and the implications of associated hormonal changes for the reinforcing properties of drugs [Mendelson et al., 1992b].

Opiate agonists decrease plasma levels of ACTH in humans [Stubbs et al., 1978], whereas opiate antagonists increase plasma ACTH levels [Volavka et al., 1979; Mendelson et al., 1986]. The opioid mixed agonist-antagonist buprenorphine had no effect on ACTH during placebo-cocaine administration, but reduced stimulation of ACTH by cocaine [Mendelson et al., 1992a]. These data suggest that buprenorphine–cocaine interactions are more characteristic of an opioid agonist than an antagonist with respect to ACTH. These data are concordant with our recent observations in rhesus monkeys that the μ antagonist naltrexone reduces buprenorphine's suppressive effects on cocaine self-administration [Mello et al., 1993c]. This suggests that μ opioid receptor activity may be important for buprenorphine's effects on cocaine self-administration.

SUMMARY AND CONCLUSIONS

Buprenorphine appears to be safe and effective for the treatment of heroin abuse as well as dual dependence on cocaine and opiates [Fudala et al., 1990; Gastfriend et al., 1992, 1993; Johnson et al., 1991; Mello and Mendelson, 1980; Mello et al., 1982]. Moreover, buprenorphine reduces needle use and needle sharing by polydrug abusers and accordingly reduces the risk for HIV infection [Gastfriend et al., 1992, 1993]. Buprenorphine has minimal side effects during daily maintenance and there appear to be no adverse medical consequences of buprenorphine alone or in combination with opiates or cocaine [Jasinski et al., 1978, 1989; Lange et al., 1990; Mello et al., 1982; Teoh et al., 1992a]. Buprenorphine did not accentuate the cardiovascular or respiratory effects of cocaine or morphine in controlled inpatient studies [Teoh et al., 1993]. These data suggest that buprenorphine maintenance should not be associated with cocaine- or heroin-induced toxicity in patients who continue to abuse these drugs. Studies of the long-term safety and effectiveness of buprenorphine treatment for opiate abuse and for dual dependence on cocaine and opiates are currently ongoing in several clinical research centers.

Traditionally, pharmacotherapies have been designed to reduce the abuse of drugs from a single class such as opioids or stimulants. Today, the concurrent abuse of drugs with different pharmacological effects (stimulants, opiates, benzodiazepines, alcohol, and marijuana) is increasingly common [Mendelson and Mello, 1991] and multiple drug abuse has been a prevailing trend for over a decade [Kosten et al., 1987; Kreek, 1978, 1987, 1991]. The diversity of combinations of abused drugs suggests that the primary reinforcer in polydrug abuse may be some perceptible change in state rather than any particular direction of change, that is, stimulation or sedation [Mello, 1977, 1983]. Most medical problems associated with chronic drug abuse are exacerbated by polydrug abuse. It is now recognized that intravenous abuse of cocaine and heroin increases the risk for HIV infection both through needle sharing and direct suppressive effects on immune function [Donahoe and Falek, 1988; Klein et al., 1988; Chaisson et al., 1989; Mendelson and Mello, 1991]. Clearly, the design of broad-spectrum pharmacotherapies will be increasingly important, since the abuse of new drug formulations and combinations is almost inevitable. Buprenorphine may become a prototype of a new approach to drug abuse treatment, since it reduces both opiate and cocaine abuse in clinical and preclinical studies.

The ways in which buprenorphine reduces cocaine self-administration by polydrug abusers and rhesus monkeys remains an enigma. Buprenorphine did not change the stimulus properties of cocaine in human polydrug abusers. All men identified cocaine but neither euphoria nor dysphoria were significantly altered during buprenorphine maintenance [Teoh et al., 1994]. Moreover, reports of cocaine-induced euphoria, dysphoria, and drug intensity were equivalent in men who subsequently seldom used cocaine (21% cocaine-positive urines) and those who frequently used cocaine (94% cocaine-positive urines) during the first 4 weeks of outpatient buprenorphine treatment [Teoh et al., 1994]. A similar dissociation between subjective responses to morphine during inpatient buprenorphine maintenance and outpatient opiate use was observed. Buprenorphine prevented morphine detection during inpatient trials, but during outpatient buprenorphine maintenance seven patients continued opiate abuse (69% opiate-positive urines) and nine patients used opiates only occasionally (17% opiate-positive urines) [Teoh et al., 1994]. Since subjective responses to cocaine and to morphine during inpatient buprenorphine maintenance did not predict outpatient drug abuse, any attempt to categorize buprenorphine's effects as increasing euphoria or dysphoria is probably too simplistic.

There is now considerable evidence that buprenorphine decreases self-administration of cocaine by man and by rhesus monkey. Interpretation of discrepant findings from well-controlled animal studies of buprenorphine's effects on cocaine discrimination and cocaine-induced conditioned place preference is limited by procedural and species differences. Moreover, data from drug self-administration paradigms may not invariably parallel data from drug discrimination paradigms. The equivalence and relative sensitivity of drug self-administration and drug discrimination methods for evaluating pharmacotherapies for drug abuse treatment remains an empirical question [Mello, 1992]. However, the potential value of preclinical models for evaluation of new pharmacotherapies for drug abuse has been persuasively demonstrated in the development of buprenorphine. Clinical trials of buprenorphine for the treatment of cocaine and heroin abuse evolved from the finding that buprenorphine significantly reduced cocaine self-administration by rhesus monkeys [Mello et al., 1989, 1990a].

Whatever the eventual resolution of questions about its mechanism of action, the apparent effectiveness of buprenorphine in reducing cocaine self-administration has far-reaching heuristic implications for the development of new pharmacotherapies. It is likely that the interactions between dopaminergic and endogenous opioid systems influence the reinforcing properties of cocaine as well as other drugs. The extent to which both μ and κ opioid receptors are important factors in buprenorphine's effects on cocaine self-administration remains to be determined. Recent evidence that cocaine-induced changes in neuroendocrine hormones may be temporally coincident with rapid mood changes adds another dimension to our evolving concepts of the biological aspects of drug reinforcement [Mendelson et al., 1992, 1993]. It is obvious that the behavioral and biological determinants of drug abuse are extraordinarily complex and defy any simple explanation. But identification and analysis of pharmacotherapies that significantly change drug-taking behavior eventually may lead to a better understanding of the neural and hormonal systems that modulate the reinforcing effects of drugs.

ACKNOWLEDGMENTS

The research summarized in this review was supported in part by grants DA 00101, DA 00064, DA 04059, DA 02519, and DA 06116 from the National Institute on Drug Abuse. We are indebted to our colleagues who made many important contributions to this research program. We especially thank Scott E. Lukas, PhD, John Drieze, MS, Mark P. Bree, and Jonathan B. Kamien, PhD, from the Behavioral Science Laboratory and David Gastfriend, MD, Siew Koon Teoh, MD, Pradit Sintavanarong, MD, MPH, and John Kuehnle, MD, from the Clinical Research and Treatment Program of the Alcohol and Drug Abuse Research Center, McLean Hospital–Harvard Medical School. We thank Dr. Edward J. Cone, Laboratory of Chemistry and Metabolism, Addiction Research Center, NIDA, for measurement of buprenorphine plasma levels and for his many helpful comments and suggestions. We are grateful to Wallis Sholar and Lynne G. Wighton for assistance in data analysis and graphic displays, and to Loretta Carvelli for preparation of the manuscript.

REFERENCES

Ascher EK, Stauffer JC, Gawasch WH (1988): Coronary artery spasm, cardiac arrest, transient electrocardiographic Q waves, and stunned myocardium in cocaine-associated acute myocardial infarction. Am J Cardiol 61:939–941.

Balster RL, Mansbach RS, Gold L, Harris LS (1992): Preclinical methods for the development of pharmacotherapies for cocaine abuse. NIDA Res Monogr 119:160–164.

Banks CD (1979): Overdose of buprenorphine: Case report. NZ Med J 89:255–256.

Bickel WK, Stitzer ML, Bigelow GE, Liebson IR, Jasinski DR, Johnson RE (1988): Buprenorphine: Dose-related blockade of opioid challenge effects in opioid dependent humans. J Pharmacol Exp Ther 247:47–53.

Blaine JD, Renault P, Levine GL, Whysner JA (1978): "Clinical Use of LAAM." Ann NY Acad Sci 311:214–231.

Blaine JD, Renault P, Thomas DB, Whysner JA (1981): Clinical status of methadyl acetate (LAAM). Ann NY Acad Sci 362:101–115.

Blumberg H, Ikeda C (1978): Naltrexone, morphine and cocaine interactions in mice and rats. J Pharmacol Exp Ther 206:303–310.

Borowsky B, Kuhn CM (1991): Monoamine mediation of cocaine-induced hypothalamo-pituitary-adrenal activation. J Pharmacol Exp Ther 256:204–210.

Bozarth MA, Wise RA (1981): Heroin reward is dependent on a diapaminergic substrate. Life Sci 29:1881.

Brady JV, Lukas SE (1984): "Testing Drugs for Physical Dependence Potential and Abuse Liability." NIDA Res Monogr 52.

Brown EE, Finlay JM, Wong JTF, Damsma G, Fibiger HC (1991): Behavioral and neurochemical interactions between cocaine and buprenorphine: Implications for the pharmacotherapy of cocaine abuse. J Pharmacol Exp Ther 256:119–126.

Chaisson RE, Bacchetti P, Osmond D, Brodie B, Sande MA, Moss AR (1989): Cocaine use and HIV infection in intravenous drug users in San Francisco. J Am Med Assoc 261:561–565.

Cone E, Holicky B, Pickworth W, Johnson RE (1992): Pharmacologic and behavioral effects of high doses of intravenous buprenorphine. NIDA Res Monogr 110:244.

Cowan A (1974): Evaluation in nonhuman primates: Evaluation of the physical dependence capacities of oripavine–thebaine partial agonists in patas monkeys. In Braude MC, Harris LS, May EL, Smith JP, Villarreal JE (eds): "Narcotic Antagonists." New York: Raven, pp 427–438.

Cregler LL, Mark H (1986): Medical complications of cocaine abuse. N Engl J Med 315:1495–1500.

Crowley TJ, Hydinger M, Stynes AJ, Feiger A (1975): Monkey motor stimulation and altered social behavior during chronic methadone administration. Psychopharmacologia 43:135–144.

Dole VP, Nyswander M (1965): A medical treatment for diacetylmorphine (heroin) addiction: A clinical trial with methadone hydrochloride. J Am Med Assoc 193:646–650.

Donohoe RM, Falek A (1988): Neuroimmunomodulation by opiates and other drugs of abuse: Relationship to HIV infection and AIDS. In Bridge TP, Mirsky FA, Goodwin FK (eds): "Psychological, Neuropsychiatric and Substance Abuse Aspects of AIDS." New York: Raven, pp 145–157.

Drieze JM, Mello NK, Lukas SE, Mendelson JH (1993): Buprenorphine attenuates cocaine's reinforcing properties in rhesus monkeys. NIDA Res Monogr 132:94.

Dykstra LA (1990): Butorphanol, levallorphan, nalbuphine and nalorphine as antagonists in the squirrel monkey. J Pharmacol Exp Ther 254:245–252.

Emrich HM, Vogt P, Herz A (1982): Possible antidepressive effects of opioids: Action of buprenorphine. Ann NY Acad Sci 398:108–112.

Fischman MW (1987): Cocaine and the amphetamines. In Meltzer HY (ed): "Psychopharmacology, The Third Generation of Progress." New York: Raven, pp 1543–1553.

Fischman MW, Schuster CR, Javaid J, Hatano Y, Davis J (1985): Acute tolerance development to the cardiovascular and subjective effects of cocaine. J Pharmacol Exp Ther 235:677–682.

Fischman MW, Foltin RW, Nestadt G, Pearlson GD (1990); Effects of desipramine maintenance on cocaine self-administration by humans. J Pharmacol Exp Ther 253:760–770.

Foltin RW, Fischman MW (1991): Smoked and intravenous cocaine in humans: Acute tolerance, cardiovascular and subjective effects. J Pharmacol Exp Ther 257:247–261.

Fudala PJ, Johnson RE, Bunker E (1989): Abrupt withdrawal of buprenorphine following chronic administration. Clin Pharmacol Ther 45:186.

Fudala PJ, Jaffe JH, Dax EM, Johnson RE (1990): Use of buprenorphine in the treatment of opioid addiction. II. Physiologic and behavioral effects of daily and alternate-day administration and abrupt withdrawal. Clin. Pharmacol. Ther. 47:525–534.

Fudala PJ, Johnson RE, Jaffe JH (1991): Outpatient comparison of buprenorphine and methadone maintenance. II. Effects on cocaine usage, retention time in study and missed clinic visits. NIDA Res Monogr 105:587–588.

Gastfriend DR, Mendelson JH, Mello NK, Teoh SK (1992): Preliminary results of an open trial of buprenorphine in the outpatient treatment of combined heroin and cocaine dependence. NIDA Res Monogr 119:461.

Gastfriend D, Mendelson JH, Mello NK, Teoh SK, Reif S (1993): Buprenorphine pharmacotherapy for concurrent heroin and cocaine dependence. Am J Addict: 269–278.

Gawin F, Ellinwood EH (1988): Cocaine and other stimulants: Actions, abuse and treatment. New Engl J Med 318:1173–1182.

Gmerek DE, Dykstra LA, Woods JH (1987): Kappa opioids in rhesus monkeys. III. Dependence associated with chronic administration. J Pharmacol Exp Ther 242:428–436.

Griffiths R, Brady J, Bradford L (1979): Predicting the abuse liability of drugs with animal drug self-administration procedure: Psychomotor stimulants and hallucinogens. In Thompson T, Dews PB (eds): "Advances in Behavioral Pharmacology." New York: Academic, pp 163–208.

Hambrook JM, Rance MJ (1976): The interaction of buprenorphine with the opiate receptor: Lipophilicity as a determining factor in drug-receptor kinetics. In Kosterlitz HW (ed): "Opiates and Endogenous Opioid Peptides." Amsterdam: North-Holland, pp 295–301.

Hammer RP (1989): Cocaine alters opiate receptor binding in critical brain reward regions. Synapse 3:55–60.

Harrigan SE, Downs DA (1981): Pharmacological evaluation of narcotic antagonist delivery systems in rhesus monkeys. NIDA Res Monogr 28:77–92.

Herz A, Shippenberg TS (1989): Neurochemical aspects of addiction: Opioids and other drugs of abuse. In Goldstein A (ed): "Molecular and Cellular Aspects of the Drug Addictions." New York: Springer-Verlag, pp 111–141.

Houde RW (1979): Analgesic effectiveness of the narcotic agonist–antagonists. Br J Clin Pharmacol 7(Suppl 3):297–308.

Ishizuka Y, Rockhold RW, Hoskins B, Ho IK (1988): Cocaine-induced changes in ^{3}H-naloxone binding in brain membranes isolated from spontaneously hypertensive and Wistar-Kyoto rats. Life Sci 43:2275–2282.

Isner JM, Estes M, Thompson PD, Costanzo-Nordin MR, Subramanian R, Miller G, Katsas G, Sweeney K, Sturner WQ (1986): Acute cardiac events temporarily related to cocaine abuse. N Engl J Med 315:1438–1443.

Jaffe JH (1990): Drug addiction and drug abuse. In Gilman AG, Rall TW, Nies AS, Taylor P (eds): "The Pharmacological Basis of Therapeutics," 8th ed. New York: Pergamon, pp 522–573.

Jaffe JH, Martin WR (1990): Opioid analgesics and antagonists. In Gilman AG, Rall TW, Nies AS, Taylor P (eds): "The Pharmacological Basis of Therapeutics," 8th ed. New York: Pergamon, pp 485–521.

Jasinski DR, Pevnick JS, Griffith JD (1978): Human pharmacology and abuse potential of the analgesic buprenorphine. Arch Gen Psychiatry 35:601–616.

Jasinski DR, Fudala PJ, Johnson RE (1989): Sublingual versus subcutaneous buprenorphine in opiate abusers. Clin Pharmacol Ther 45:513–519.

Johanson CE, Fischman MW (1989): The pharmacology of cocaine related to its abuse. Pharmacol Rev 41:3–52.

Johanson CE, Balster RL, Bonese K (1976): Self-administration of psychomotor stimulant drugs: The effects of unlimited access. Pharmacol Biochem Behav 4:45–51.

Johnson RE, Fudala PJ, Jaffe JH (1991): Outpatient comparison of buprenorphine and methadone maintenance. 1. Effects on opiate use and self-reported adverse effects and withdrawal symptomatology. NIDA Res Monogr 105:585–586.

Johnson RE, Jaffe JH, Fudala PJ (1992): A controlled trial of buprenorphine treatment for opioid dependence. J Am Med Assoc 267:2750–2755.

Jones BE, Prada JA (1975): Drug seeking behavior during methadone maintenance. Psychopharmacologia 41:7–10.

Jones BE, Prada JA (1977): Effects of methadone and morphine maintenance on drug-seeking behavior in the dog. Psychopharmacology 54:109–112.

Julius D, Renault P (eds) (1976): "Narcotic Antagonists: Naltrexone Progress Report." NIDA Res Monogr 9.

Kamien JB, Spealman RD (1991): Modulation of the discriminative-stimulus effects of cocaine by buprenorphine. Behav Pharmacol 2:517–520.

Kay DC, Eisenstein RB, Jasinski DR (1969): Morphine effects on human REM state, waking state and NREM sleep. Psychopharmacologia 14:404–416.

Kiritsky-Roy JA, Standish SM, Whitmore RD, Smith M, Halter JB, Terry LC (1988): Cocaine-induced cardiovascular (CV) and pituitary–adrenal stimulation: Role of dopamine (DA) and corticotropin releasing hormone (CRH). Soc Neurosci Abstr 14:445.

Kiritsky-Roy JA, Standish SM, Terry LC (1989): Dopamine D-1 and D-2 receptor antagonists potentiate analgesic and motor effects of morphine. Pharmacol Biochem Behav 32:717–721.

Klein TW, Newton CA, Freidman H (1988): Suppression of human and mouse lymphocyte proliferation by cocaine. In Bridge TP, Mirsky FA, Goodwin FK (eds): "Psychological, Neuropsychiatric and Substance Abuse Aspects of AIDS." New York: Raven, pp 139–143.

Kosten TA, Marby DW, Nestler EJ (1991): Cocaine conditioned place preference is attenuated by chronic buprenorphine treatment. Life Sci 49:PL-201–PL-206.

Kosten TR, Rounsaville BJ, Kleber HD (1987): A 2.5 year follow-up of cocaine use among treated opioid addicts. Arch Gen Psychiatry 44:281–284.

Kosten TR, Kleber HD, Morgan C (1989a): Role of opioid antagonists in treating intravenous cocaine abuse. Life Sci 44:887–892.

Kosten TR, Kleber HD, Morgan C (1989b): Treatment of cocaine abuse with buprenorphine. Biol Psychiatry 26:637–639.

Kosten TR, Morgan C, Kosten TA (1990): Depressive symptoms during buprenorphine treatment of opioid abusers. J Subst Abuse Treatment 7:51–54.

Kreek MJ (1978): Medical complications in methadone patients. Ann NY Acad Sci 311:110–134.

Kreek MJ (1987): Multiple drug abuse patterns and medical consequences. In Meltzer H (ed): "Psychopharmacology, The Third Generation of Progress." New York: Raven, pp 1597–1604.

Kreek MJ (1991): Multiple drug abuse patterns: Recent trends and associated medical consequences. In Mello NK (ed): "Advances in Substance Abuse, Behavioral and Biological Research" (Vol. 4). London: Jessica Kingsley, pp 91–112.

Kuhar MJ, Ritz MC, Sharkey J (1988): Cocaine receptors on dopamine transporters mediate cocaine reinforced behavior. NIDA Res Monogr 88:14–22.

Kuljis R, Advis J (1989): Immunocytochemical and physiological evidence of a synapse between dopamine and luteinizing hormone releasing hormone containing neurons in the ewe median eminence. Endocrinology 124:1579–1581.

Kumor KM, Sherer MA, Thompson L, Cone E, Mahaffey J, Jaffe JH (1988): Lack of cardiovascular tolerance during intravenous cocaine infusions in human volunteers. Life Sci 42:2063–2071.

Lange WR, Fudala PJ, Dax EM, Johnson RE (1990): Safety and side-effects of buprenorphine in the clinical management of heroin addiction. Drug Alcohol Depend 26:19–28.

Lewis JW (1974): Ring C-bridged derivatives of thebaine and oripavine. In Braude MC, Harris LS, May EL, Smith JP, Villarreal JE (eds): "Narcotic Antagonists." New York: Raven, pp 123–136.

Lewis JW, Rance MJ, Sanger DJ (1983): The pharmacology and abuse potential of buprenorphine: A new antagonist analgesic. In Mello NK (ed): "Advances in Substance Abuse: Behavioral and Biological Research" (Vol. 3). Greenwich, CT: JAI Press, pp 103–154.

Lukas SE, Mendelson JH (1988): Electroencephalographic activity and plasma ACTH during ethanol-induced euphoria. Biol Psychiatry 23:141–148.

Lukas SE, Mello NK, Bree MP, Mendelson JH (1988): Differential tolerance development to buprenorphine-, diprenorphine-, and heroin-induced disruption of food-maintained responding in macaque monkeys. Pharmacol Biochem Behav 30:977–982.

Lukas SE, Dorsey CM, Abdulali A, Fortin M, Abdulali S, Mendelson JH, Mello NK, Cunningham S (1992): EEG sleep architecture in cocaine- and heroin-dependent subjects during buprenorphine maintenance. NIDA Res Monogr 119:462.

Lukas SE, Mello NK, Drieze JM, Mendelson JH (submitted): Modulation of the reinforcing effects of cocaine by buprenorphine in rhesus monkeys.

Mansky PA (1978): Opiates: Human psychopharmacology. In Iversen LL, Iversen SD, Snyder SH (eds): "Drugs of Abuse." New York: Plenum, pp 95–185.

Mansour A, Lewis ME, Khachaturian H, Akil H, Watson SJ (1986): Multiple opioid receptor subtypes in the pituitary-adrenal axis: A cross-species study. NIDA Res Monogr 75:311–314.

Martin WR, Jasinski DR, Haertzen CA, Kay DC, Jones BE, Mansky PA, Carpenter RW (1973): Methadone—A reevaluation. Arch Gen Psychiatry 28:286–295.

Mello NK (1977): Stimulus self-administration: Some implications for the prediction of drug abuse liability. In Thompson T, Unna K (eds): "Predicting Dependence Liability of Stimulant and Depressant Drugs." Baltimore: University Park Press, pp 243–260.

Mello NK (1983): A behavioral analysis of the reinforcing properties of alcohol and other drugs in man. In Kissin B, Begleiter H (eds): "The Pathogenesis of Alcoholism, Biological Factors." New York: Plenum, pp 133–198.

Mello NK (1991): Preclinical evaluation of the effects of buprenorphine, naltrexone and desipramine on cocaine self-administration. NIDA Res Monogr 105:189–195.

Mello NK (1992): Behavioral strategies for the evaluation of new pharmacotherapies for drug abuse treatment. NIDA Res Monogr 119:150–154.

Mello NK, Mendelson JH (1965): Operant analysis of drinking patterns of chronic alcoholics. Nature 206:43–46.

Mello NK, Mendelson JH (1972): Drinking patterns during work-contingent and noncontingent alcohol acquisition. Psychosom Med 34:139–164.

Mello NK, Mendelson JH (1980): Buprenorphine suppresses heroin use by heroin addicts. Science 27:657–659.

Mello NK, Mendelson JH (1985): Behavioral pharmacology of buprenorphine. Drug Alcohol Depend 14:283–303.

Mello NK, Mendelson JH (1992): Primate studies of the behavioral pharmacology of buprenorphine. NIDA Res Monogr 121:61–100.

Mello NK, Mendelson JH (1993): Buprenorphine's effects on cocaine and heroin abuse. In Korenman S, Barchas J (eds): "The Biological Basis of Substance Abuse." New York: Oxford University Press, pp 463–485.

Mello NK, Mendelson JH, Kuehnle JC, Sellers M (1978): Human polydrug use: Marihuana and alcohol. J Pharmacol Exp Ther 207:922–935.

Mello NK, Mendelson JH, Kuehnle JC, Sellers ML (1981): Operant analysis of human heroin self-administration and the effects of naltrexone. J Pharmacol Exp Ther 216:45–54.

Mello NK, Mendelson JH, Kuehnle JC (1982): Buprenorphine effects on human heroin self-administration: An operant analysis. J Pharmacol Exp Ther 223:30–39.

Mello NK, Bree MP, Mendelson JH (1983): Comparison of buprenorphine and methadone effects on opiate self-administration in primates. J Pharmacol Exp Ther 225:378–386.

Mello NK, Lukas SE, Bree MP, Mendelson JH (1988): Progressive ratio performance maintained by buprenorphine, heroin and methadone in macaque monkeys. Drug Alcohol Depend 21:81–97.

Mello NK, Mendelson JH, Bree MP, Lukas SE (1989): Buprenorphine suppresses cocaine self-administration by rhesus monkeys. Science 245:859–862.

Mello NK, Mendelson JH, Bree MP, Lukas SE (1990a): Buprenorphine and naltrexone effects on cocaine self-administration by rhesus monkeys. J Pharmacol Exp Ther 254:926–939.

Mello NK, Mendelson JH, Drieze J, Kelly M (1990b): Acute effects of cocaine on prolactin and gonadotropins in female rhesus monkey during the follicular phase of the menstrual cycle. J Pharmacol Exp Ther 254:815–823.

Mello NK, Mendelson JH, Drieze J, Kelly M (1990c): Cocaine effects on luteinizing hormone-releasing hormone-stimulated anterior pituitary hormones in female rhesus monkey. J Clin Endocrinol Metab 71:1434–1441.

Mello NK, Lukas SE, Kamien JB, Mendelson JH, Drieze J, Cone EJ (1992): The effects of chronic buprenorphine treatment on cocaine and food self-administration by rhesus monkeys. J Pharmacol Exp Ther 260:1185–1193.

Mello NK, Kamien JB, Lukas SE, Drieze J, Mendelson JH (1993a): The effects of nalbuphine and butorphanol treatment on cocaine and food self-administration by rhesus monkeys. Neuropsychopharmacology 8:45–55.

Mello NK, Kamien JB, Lukas SE, Mendelson JH, Drieze JM, Sholar JW (1993b): Effects of intermittent buprenorphine administration on cocaine self-administration by rhesus monkeys. J Pharmacol Exp Ther 264:530–541.

Mello NK, Mendelson JH, Lukas SE, Drieze JM (1993c): Naltrexone attenuates buprenorphine's reduction of cocaine self-administration in rhesus monkeys. NIDA Res Monogr 132:236.

Mello NK, Lukas SE, Mendelson JH, Drieze J (1993d): Naltrexone–buprenorphine interaction: Effects on cocaine self-administration. Neuropsychopharmalogy 9:211–224.

Mendelson JH, Mello NK (1991): Commonly abused drugs. In Wilson JD, Braunwald E, Isselbacher JJ, Petersdorf RG, Martin JB, Fauci AS, Root RK (eds): "12th Edition of Harrison's Principles of Internal Medicine." New York: McGraw-Hill, pp 2155–2161.

Mendelson JH, Mello NK (1992): Human laboratory studies of buprenorphine. NIDA Res Monogr 121:38–60.

Mendelson JH, Rossi AM, Meyer RE (eds) (1974): "The Use of Marihuana: A Psychological and Physiological Inquiry." New York: Plenum.

Mendelson JH, Mello NK, Cristofaro P, Skupny A, Ellingboe J (1986): Use of naltrexone as a provocative test for hypothalamic–pituitary hormone function. Pharmacol Biochem Behav 24:309–313.

Mendelson JH, Teoh SK, Lange U, Mello NK, Weiss R, Skupny A, Ellingboe J (1988): Anterior pituitary, adrenal and gonadal hormones during cocaine withdrawal. Am J Psychiatry 145:1094–1098.

Mendelson JH, Mello NK, Teoh SK, Kuehnle J, Sintavanarong P, Dooley-Coufos K (1991a): Buprenorphine treatment for concurrent heroin and cocaine dependence: Phase I study. NIDA Res Monogr 105:196–202.

Mendelson JH, Mello NK, Teoh SK, Ellingboe J (1991b): Use of naltrexone for the diagnosis and treatment of reproductive hormone disorders in women. NIDA Res Monogr 105:161–167.

Mendelson JH, Mello NK, Teoh SK, Lukas SE, Phipps W, Ellingboe J, Palmieri SL, Shiff I (1992): Human studies on the biological basis of reinforcement. In O'Brien CP, Jaffe JH (eds): "Addictive States." New York: Raven, pp 131–155.

Mendelson JH, Teoh SK, Mello NK, Ellingboe J (1993): Buprenorphine attenuates the effects of cocaine on adrenocorticotropin (ACTH) secretion and mood states in man. Neuropsychopharmacology 7:157–162.

Meyer RE, Mirin SM (1979): "The Heroin Stimulus." New York: Plenum.

Moldow RI, Fischman AK (1987): Cocaine induced secretion of ACTH, beta-endorphin and corticosterone. Peptides 8:819–822.

Negus SS, Dykstra LA (1988): Kappa antagonist properties of buprenorphine in the shock titration procedure. Eur J Pharmacol 56:77–86.

Negus SS, Picker MJ, Dykstra LA (1990): Interactions between mu and kappa opioid agonists in the rat drug discrimination procedure. Psychopharmacology 102:465–473.

Negus SS, Picker MJ, Dykstra LA (1991): Inreractions between the discriminative stimulus effects of mu and kappa opioids in the squirrel monkey. J Pharmacol Exp Ther 256:149–158.

O'Connor JJ, Moloney E, Travers R, Campbell A (1988): Buprenorphine abuse among opiate addicts. Br J Addiction 83:1085–1087.

Oswald I, Evans JI, Lewis SA (1969): Addictive drugs cause suppression of paradoxical sleep with rebound. In "Scientific Basis of Drug Dependence." London: Churchill, pp 243–247.

Richards ML, Sadee, W (1985): Buprenorphine is an antagonist at the kappa opioid receptor. Pharm Res 2:178–181.

Ritz MC, Lamb RJ, Goldberg SR, Kuhar MJ (1987): Cocaine receptors on dopamine transporters are related to self-administration of cocaine. Science 237:1219–1223.

Rivier C, Vale W (1987): Cocaine stimulated adrenocorticotropin (ACTH) secretion through a corticotropin-releasing factor (CRF)-mediated mechanism. Brain Res 422:403–406.

Rosen MK, Pearsall HR, McDougle CJ, Price LH, Woods, SW, Kosten TR (1992): Effects of acute buprenorphine on responses to intranasal cocaine. Personal communication.

San L, Camí J, Fernández T, Ollé JM, Peri JM, Torrens M (1992): Assessment and management of opioid withdrawal symptoms in buprenorphine-dependent subjects. Br J Addict 87:55–62.

Schecter A (1980): The role of narcotic antagonists in the rehabilitation of opiate addicts: A review of naltrexone. Am J Drug Alcohol Abuse 7:1–18.

Schmidt WK, Tam SW, Shotzberger GS, Smith Jr DH, Clark R, Vernier VG (1985): Nalbuphine. Drug Alcohol Depend 14:339–362.

Shippenberg TS, Herz A (1987): Place preference conditioning reveals the involvement of D-1 dopamine receptors in the motivational properties of μ- and κ-opioid agonists. Brain Res 436:169–172.

Snyder EW, Dustman RE, Straight RC, Wayne AW, Beck EC (1977): Sudden toxocity of methadone in monkeys: Behavioral and electrophysiological evidence. Pharmacol Biochem Behav 6:87–92.

Spealman RD (1992): Use of cocaine-discrimination techniques for preclinical evaluation of candidate therapeutics for cocaine dependence. NIDA Res Monogr 119:175–179.

Spealman RD, Bergman J (1992): Modulation of the discriminative stimulus effects of cocaine by mu and kappa opioids. J Pharmacol Exp Ther 261:607–615.

Stretch R (1977): Discrete-trial control of cocaine self-injection behaviour in squirrel monkeys: Effects of morphine, naloxone, and chlorpromazine. Can J Physiol Pharmacol 55:778–790.

Stubbs WA, Delitala G, Nones A, Jeffcoate WJ, Edwards CRW, Ratter S, Besser GM, Bloom SR, Alberti KGM (1978): Hormonal and metabolic responses to an enkephalin analogue in normal man. Lancet 2:25–27.

Tallarida RJ, Murray RB (1981): "Manual of Pharmacologic Calculations." New York: Springer-Verlag.

Teoh SK, Mendelson JH, Mello NK, Kuehnle J, Sintavanarong P, Rhoades EM (1993): Acute interactions of buprenorphine with intravenous cocaine and morphine: An investigational new drug phase I safety evaluation. J Clin Psychopharm 13:87–99.

Teoh SK, Mendelson JH, Mello NK, Kuehnle J, Rhoades E, Sholar W, Gastfriend D (1994): Buprenorphine's effects on morphine and cocaine induced subjective responses by drug-dependent men. J Clin Psychopharm 14:15–27.

Thompson T, Unna KR (1977): "Predicting Dependence Liability of Stimulant and Depressant Drugs." Baltimore: University Park Press.

Vale W, Rivier C, Brown MR, Spiess J, Koob G, Swanson L, Bilezikjian L, Bloom F, Rivier J (1983): Chemical and biological characterization of corticotropin releasing factor. Recent Prog Horm Res 39:245–270.

Verebey K, Volavka J, Mule SJ, Resnick RB (1976): Naltrexone: Disposition, metabolism and effects after acute and chronic dosing. Clin Pharmacol Ther 20:315–328.

Volavka J, Cho D, Mallya A, Danman J (1979): Naloxone increases ACTH and cortisol levels in man. New Engl J Med 300:1056–1057.

Weddington WW, Brown BS, Haertzen CA, Cone EJ, Dax EM, Herning RI, Michaelson BS (1990): Changes in mood, craving and sleep during short-term abstinence reported by male cocaine addicts. A controlled, residential study. Arch Gen Psychiatry 47:861–868.

Winger G, Skjoldager P, Woods, JH (1992): Effects of buprenorphine and other opioid agonists and antagonists on alfentanil- and cocaine-reinforced responding in rhesus monkeys. J Pharmacol Exp Ther 261:311–317.

Witkin JM Johnson RE, Jaffe JH, Goldberg SR, Grayson NA, Rice KC, Katz JL (1991): The partial opioid agonist, buprenorphine, protects against lethal effects of cocaine. Drug Alcohol Depend 27:177–184.

Woolverton WL, Kleven MS (1992): Assessment of new medications for stimulant abuse treatment. NIDA Res Monogr 119:155–159.

Yanagita T, Katoh S, Wakasa Y, Oinuma N (1982): Dependence potential of buprenorphine studied in rhesus monkeys. NIDA Res Monogr 41:208–214.

Yen SSC (1986): Neuroendocrine control of hypophyseal function. In Yen SSC, Jaffe RB (eds): "Reproductive Endocrinology." Philadelphia: Saunders, pp 33–74.

Yen SSC, Quigley ME, Reid RL, Ropert JF, Cetel NS (1985): Neuroendocrinology of opioid peptides and their role in the control of gonadotropin and prolactin secretion. Am J Obstet Gynecol 152:485–493.

DETOXIFICATION AND INDUCTION ONTO NALTREXONE

MARC ROSEN and THOMAS R. KOSTEN
Department of Psychiatry, Division of Substance Abuse, Yale University School of Medicine, New Haven, CT 06519

INTRODUCTION

There are several drawbacks to current methods of detoxification from opiates. Methadone stabilization and then taper requires the patient to tolerate withdrawal symptoms for 2 weeks or longer, leading to intense craving and potential illicit opiate use by outpatients during this time [Cushman and Dole, 1973]. Clonidine detoxification [Gold et al., 1978] and clonidine–naltrexone detoxification [Vining et al., 1988] require the patient to tolerate moderately severe symptoms for several days. Clonidine itself produces an unpleasant sedation and lightheadedness, and neither clonidine nor naltrexone is reinforcing. Clonidine–naltrexone has been a significant advance, because it has reduced opiate detoxification from 2 weeks to only 3 days. Similar rapid detoxification approaches may hold promise in combination with buprenorphine, an opioid partial agonist.

Transition to antagonists, such as naltrexone, has been clinically difficult, because naltrexone introduction must be delayed to allow for an adequate opioid-free interval during which physical dependence will be lost, in order to avoid precipitating withdrawal [Kosten and Kleber, 1984]. Previous work has shown that naltrexone, even at very low doses (1 mg), will precipitate substantial withdrawal symptoms in patients discontinued from methadone 18 hr earlier [Charney et al., 1984]. Typically, 10–14 days must intervene between the last methadone dose and the first naltrexone dose in order to avoid precipitating withdrawal [Kosten and Kleber, 1984].

As an alternative to this long opioid-free period, during which relapse to drug abuse is likely, we have explored the early introduction of naltrexone [Charney et al., 1981; Charney et al., 1982; Kleber et al., 1987; Vining et al., 1988]. In the

---

Buprenorphine: Combatting Drug Abuse With a Unique Opioid, pages 289–305
© 1995 Wiley-Liss, Inc.

initial studies, an attempt was made to introduce very small doses of antagonist directly after stopping the agonist [Charney et al., 1982; Kleber et al., 1987]. Conceptually, this was an attempt to introduce opioid antagonism gradually in order to avoid precipitating withdrawal. Because even low naltrexone doses precipitated withdrawal, it was necessary to administer clonidine to suppress these symptoms. Opioid-dependent patients detoxified with clonidine and naltrexone exhibited approximately a 50% reduction in the duration of their acute withdrawal symptoms [Charney et al., 1982; Kleber et al., 1987; Vining et al., 1988]. These studies suggested that recovery from opioid physical dependency might be accelerated by the introduction of an opioid antagonist.

Another strategy to accelerate reduction in opioid dependency would be to introduce an opioid partial agonist that could precipitate mild withdrawal, and at the same time minimize these symptoms by its agonist activity. This strategy has not been feasible with previously available partial agonists such as pentazocine, cyclazocine, butorphanol, or nalorphine, since they precipitate significant withdrawal in morphine-dependent patients and may cause psychosis in some patients [Martin et al., 1966; Martin, 1967; Pircio et al., 1976; Jacob et al., 1979; Woods and Gmerek, 1985]. However, a new partial agonist, buprenorphine, showed therapeutic promise because of early work indicating that it did not precipitate significant withdrawal in methadone- or morphine-dependent patients [Jasinski et al., 1978, 1984; Mello and Mendelson, 1980; Mello et al., 1982].

Buprenorphine is an opioid partial agonist or mixed agonist–antagonist that addicts report does not produce a heroin-like rush [Jasinski et al., 1984]. Buprenorphine doses of 2–4 mg have been substituted for 20–30 mg of methadone without precipitating substantial withdrawal symptoms, although buprenorphine may act as an opioid antagonist at doses as low as 8 mg [Jasinski et al., 1978; Mello and Mendelson, 1980; Mello et al., 1982]. After chronic administration, buprenorphine does not produce significant physical dependence, as suggested by the minimal withdrawal symptoms that occur when it is stopped [Dum et al., 1981; Jasinski et al., 1984; Lewis, 1985]. Because of these properties, buprenorphine was examined to determine whether it might facilitate the transition from opioid agonists to antagonists in a three-step process: (1) buprenorphine substitution for agonists such as methadone, (2) buprenorphine-induced reduction in physical dependency, and (3) discontinuation of buprenorphine with rapid introduction of naltrexone.

The design for this study with opioid-dependent patients involved an initial outpatient protocol, which was followed by an inpatient protocol for patients in the last two thirds of this study. The outpatient protocol included discontinuation of either heroin or methadone followed by a 30-day trial on a range of buprenorphine dosages (Protocol A). Induction onto naltrexone was attempted in all of those patients who completed 30 days on buprenorphine. For the first third of the subjects, there was an attempt to complete their induction as outpatients, but because this was generally unsuccessful, a second inpatient protocol was developed. The inpatient protocol (Protocol B) began after the 30-day outpatient buprenorphine maintenance and included blinded discontinuation of the buprenorphine followed by double-blind placebo-controlled challenges with either low-dose naltrexone or high-dose naloxone. These two inpatient challenges carefully compared the capacity of either low-dose naltrexone or high-dose naloxone to precipitate withdrawal in buprenorphine-maintained patients.

Although the initial challenge studies were done with low-dose naltrexone (1 mg oral) in order to parallel earlier work with naltrexone in methadone-maintained patients [Charney et al., 1984], the subsequent studies used high-dose naloxone rather than naltrexone for three reasons. First, previous work with partial agonists such as butorphanol, nalbuphine, and pentazocine had used high doses of naloxone to precipitate a withdrawal syndrome after chronic treatment with these agents, thereby providing some guidelines for its use with buprenorphine. In these studies with nalbuphine and butorphanol, a withdrawal syndrome could be precipitated by 4 mg of naloxone [Jasinski et al., 1968; Pircio et al., 1976; Jacob et al., 1979; Woods and Gmerek, 1985]. With pentazocine, 10–15 mg of naloxone was necessary to reverse its agonist as well as its dysphoric effects, while up to 16 mg of naloxone did not reverse the respiratory depression associated with buprenorphine [Kallos and Smith, 1968; Jasinski et al., 1978; Dum et al., 1981; Quigley et al., 1984; Lewis, 1985]. Thus, high doses of naloxone had been given previously to patients treated with these partial agonists, and it appeared that more than 16 mg of naloxone might be needed with buprenorphine in order to precipitate any withdrawal.

Second, translating this naloxone dose into an equivalent dose of naltrexone is not straightforward, because the half-life is markedly longer for naltrexone than for naloxone and because two indicators of antagonist potency—precipitation of withdrawal and blocking of exogenous opioids—suggest different relative potencies of these two medications [Martin, 1967; Kosten and Kleber, 1984]. Precipitation of withdrawal in opioid-dependent patients can be induced by as little as 0.1 mg of naloxone or 1 mg of naltrexone, while blocking a 25 mg injection of heroin after antagonist administration requires 1 mg of intravenous naloxone or 50 mg of oral naltrexone [Charney et al., 1982, 1984; Kleber et al., 1987; Vining et al., 1988]. These two assessments indicate a 10- to 50-fold relative potency of naloxone over naltrexone, suggesting that unacceptably large naltrexone doses, theoretically over 800 mg (50 times 16 mg naloxone), would be required to precipitate significant withdrawal in buprenorphine-maintained patients. This high dose of naltrexone (above 200 mg daily) has been associated with liver toxicity.

Third, any substantial withdrawal syndrome precipitated by the high-dose antagonist would last substantially longer with naltrexone than with naloxone [Resnick et al., 1977; Charney et al., 1982; Kosten and Kleber, 1984; Kleber et al, 1987]. In addition to behavioral ratings of withdrawal symptoms, physiological (blood pressure) and biochemical (3-methoxy-4-hydroxyphenethylamine glycol or MHPG) responses to antagonist challenge were monitored. Previous work had demonstrated increases in both blood pressure and plasma-free MHPG, an index of norepinephrine turnover, during withdrawal precipitated by low-dose naltrexone in methadone-maintained patients [Charney, et al., 1984]. Thus, these two other measurements could be used to provide important objective correlates of any opioid withdrawal precipitated by naltrexone in buprenorphine-maintained patients.

METHODS

Subjects

Forty-one opioid-dependent patients were entered into the month-long outpatient Protocol A. The patients included 31 males and 10 females whose mean age was 31 (±1 SEM) years. Fourteen patients came from methadone maintenance at a dose of 25 mg/day, and the other 27 patients were using street heroin. For those using heroin, opioid addiction was confirmed by urine toxicology and challenge with naloxone at 0.8 mg intramuscularly [Wang et al., 1974; Kleber et al., 1985]. To qualify for inclusion, these heroin abusers had to attain a withdrawal score above 35 on our clinician-rated scale within 15 min of naloxone injection. The outpatient withdrawal rating scale includes 24 items with severity ratings that range from 0 to 3 points and has a score range of 0 to 72 [Kosten et al., 1985]. An item score of 3 indicates "severe" withdrawal, and total scores less than 20 indicate minimal withdrawal. For the inpatient challenges, a 15-item subscale with a range from 0 to 45 was used. This shorter subscale enabled more rapid administration and used items considered more responsive to acute change over the course of the antagonist challenges.

Of the 41 patients, 18 entered the inpatient Protocol B after completion of the 30-day outpatient Protocol A. The inpatient protocol included double-blind challenges with either low-dose naltrexone (in 13 patients) or high-dose naloxone (in 5 patients). The other 23 patients either dropped out of the outpatient Protocol A before 30 days ($n = 13$) or completed the open outpatient Protocol A and attempted outpatient induction onto naltrexone from buprenorphine ($n = 10$). The 18 patients who entered the inpatient protocol and the 23 other patients were not significantly different in demographics or percentage from methadone maintenance patients (26% inpatient vs 39% outpatient).

The findings from the inpatient protocol were compared with those of a previously published study in which 15 methadone-maintained patients (mean dose 35 mg daily, range 20–65 mg) were given low-dose (1 mg) oral naltrexone challenges [Charney et al., 1984]. The latter patients included 10 males and 5 females and had a mean age of 31 (±1) years. The mean duration of methadone treatment was 3 (±1) years. Eight other methadone-maintained patients were given a blinded challenge with placebo, but data were incomplete on one of them. The 7 placebo patients included 5 males and 2 females and had a mean age of 30 years.

Study Design

Protocol A (Outpatient). The patients who started buprenorphine treatment were discontinued from either methadone maintenance or street heroin, and within 24 hr of their last dose they started on sublingual buprenorphine at 2 mg, except for four patients. Two of the four started at 4 mg and the other two started at 8 mg in a dosage induction experiment that examined whether opioid withdrawal would be precipitated by the higher buprenorphine dosages. Among the 37 patients who started at 2 mg, a wider range of maintenance dosages was examined, and after the first 5 days the maintenance dosages were 2 mg ($n = 16$), 3 mg ($n = 14$), 4 mg ($n = 4$), and 6 mg ($n = 3$). This was an open trial with once-a-day dosing 7 days a week. After the first 5 days, the patients remained on fixed dosages of buprenorphine for Days 6–30 as outpatients and then either stopped the buprenorphine as outpatients or entered Protocol B. During this 30-day outpatient trial, withdrawal symptoms were rated daily by a clinician, and urine toxicologies were obtained twice weekly on a randomized schedule. For the 12 patients who had been transferred from methadone maintenance to buprenorphine, urine toxicologies for the 2 months before the start of buprenorphine were also obtained. Two months of urines were used, because only random weekly urines, rather than twice weekly urines, were obtained in the methadone program. The rate of cocaine- and opioid-positive urines were compared for methadone and buprenorphine treatments.

Buprenorphine was discontinued abruptly in an open trial with 10 outpatients who took buprenorphine for 30 days. Maintenance dosages for these patients were 2 mg ($n = 3$), 3 mg ($n = 2$), 4 mg ($n = 3$), 6 mg ($n = 1$), and 8 mg ($n = 1$). They reported few withdrawal symptoms during the 3–5 days that we were able to follow them after discontinuation of buprenorphine, but only two (at 2 mg and 4 mg) took any naltrexone, and they all returned to methadone maintenance or illicit opioid use.

Protocol B (Inpatient). Following the 30 days on buprenorphine, 18 patients were hospitalized at the Connecticut Mental Health Center Clinical Neuroscience Research Unit for 4–7 days. Upon admission to the hospital, each patient received a maintenance dose of buprenorphine once daily at 5 PM for 3 days. The maintenance doses were as follows: 2 mg ($n = 7$), 3 mg ($n = 8$), 4 mg ($n = 2$), 6 mg ($n = 1$). The buprenorphine was then discontinued abruptly by blinded substitution of placebo on Day 3, after each patient completed a placebo antagonist challenge. In the naltrexone challenge, an oral placebo was given at 9 AM on that day. In the naloxone challenge, an intravenous placebo was given instead. The day after buprenorphine placebo substitution, patients were given a challenge of either active naltrexone (1 mg p.o.) or active naloxone (0.5 mg/kg i.v.) at 9 AM. The intravenous naloxone infusion was given over a 20-min period using a 10-mg/ml naloxone solution. Throughout the hospitalization, all patients received a vanillymandelic acid exclusion diet [Charney et al., 1984].

Prior to the naltrexone or naloxone challenge procedure each patient fasted overnight for 10 hr and remained in the fasting state during the procedure until approximately 3 PM. An intravenous catheter was placed in the patient's arm to obtain two blood samples during the hour before receipt of naltrexone or naloxone (baseline), and then every 30 min for the next 3.5 hr after naltrexone or naloxone administration. A separate intravenous injection site was used for the i.v. naloxone infusions. Blood pressure measurements and opioid withdrawal ratings were obtained at the same time points. Withdrawal during the inpatient protocols was rated with a 15-item subscale of our 24-item scale with items scored from 0 to 3, giving a score range from 0 to 45 [Charney et al., 1981, 1984]. This shorter subscale had been developed in our earlier methadone–naltrexone challenge study and was adopted to facilitate comparison between the current study and previous work [Charney et al., 1984]. The comparison methadone maintenance patients had been continued on a stable dose of methadone before admission to the Research Unit and then participated in a procedure identical to the 1-mg oral naltrexone challenges following abrupt discontinuation of their methadone dose [Charney et al., 1984]. Raters and patients were blind as to whether placebo or naltrexone was administered.

Biochemical Methods (Protocol B)

Two 1-ml aliquots of plasma were taken from iced blood samples that were centrifuged within 2 hr. Assays for MHPG then were conducted on these duplicate samples by selected ion monitoring with a gas chromatograph—mass spectrometer (Finnegan Model 3300 series) [Elsworth et al., 1982]. Because of difficulties in finding adequate veins for blood sampling, MHPG determinations could be made at antagonist challenges for only eight of the naltrexone-challenged and four of the naloxone-challenged buprenorphine patients.

Data Analysis

For Protocols A and B, data analysis included simple descriptive measures of treatment retention, withdrawal symptoms, and illicit drug use. Comparisons were made across buprenorphine dosages as well as between dropouts and the remaining sample by means of contingency tables or repeated measures analysis of variance (ANOVA-R), as appropriate. To facilitate data analyses, ANOVA-R of withdrawal ratings in the outpatient trial (Protocol A) used ratings from days 2, 5, 8, 11, 14, 17, 20, 23, 26, and 29. For Protocol B, withdrawal ratings over a 3-hr period were compared for the naltrexone- and the naloxone-challenged buprenorphine patients and the naltrexone-challenged methadone patients (Treatment Type), as well as between the placebo and naltrexone challenges (Challenge Type), by means of a three-way ANOVA-R (e.g., Treatment Type by Challenge Type by Time Point, repeated measure). The naltrexone-challenged methadone patients were used as a further historical comparison group. Mean blood pressures were calculated as [2(systolic − diastolic)/3] + diastolic pressure. Blood pressures and plasma levels of MHPG were analyzed by determining the peak change in blood pressure or MHPG for each patient, because of considerable variability in the time course of withdrawal symptoms induced by naloxone in the buprenorphine patients and by naltrexone in the methadone-maintained patients. The peak changes in blood pressure and in MHPG were compared for the various groups with covariance adjustment for baseline differences (ANCOVA).

RESULTS

Outpatient Buprenorphine Treatment and Overall Outcome (Protocol A)

The 41 opioid-dependent patients generally had minimal withdrawal symptoms while maintained on buprenorphine. When started on buprenorphine, the patients had mild withdrawal symptoms, but which declined over the first 2 weeks on buprenorphine. The mean score (on the 24-item, 72-point scale) was 18 ± 15 (SD) at Day 2 and had declined to 11 ± 9 by Day 14 and to 9 ± 8 by Day 21. Maintenance doses of buprenorphine during the course of the trial were examined only for those patients started at 2 mg (four were started at higher doses) and the patients were categorized into three groups: 2 mg ($n = 16$), 3 mg ($n = 14$), 4 or 6 mg ($n = 7$). Because all three groups were on 2 mg during the first 5 days, we ran analyses only for Days 8–29. Withdrawal symptoms for the three Dose Groups were quite similar for Weeks 2–4 of the trial, and no significant Dose effect was seen. Withdrawal symptoms at Day 2 were higher among patients getting the 2-mg standard induction dose (18.6 ± 15) than among the four patients started at either 4 mg or 8 mg of buprenorphine (8 ± 8) ($t = 2.3, p < 0.05$, one-tail), which suggested that higher buprenorphine doses did not precipitate withdrawal. Instead, the starting dose of 2 mg may have been somewhat low for the patients coming from methadone maintenance (all at 25 mg), because the 10 methadone patients who started at 2 mg had fairly sustained mild withdrawal symptoms over the first 2 weeks, while the 27 street heroin addicts showed a decline in symptom levels (repeated-measures ANCOVA for Time: $F = 2.9; df = 4, 35; p < 0.03$; for Treatment: $F = 2.1; df = 1, 35; p < 0.1$). At Day 2 the symptom ratings of the methadone and street groups were equivalent (18 vs 19), but by Day 8 the methadone group's ratings remained at 18 ± 15, while the street group's dropped to 11 ± 8 ($t = 2.1; df = 35; p < 0.05$). The methadone group remained above the street group at Day 11 (18 vs 11) ($t = 2.2; p < 0.03$) and Day 14 (16 vs 9) ($t = 2.3; p < 0.03$). Interestingly, the four patients who started above 2 mg and who had lower levels of withdrawal than the 2-mg patients (see above), had all come from methadone maintenance, and they had somewhat lower withdrawal levels throughout the first (mean = 13) and second (mean = 8) weeks of buprenorphine treatment. Thus, although these levels of withdrawal symptoms were mild and generally not related to maintenance doses of buprenorphine, the patients coming from methadone maintenance appeared to have a more sustained period of withdrawal adjustment and might have benefited from a starting dose higher than 2 mg.

During the 30-day outpatient protocol, patients showed good retention and reduced illicit opioid use. Of the 41 entrants, 29 patients (71%) reported in daily and completed this protocol. The mean stay was 25 ± 8 days, and several of the 12 dropouts left because of circumstances unrelated to the medication (e.g., unexpected job transfer) or to illicit drug abuse. Illicit opioid use for the patients who completed treatment declined from 33% of urines in Week 1 to 19% in Week 4, and was not related to dosage of buprenorphine. For the dropouts, illicit opioid use remained at 50% of urines through Week 3, and both dropouts in Week 4 were using illicit opioids. For the whole trial, the percentage of illicit opioid urines was greater among dropouts (51%) than among those remaining in treatment (27%) ($t = 2.3, p < 0.03$), and there was an inverse correlation between days in treatment and number of illicit urines ($r = -0.34, p < 0.03$) (more illicit urines with fewer days in treatment).

Demographic comparisons and overall outcome for the various maintenance dosages of buprenorphine are shown in Table I. None of the outcomes, including

TABLE I. Sample Characteristics and Global Outcome by Maintenance Dose of Buprenorphine ($n = 41$)[a]

	Buprenorphine dose			
	2 mg	3 mg	4, 6, 8 mg	All
Characteristic				
Sample size	16	14	11 (6, 3, 2)	41
Males (%)	69	79	91	77
Age (year ± SD)	31 ± 7	29 ± 6	33 ± 7	31 ± 7
From "street" (%)	69	70	55	68
Outcomes				
Stay 30 days (%)	63	70	82	71
Opiate use (%)	37	27	37	33
Take naltrexone (%)	50	50	46	49
Naltrexone >2 weeks (%)	6	22	0	10

[a]No differences were statistically significant.

retention for 30 days, percentage of urines positive for illicit opioid use, taking at least one dose of naltrexone, and being maintained on naltrexone for at least 2 weeks were significantly different among the dosage groups. Although the rate of successful naltrexone maintenance appears to be best with a 3-mg dose of buprenorphine (22% vs. 7% and 0%), this is an artifact of the allocation of patients to naloxone challenges compared with naltrexone challenges. All naloxone-challenge patients had been maintained at 3 mg of buprenorphine, and this naloxone procedure, rather than dose of buprenorphine, seemed generally more effective at eventual naltrexone induction. The patients who received a high-dose naloxone challenge were quite successful at being maintained on naltrexone. This success appeared to result from the tolerance by these patients of a rapid increase in naltrexone dosage from 6 mg to a full 50 mg over the 24–36 hr after the withdrawal from high-dose naloxone had resolved (within 4–5 hr). During this rapid induction onto naltrexone, patients had trouble sleeping and one had some vague muscle aches, but none showed severe signs of withdrawal. Thus, the high-dose naloxone enabled a very rapid detoxification from buprenorphine. These challenges are addressed in more detail below.

Inpatient Antagonist Challenges (Protocol B)

Of the 18 inpatients given antagonist challenges, 13 received naltrexone and 5 received naloxone. None of the 13 patients on buprenorphine had marked differences in their response to 1 mg of oral naltrexone, compared with their responses to placebo. In contrast, in an earlier study [Charney et al., 1984] naltrexone (1 mg) induced substantial withdrawal symptoms in 13 of 15 methadone-maintained patients. Significant increases in withdrawal symptoms were induced in five

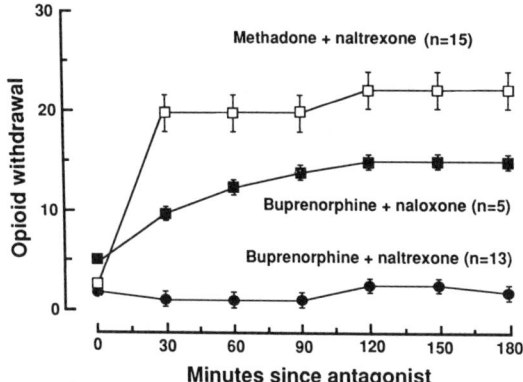

Fig. 1. Plot of patients' ratings of opioid withdrawal symptoms for patients maintained on buprenorphine and given either naltrexone (1 mg p.o.) or naloxone (0.5 mg/kg [mean = 35 mg] i.v.) and for methadone-maintained patients given naltrexone (1 mg p.o.). Placebo challenge for all three groups did not differ from buprenorphine + naltrexone. Mean scores ± SEM are plotted.

buprenorphine patients by high-dose naloxone infusions (0.5 mg/kg i.v.) (mean weight = 74 ± 5 kg). This naloxone dose (mean = 35 mg) is about 100 times the dose usually needed to precipitate withdrawal in methadone- or heroin-dependent subjects and was about 50 times greater than the 0.8-mg dose that precipitated withdrawal in the heroin-dependent patients before they started on buprenorphine 30 days earlier.

The patients' ratings of withdrawal symptoms (and standard errors) for the two buprenorphine groups are shown in Figure 1 along with the methadone group response to 1 mg naltrexone for comparison. The placebo responses for the three different conditions were indistinguishable from the 1-mg naltrexone response in the buprenorphine group ($n = 13$) and were omitted for clarity.

The withdrawal symptoms among the buprenorphine patients were significantly greater for the naloxone (0.5 mg/kg) than for the naltrexone (1 mg) challenge (Treatment Type), as shown by main effects (Treatment: $F = 10.5$; $df = 1, 32$; $p < 0.003$; Challenge: $F = 2.7$; $df = 1, 32$; $p < 0.05$, one-tail) and interaction with Time or the repeated measure (Treatment × Challenge × Time: $F = 7.0$; $df = 5, 160$; $p < 0.0001$). The comparison of withdrawal severity scores among the methadone and two buprenorphine groups (Treatment Type) was also highly significant for main effects (Treatment: $F = 18.7$; $df = 2, 52$; $p < 0.0001$; Challenge: $F = 20.5$; $df = 1, 52$; $p < 0.001$) and for Time interaction (Treatment × Challenge × Time: $F = 5.8$; $df = 10, 260$; $p < 0.0001$). As shown in Figure 1, the withdrawal response for the naloxone-challenged buprenorphine patients was substantially less than that for the naltrexone-challenged methadone patients. The placebo challenge responses in the methadone and both buprenorphine groups were not significantly different from each other or from the naltrexone challenge in the buprenorphine

TABLE II. Effect of Naltrexone- or Naloxone-Precipitated Opiate Withdrawal on Plasma-Free MHPG Levels in Buprenorphine- and Methadone-Maintained Patients[a]

Treatment group	Number	MHPG (ng/ml ± SD)	
		Baseline	Peak
Buprenorphine			
Naltrexone	8	3.3 ± 0.6	3.5 ± 0.7
Placebo	8	3.2 ± 0.9	3.6 ± 0.8
Naloxone	4	3.6 ± 0.7	3.7 ± 0.7
Placebo	4	3.5 ± 0.8	3.9 ± 1.0
Methadone			
Naltrexone	15	3.1 ± 0.9	4.0 ± 1.2
Placebo	7	3.9 ± 0.6	4.4 ± 0.7

[a]Significant differences are indicated in text.

patients. The withdrawal symptoms severity was not related to either dose of buprenorphine or to methadone versus "street" induction onto buprenorphine.

Before administration of the active naltrexone challenge, the baseline plasma-free MHPG levels were not significantly different among the two buprenorphine (3.3 ± 0.6 ng/ml and 3.6 ± 0.7 ng/ml) and methadone groups (3.1 ± 0.9 ng/ml) (Table II). The peak increase in plasma MHPG, however, was significantly greater for the methadone (0.9 ng/ml) than for either the naltrexone-challenged (0.2 ng/ml) or naloxone-challenged (0.1 ng/ml) buprenorphine groups. Using ANCOVA (covariance) to adjust for baseline levels, the overall F ratio was 54 ($df = 6, 45; p < 0.0001$), with significant Challenge ($F = 4.6; df = 1, 45; p < 0.04$), Treatment Type (methadone and two buprenorphine groups) ($F = 8.6; df = 2, 45; p < 0.001$), and interaction effects ($F = 5.0; df = 2, 45; p < 0.01$). The major source of this MHPG interaction was that the buprenorphine patients showed lesser MHPG responses to active challenge than to placebo, while the methadone patients showed greater responses to active challenge than to placebo.

Among the buprenorphine-maintained patients, a significant change in blood pressure was induced by the naloxone challenge, but not by the naltrexone challenge, compared to placebo challenge. Because the baseline mean blood pressures ranged from 93 to 104 mm Hg, as shown in Table III, covariance adjustments were used for comparisons. Using ANCOVA to compare the two challenges for the three treatments gave an overall F ratio of 12.4 ($df = 6, 57; p < 0.0001$) with significant Treatment ($F = 11.4; df = 2, 57$) and Challenge ($F = 11.5; df = 1, 57$) effects but no significant interaction. The difference between the placebo and naloxone challenges for the buprenorphine patients (2.4 mm Hg drop for placebo and 10.5 mm Hg rise for naloxone) was substantially greater than the difference for the naltrexone-challenged buprenorphine patients (8.7 mm Hg rise for placebo and 9.9 mm Hg rise for naltrexone) and was equivalent to the difference for the methadone patients (4.6 mm Hg rise for placebo and 13.6 mm rise for naltrexone). Thus, the

TABLE III. Effect of Naltrexone- or Naloxone-Precipitated Opiate Withdrawal on Mean Standing Blood Pressure in Buprenorphine- and Methadone-Maintained Patients[a]

Treatment group	Number	Mean blood pressure (±SD, mm Hg)		
		Baseline	Peak	Difference
Buprenorphine				
Naltrexone	13	96.7 ± 10.5	106.6 ± 10.5	9.9
Placebo	13	93.3 ± 8.9	102.0 ± 10.3	8.7
Naloxone	5	94.8 ± 4.6	105.3 ± 7.0	10.5
Placebo	5	104.7 ± 14.0	102.3 ± 8.7	−2.4
Methadone				
Naltrexone	15	103.5 ± 11.7	117.1 ± 8.6	13.6
Placebo	7	101.3 ± 9.1	105.9 ± 9.2	4.6

[a]Significant differences are indicated in text.

blood pressure changes were consistent with the differences in withdrawal symptoms among the three treatment groups.

DISCUSSION

Extrapolating from these studies, induction onto naltrexone appears to be easier from buprenorphine than from methadone. Whereas 1 mg of naltrexone precipitates no detectable withdrawal in buprenorphine-maintained patients, it causes severe withdrawal in methadone-maintained patients. Extremely high doses of intravenous naloxone precipitate moderate withdrawal in patients maintained on 3 mg sublingually of buprenorphine, whereas they undoubtedly would precipitate severe withdrawal in methadone-maintained patients. These studies are uncontrolled, and placebo-controlled random assignment studies remain to be done. Nevertheless, these studies suggest that therapeutic doses of buprenorphine induce less dependence than full μ agonists such as methadone and that detoxification can be rapidly and effectively completed.

Although gradual discontinuation of buprenorphine produces the clinically undesirable side effect of a prolonged abstinence syndrome, it is relatively mild, and this mildness has facilitated antagonist-precipitated detoxification approaches. There are two possible hypotheses to explain the mildness of the buprenorphine withdrawal syndrome. First is that buprenorphine, being a μ partial agonist, induces only partial tolerance and dependence. This hypothesis is supported by studies showing milder antagonist-precipitated withdrawal from buprenorphine than from full μ agonists. However, a caveat here is that antagonist-precipitated withdrawal depends on relative affinities of the antagonist and buprenorphine for receptors, as well as the level of tolerance and dependence. Thus, antagonists may precipitate a mild withdrawal syndrome in buprenorphine-treated addicts because the antagonist does not effectively displace the buprenorphine from opiate receptors. A second hypotheses is that buprenorphine's long duration of action and slow dissociation from receptors [Walsh et al., 1993] produces a self-taper effect. Either explanation for our success with rapid detoxification approaches leaves many details of the pharmacology of buprenorphine and the neurobiology of opioid agonist–antagonists to be explored. From this neurobiology we hope to develop new treatments to reduce relapse to heroin dependence by controlling protracted withdrawal and other neurobiological precipitants of relapse in this chronic psychiatric disorder.

Heroin addicts or methadone-maintained patients can be transferred onto the opioid partial agonist buprenorphine for a 1-month outpatient program with good retention, minimal withdrawal symptoms, and a reduction in illicit opioid and cocaine use. The reduction in cocaine abuse was particularly striking and may offer a potential new treatment for this serious addiction. The optimal dose of buprenorphine for outpatient treatment or for the transition to naltrexone appears to be quite flexible within the sublingual range of 2–8 mg, and dosing may be quite similar to that with methadone maintenance, in which wide individual variations are common. On the basis of its association with good retention and limited illicit drug use, buprenorphine clearly holds promise as a treatment agent for opioid addicts.

Following a month on buprenorphine, patients can be given low doses of the opioid antagonist naltrexone (1 mg) without precipitating withdrawal symptoms or increases in blood pressure and norepinephrine turnover, as reflected by plasma MHPG levels. When given to patients maintained on the pure agonist methadone, the same dose of naltrexone precipitated substantial withdrawal and increases in blood pressure and MHPG levels. Withdrawal can be precipitated in buprenorphine patients with high-dose i.v. naloxone (0.5 mg/kg), but this withdrawal syndrome is less intense than that produced by even low-dose naltrexone in methadone-maintained patients. More importantly, naltrexone maintenance can be rapidly initiated after the naloxone-precipitated withdrawal without precipitating further withdrawal symptoms. These findings suggest an attenuation of opioid physical dependence by the limited antagonist activity of buprenorphine, since these buprenorphine patients had been dependent on the opioid agonists methadone or heroin before starting buprenorphine. In several previous reports, our group and others have shown that opioid-dependent patients can be switched from the pure agonists to buprenorphine with minimal withdrawal symptoms [Jasinski et al., 1978, 1984; Kosten and Kleber, 1988], but systematic examination of the transition from buprenorphine to a pure antagonist has not been reported previously.

Two concepts seem important in explaining the authors' findings concerning antagonist challenge in the buprenorphine patients—the higher opioid receptor affinity of buprenorphine compared with that of commonly prescribed antagonists, and antagonist resetting of receptor mechanisms from an opioid-dependent to an opioid-naive state.

Whereas naltrexone has a greater affinity for the μ receptor than does methadone or heroin, buprenorphine is an unusual partial agonist in apparently binding more tightly than naltrexone to these receptors [Neil, 1984; Lewis, 1985]. This difference in affinity has been offered as an explanation for naloxone's inability to precipitate withdrawal in buprenorphine-maintained animals [Lewis, 1985; Kosten et al.,

1988]. Thus, one reason low-dose naltrexone produced minimal withdrawal was simply the inability of this pure antagonist to displace buprenorphine, while the high-dose naloxone worked by the law of mass action to occupy enough of the receptors for a long enough time to precipitate withdrawal.

The capacity of opioid antagonists to actively reset receptor mechanisms, thereby decreasing physical dependence, also may contribute to the minimal withdrawal response exhibited by the naloxone-challenged patients when they were rapidly inducted onto naltrexone over a 36-hr period. Previous studies have shown that coadministration of opioid agonists with antagonists inhibits the development of physical dependency or accelerates recovery [Cochin and Mushlin, 1976]. Also, clinical studies with rapid clonidine naltrexone detoxification found that giving naltrexone to opioid-dependent patients can compress the abstinence syndrome into a relatively brief period [Charney et al., 1982; Kleber et al., 1987; Vining et al., 1988]. Antagonist exposure appears to actively reset relevant receptor mechanisms, not only attenuating the development of physical dependence, but also reversing receptor changes and receptor coupling to second messengers induced during physical dependence on opioids [Cochin and Mushlin, 1976; Aceto et al., 1977; Bardo et al., 1983; Collier et al., 1983; Rothman et al., 1986; Zukin and Tempel, 1986; Krystal et al., 1989]. Buprenorphine may have induced some receptor resetting during the month of treatment. Clearly, some mild withdrawal was produced by buprenorphine over the first 1–2 weeks after the transition from the pure agonists, and less severe withdrawal, minimal MHPG elevation, and a relatively small blood pressure increase compared with that found in the methadone patients were precipitated by the high-dose naloxone.

In this dosage range, however, buprenorphine has predominantly agonist activity, as suggested by neuroendocrine assessments [Mendelson et al., 1982; Rolande et al., 1983]. By using cortisol, growth hormone, prolactin, and luteinizing hormone as markers, it has been concluded that the pattern of hormonal response to acute and repeated dosing of buprenorphine is consistent with an agonist rather than antagonist profile [Brown et al., 1978; Mendelson et al., 1982; Rolande et al., 1983]. This agonist profile is consistent with the easy transition from methadone to buprenorphine. Further work may explore the neuroendocrine profile of buprenorphine after a month of treatment, since previous studies have involved only several days of treatment. Perhaps with longer treatment, buprenorphine accumulates, resulting in more antagonist activity, as has been shown acutely in animals given much higher dosages of buprenorphine [Cowan et al., 1977; Lewis, 1985]. A definitive test of the antagonist hypothesis could best be obtained by using an antagonist with a higher receptor affinity than buprenorphine, but such a drug is not available for testing in humans.

Clinically, previous studies have shown that buprenorphine withdrawal may be substantially less severe than withdrawal from pure agonists, such as methadone [Jasinski et al., 1984]. Since a major problem with methadone maintenance treatment for opioid abuse has been the continuation of substantial withdrawal symptoms following attainment of a drug-free state, buprenorphine may offer a method for minimizing the withdrawal symptoms that follow detoxification from maintenance treatment [Cushman and Dole, 1973; Kosten and Kleber, 1988]. Furthermore, patients may start naltrexone more readily after buprenorphine and thereby have a greater chance of remaining drug-free [Kosten and Kleber, 1984]. The endogenous opioid system may, indeed, be set closer to normal baseline functioning when chronically exposed to the partial agonist buprenorphine than to the pure agonist methadone. The rapid detoxification from buprenorphine with high-dose naloxone seems to offer an exciting possibility of stabilizing buprenorphine-treated patients on naltrexone within a day. In a recently completed comparison of buprenorphine with clonidine alone for outpatient detoxification of heroin addicts, our colleagues found a 92% success rate ($n = 12$) with buprenorphine compared with 77% ($n = 13$) with clonidine alone [Shi et al., 1993]. An interesting aspect of these studies was stabilization on buprenorphine for only 3–7 days rather than 1 month, thereby greatly simplifying this outpatient approach. An ongoing randomized clinical trial is examining this buprenorphine procedure further in comparison with the standard clonidine approaches without using buprenorphine, and SPECT (single photon emission computed tomography) studies of naltrexone-precipitated buprenorphine withdrawal have confirmed the relative mildness of this precipitated withdrawal [Thomas et al., 1992]. In these SPECT studies of brain blood flow, less perturbation of cortical blood flow was shown with precipitated withdrawal from buprenorphine than from methadone (JH Krystal, personal communication). Thus, a series of clinical and neurobiological studies have supported the potential utility of buprenorphine for opiate detoxification and further studies are clearly justified.

ACKNOWLEDGMENTS

Support was provided by the National Institute on Drug Abuse Career Scientist Development Award KO2-DA00112 to TRK and Center grant P50-DA04060.

REFERENCES

Aceto MD, Flora RE, Harris LS (1977): The effects of naloxone and nalorphine during the development of morphine dependence in rhesus monkeys. Pharmacology 15:1–9.

Bardo MT, Bhatnagar RK, Gebhart GF (1983): Chronic naltrexone increases opiate binding in brain and produces supersensitivity to morphine in the locus coeruleus of the rat. Brain Res 289:223–234.

Brown B, Dettmar PW, Dobson PR, Lynn AG, Metcalf G, Morgan BA (1978): Opiate analgesics: The effect of agonist–antagonist character on prolactin secretion. J Pharm Pharmacol 30:644–645.

Charney DS, Redmond DE, Galloway MP, Kleber HD, Henninger GR, Murberg M, Roth RH (1984): Naltrexone precipitated opiate withdrawal in methadone addicted human subjects: Evidence for noradrenergic hyperactivity. Life Sci 35:1263–1272.

Charney DS, Sternberg DE, Kleber HD, Heninger GR, Redmond DE Jr (1981): The clinical use of clonidine in the abrupt withdrawal from methadone. Arch Gen Psychiatry 38:1273–1277.

Charney DS, Riordan CE, Kleber HD, Murburg M, Braverman P, Sternberg DE, Heninger GR, Redmond DE (1982): Clonidine and naltrexone: A safe, effective and rapid treatment of abrupt withdrawal from methadone therapy. Arch Gen Psychiatry 39:1327–1332.

Cochin J, Mushlin BE (1976): Effect of agonist–antagonist interaction of the development of tolerance and dependence. Ann NY Acad Sci 281:244–251.

Collier HOJ, Plant NT, Tucker JF (1983): Pertussis vaccine inhibits the chronic but not acute action of normorphine on the myenteric plexus of guinea-pig ileum. Eur J Pharmacol 91:325–326.

Cowan A, Lewis JW, Macfarlane IR (1977): Agonist and antagonist properties of buprenorphine, a new antinociceptive agent. Br J Pharmacol 60:537–545.

Cushman P, Dole VP (1973): Detoxification of methadone maintenance patients. J Am Med Assoc 226:747–751.

Dum J, Bläsig J, Herz A (1981): Buprenorphine: Demonstration of physical dependence liability. Eur J Pharmacol 70:293–300.

Elsworth JD, Redmond DE Jr, Roth RH (1982): Plasma and CSF 3-methoxy-4-hydroxy phenylethylene glycol (MHPG) as indices of brain norepinephrine metabolites in primates. Brain Res 235:115–124.

Gold MS, Redmond DE, Kleber HD (1978): Clonidine for opiate withdrawal. Lancet 1:929–930.

Jacob JJC, Michaud GM, Tremblay EC (1979): Mixed agonist–antagonist opiates and physical dependence. Br J Clin Pharmacol 7(Suppl 3):291S-296S.

Jasinski DR, Martin WR, Sapira JD (1968): Antagonism of the subjective, behavioral, pupillary and respiratory depressant effects of cyclazocine by naloxone. Clin Pharmacol Ther 9:215–222.

Jasinski DR, Pevnick JS, Griffith JD (1978): Human pharmacology and abuse potential of the analgesic buprenorphine. Arch Gen Psychiatry 35:510–516.

Jasinski DR, Boren JJ, Henningfield JE, Johnson RE, Lange WR, Lukas SE (1984): Progress report from the NIDA Addiction Research Center, Baltimore, Maryland.

Kallos T, Smith TC (1968): Naloxone reversal of pentazocine induced respiratory depression (letter). J Am Med Assoc 204:932.

Kleber HD, Riordan CE, Rounsaville BJ, Kosten TR, Charney D, Gaspari J, Hogan I, O'Connor C (1985): Clonidine in outpatient detoxification from methadone maintenance. Arch Gen Psychiatry 42:391–398.

Kleber HD, Topazian M, Gaspari J, Riordan CE, Kosten, TR (1987): Clonidine and naltrexone in the outpatient treatment of heroin withdrawal. Am J Drug Alcohol Abuse 13:1–18.

Kosten TR, Kleber HD (1984): Strategies to improve compliance with narcotic antagonists. Am J Drug Alcohol Abuse 10:249–266.

Kosten TR, Kleber HD (1988): Buprenorphine detoxification from opioid dependence: A pilot study. Life Sci 42:635–641.

Kosten TR, Rounsaville BJ, Kleber HD (1985): Comparison of clinician ratings to self reports of withdrawal during clonidine detoxification of opiate addicts. Am J Drug Alcohol Abuse 11:1–10.

Kosten TR, Krystal J, Morgan C, Charney D, Price L, Kleber H (1988): Opioid detoxification using buprenorphine. NIDA Res Monogr 90:68.

Krystal, JH, Walker MW, Heninger GR (1989): Intermittent naloxone attenuates the development of physical dependence on methadone in rhesus monkeys. Eur J Pharmacol 160:331–338.

Lewis JW (1985): Buprenorphine. Drug Alcohol Depend 14:363–372.

Martin WR (1967): Opioid antagonists. Pharmacol Rev 19:463–521.

Martin WR, Gorodetzky CW, McClane TK (1966): An experimental study in the treatment of narcotic addicts with cyclazocine. Clin Pharmacol Ther 7:455–485.

Mello NK, Mendelson JH (1980): Buprenorphine suppresses heroin use by heroin addicts. Science 207:657–659.

Mello NK, Mendelson JH, Kuehnle JC (1982): Buprenorphine effects on human heroin self-administration: An operant analysis. J Pharmacol Exp Ther 223:30–39.

Mendelson JH, Ellingboe J, Mello NK, Kuehnle J (1982): Buprenorphine effects on plasma luteinizing hormone and prolactin in male heroin addicts. J Pharmacol Exp Ther 220:252–255.

Neil A (1984): Affinities of some common opioid analgesics towards four binding sites in mouse brain. Naunyn-Schmeidebergs Arch Pharmacol 328:24–29.

Pircio AW, Gylys JA, Cavanagh RL, Buyniski JP, Bierwagen ME (1976): The pharmacology of butorphanol, a 3,14-dihydroxymorphinan narcotic antagonist analgesic. Arch Int Pharmacodyn Ther 220:231–257.

Quigley AJ, Bredemeyer DE, Seow SS (1984): A case of buprenorphine abuse. Med J Australia 140:425–426.

Resnick RB, Kestenbaum RS, Washton A, Poole D (1977): Naloxone precipitated withdrawal: A method for rapid induction onto naltrexone. Clin Pharmacol Ther 21:409–411.

Rolande E, Marabini A, Franceschini R, Messina V, Bongera P, Barreca, T (1983): Changes in pituitary secretion induced by an agonist antagonist opioid drug, buprenorphine. Acta Endocrinol 104:257–260.

Rothman RB, Danks JA, Jacobson AE, Burke TR, Rice KC, Tortella FC, Holaday JW (1986): Morphine tolerance increases mu-noncompetitive delta binding sites. Eur J Pharmacol 124:113–119.

Shi JM, O'Connor PG, Constantino JA, Carroll KM, Schottenfeld RS, Rounsaville BJ (1993): Three methods of ambulatory opiate detoxification: Preliminary results of a randomized clinical trial. NIDA Res Monogr 132:309.

Thomas HM, van Dyck CH, Rosen MI, Waugh ME, O'Connor PG, Pearsall HR, Hoffer PB, Woods SW, Kosten TR (1992): Regional cerebral blood flow (rCBF) changes in naltrexone precipitated withdrawal from buprenorphine. NIDA Res Monogr 119:459.

Vining E, Kosten TR, Kleber HD (1988): Clinical utility of rapid clonidine–naltrexone detoxification for opioid abusers. Br J Addict 83:567 575.

Walsh SL, Preston KL, Stitzer ML, Liebson IA, Bigelow GE (1993): Comparison of the acute effects of buprenorphine and methadone in non-dependent humans. NIDA Res Monogr 132:333.

Wang RIH, Weisen RL, Lamid S, Roh BL (1974): Rating the presence and severity of opiate dependence. Clin Pharmacol Ther 16:653–658.

Woods JH, Gmerek DE (1985): Substitution and primary dependence studies in animals. Drug Alcohol Depend 14:233–247.

Zukin R, Tempel A (1986): Neurochemical correlates of opiate receptor regulation. Biochem Pharmacol 35:1623–1627.

PERSPECTIVE

BUPRENORPHINE: WHAT INTERESTS THE NATIONAL INSTITUTE ON DRUG ABUSE?

DORALIE L. SEGAL
National Institute on Drug Abuse, Medications Development Division, Rockville, MD 20857

CHARLES R. SCHUSTER
National Institute on Drug Abuse, Addiction Research Center, Baltimore, MD 21224

THE PARENTAL NARCOTICS PROBLEM: A PUBLIC HEALTH EMERGENCY

Substance abuse, including opioid abuse, is one of the most significant public health problems facing our society today. The use of illicit drugs by the parenteral route, in addition to posing all of the dangers of pharmacological toxicity, has the added risks of infection. According to the latest statistics gathered by the National Institute on Drug Abuse (NIDA), approximately one third of all the reported AIDS cases in the United States were associated with parenteral drug abuse. When the cases of asymptomatic patients infected with the human immunodeficiency virus (HIV) are included in the above figures, the percentage becomes even greater. In fact, drug abuse is the second leading risk factor in adults for acquiring the HIV, and the primary risk factor in infants [National Commission on AIDS, 1991].

Although contaminated blood from shared needles is the primary vector of HIV into the population, the virus is also spread by sexual transmission between infected drug abusers and their partners—both non-drug-users and drug abusers. Infants become HIV-infected perinatally from infected mothers [National Commission on AIDS, 1991].

It is also believed that drug use, whether by injection or other routes, predisposes the user to participate in high-risk sexual behaviors that increase the likelihood of HIV infection. The exact mechanisms underlying the association between illicit

---

Buprenorphine: Combatting Drug Abuse With a Unique Opioid, pages 309–320
© 1995 Wiley-Liss, Inc.

drug use and high-risk sexual practices have not yet been clearly identified, but it is thought that drug-induced loss of inhibitions and perhaps a "social community bonding" play a role [National Commission on AIDS, 1991].

Needle sharing among narcotic addicts has greatly increased the incidence of viral hepatitis B in the United States. This is of special concern, since drug abusers are known to have biochemical abnormalities, whether caused solely by the abused substances (including alcohol), by contaminants injected with abused drugs, or in combination with other factors prevalent in an addict's life-style. Thus, the narcotic addicts' livers are particularly vulnerable to infective organisms that either cause or exacerbate hepatic disease. The narcotic addict also appears to be more susceptible than the general population to developing endocarditis and abscess formation [Kreek, 1978].

The social consequences of drug abuse are well known today to everyone who reads a newspaper or watches television. Hardly a day goes by without media reports of violent drug-related crimes. Concern becomes especially high when innocent citizens become victims, caught in the crossfire between drug dealers struggling over territory. The devastation is also felt acutely at the personal level. The drug addict is often riddled with pain, feels helpless to control his or her problem, loses self-respect—and then self-esteem. With that gone, close personal relationships become strained and then disintegrate. Thus, the family unit suffers and often breaks apart, leaving children with poor examples, lack of a role model, exposure to criminal activities, and feelings of rejection. The cycle continues, and the consequences spiral out of control.

It is now well established that treatment for substance abuse can help certain individuals to attain a drug-free life-style. However, despite our success with some, the majority of drug addicts still are without effective treatment. Methadone maintenance has been sufficiently encouraging in its effectiveness for the treatment of heroin dependence to prompt the search for other medications to offer to heroin addicts entering treatment. In this chapter we will briefly discuss medications for the treatment of heroin dependence that are currently available, as well as one of the most promising new medications, buprenorphine.

EXISTING OPIATE THERAPIES

Methadone

Methadone is by far the most widely used effective treatment for narcotic dependence in the United States. Methadone maintenance has been shown to decrease illicit drug use and criminal activity as well as curb the transmission of the HIV infection among intravenous drug users [Gerstein and Harwood, 1990]. The longer the retention in treatment, the less likely are these persons to relapse into drug use [D'Aunno and Vaughn, 1992]. But methadone, like most other medications, has its drawbacks.

When administered orally methadone is effective for most patients for a period of

24 hr; yet, for some patients, the drug does not completely suppress their narcotic craving for this period of time. In order to provide sustained relief of withdrawal symptoms for these patients over the entire 24 hr, larger doses of methadone are required, which may cause unwanted sedation.

A patient cannot "earn" privileges for more than one take-home dose per week until he or she has been in a program for 3 months under staff supervision [Federal Register, 1989]. But because of take-home privileges, methadone is susceptible to diversion and subsequent illicit use. Because methadone is classified as a controlled substance and must be administered according to treatment regulations, it can be dispensed only in a licensed clinic or program setting. These restrictions create a burden in terms of time and travel demands for many patients, especially for those who are trying to coordinate their own treatment needs with a family life, job, and possible educational opportunities. Furthermore, in some locations or within some groups, a stigma associated with methadone maintenance treatment has developed that limits its acceptance.

At appropriate doses, given in conjunction with a full range of counseling and rehabilitative services, methadone maintenance treatment has allowed many addicts to attain a normalcy in their vocational, educational, and social lives. Unfortunately, many methadone clinics are using suboptimal doses that are not sufficient to prevent withdrawal symptoms or to provide blockade of intravenous heroin use during the 24-hour period between methadone doses [Schuster, 1989; Ball and Ross, 1991]. Also, many clinics do not provide the full range of services necessary for successful treatment.

When higher doses and overdoses of methadone produce toxicity, respiratory depression seems to be the major acute effect [Gay and Inaba, 1976; Senay, 1988]. Methadone is an opiate agonist; it therefore causes a "severe and protracted withdrawal" [Jasinski et al., 1978; Mello and Mendelson, 1980]. This can be a significant problem for patients who are being withdrawn from methadone.

When methadone is diverted and nontolerant persons, especially children, ingest high doses of the drug, toxicity and overdose fatalities can result. Also, a market has developed for illicit sales and redistribution of methadone to heroin addicts who will not, or cannot, enter treatment but use methadone when they cannot obtain heroin. This type of behavior helps to stigmatize methadone as a drug of abuse.

LAAM

LAAM (levo-alpha-acetylmethadol or levomethadylacetate hydrochloride), a congener of methadone, is a newly FDA-approved medication for the treatment of opiate addiction and an alternative to methadone. Because LAAM is administered orally every other day or three times a week, it affords the patient a greater freedom from daily clinic visits—thus breaking the routine of daily drug-seeking behavior. This more liberal schedule should also improve patient compliance, ease the hassle and consequences of take-home dosing with its attendant concerns with diversion, as well as offer some relief to clinic personnel.

Patients describe LAAM as having a slower onset, lesser peak, and slower offset than methadone, which is consistent with LAAM's biotransformation to active metabolites in the liver. Since the active metabolites take several days to weeks to reach steady state, it is critical during this induction period to support patients who feel undermedicated. Patient dropout occurs in greatest number during the first 4 weeks of therapy.

Although LAAM's approval is a significant contribution within the field of addiction medicine and will offer an additional treatment option to many opiate-addicted patients, the drug is "not for everyone." Treatment providers must be very attentive to individualizing dosing for each patient. Should an addict who is being inducted with long-acting LAAM also use street drugs, the potential for overdosing—and respiratory depression—is very real. Patients are warned not to take illicit drugs or alcohol, especially during induction.

LAAM, which is marketed under the tradename ORLAAM®, was made available on a state-by-state basis in the spring of 1994.

Naltrexone

Naltrexone (tradename: Trexan®), an opiate antagonist, was approved by the FDA for marketing in 1984, as a treatment for opiate dependence. Naltrexone has a number of advantages for the treatment of opiate dependence: It blocks all μ agonist effects and it does not produce physical dependence or other agonist-like side effects. Unfortunately, it is unacceptable to the vast majority of persons who abuse, and are dependent upon, opiates. Before a heroin, or methadone, patient can be transferred to naltrexone, a drug-free period of 1 week, or 2 weeks, respectively, is required. This is necessary to prevent precipitation of withdrawal symptoms [Ling and Wesson, 1990]. During this interval, the addict is particularly susceptible to using street drugs. Naltrexone blocks the euphoria that opiate abusers seek, and by some accounts dysphoria is produced. These factors limit its use to only the most highly motivated patients [Altman et al., 1976; Judson et al., 1981]. Naltrexone has been shown to be of value as an adjunct in the treatment of opiate-dependent persons who are under the stringent contingencies of probation from the criminal justice system [Metzger et al., 1990].

Clonidine

Clonidine (tradename: Catapres®) is an alpha-adrenergic agonist that is approved by the FDA only for the treatment of hypertension. However, it is in limited use for treating opiate withdrawal during the initial phase of detoxification. It may be used in combination with an antagonist, such as naltrexone, to minimize precipitated withdrawal symptoms in opioid-dependent patients to allow them to become drug-free in as little as 3–4 days [Charney et al., 1982; Kleber et al., 1985; Vining et al., 1988]. Clonidine does not produce euphoria and therefore does not maintain continued drug-seeking behavior. Clonidine does not cause physical dependence of the opiate type, and since it is used only during the acute withdrawal period, it can be discontinued without incident. The major side effects of clonidine are orthostatic hypotension and drowsiness/sedation. However, clonidine only partially relieves opiate withdrawal symptoms; therefore, relapse rates are high [Jasinski et al., 1985; Smith et al., 1989; Ling and Wesson, 1990].

Other Therapies

NIDA is a very strong advocate of combining treatment modalities. Pharmacological treatment alone addresses only one of the problems associated with drug dependence and when it is employed alone, relapse rates are very high. In fact, NIDA mandates that its clinical trials of new potential medications combine behavioral and psychosocial treatment and rehabilitative interventions in the research design.

Many clinics offer only drug-free, psychological services for the treatment of substance abuse. These range from intensive inpatient programs with psychiatric counseling to rather casual outpatient services that might include individual or group therapy, or self-help groups such as Narcotics Anonymous, which is based on the 12-step program endorsed by Alcoholics Anonymous. Obviously, this type of treatment is not for everyone; only the most motivated of patients seem to respond, and the relapse rate is very high.

BUPRENORPHINE

Background

NIDA is aware that, despite the efficacy of methadone and naltrexone in selected portions of the heroin-dependent population, we are in need of alternative pharmacotherapies. One of the most promising alternative medications for the treatment of heroin dependence is buprenorphine.

Pharmacology

Buprenorphine has been of interest to drug abuse research since 1976, when Martin first described it as a partial agonist of the morphine type [Martin et al., 1976]. Jasinski, in a clinical study, recognized the drug's capacity to block the euphoria produced by opiates [Jasinski et al., 1978], and Mello and associates reported that buprenorphine decreased heroin self-administration in opiate abusers [Mello and Mendelson, 1980; Mello et al., 1982]. The drug is currently marketed in the United States in parenteral form as Buprenex®, approved by the FDA as an analgesic for relief of moderate to severe pain.

Buprenorphine has pharmacological properties that are highly desirable for the treatment of narcotic addiction [Lewis et al., 1983], and hence the drug is of great interest to NIDA's Medication Development Program. Buprenorphine produces morphine-like subjective effects, but with less intensity at doses producing peak effects. As a μ partial agonist, the drug produces only mild physical dependence with limited withdrawal signs and symptoms, which allows patients to undergo a relatively comfortable detoxification regimen [Jasinski et al., 1978; Mello et al., 1982; Lukas et al., 1984].

When heroin-dependent subjects, maintained on 8 mg sublingual buprenorphine for 37 days were abruptly terminated from this drug, most were able to complete the withdrawal phase. The subjects reported mild to moderate opioid withdrawal symptoms that peaked at 3–5 days and were over by 8–10 days. About half of the subjects required no medication for symptomatic relief; the others received varying amounts of temazepam for insomnia, and clonidine or propoxyphene for other withdrawal symptoms [Fudala et al., 1990].

The most important action of buprenorphine for the treatment of opioid dependence is that it antagonizes the euphoria produced by opiates, and its duration of action is long, some 29.5 hr —similar to that of methadone [Jasinski et al., 1978]. This is probably due to its very slow dissociation from opioid receptors [Lewis et al., 1983] and its high degree of lipophilicity [Hambrook and Rance, 1976]—in contrast to methadone's long action, which is primarily due to its long plasma half-life [Jaffe and Martin, 1985].

It would appear that buprenorphine combines the desirable features of naltrexone and methadone in one medication. Because of its partial agonist properties at the μ receptor buprenorphine will, like methadone, be taken readily by recovering heroin addicts and will, like naltrexone, antagonize the pharmacological effects of heroin.

Safety

NIDA is especially impressed with buprenorphine's low toxicity, demonstrated in preclinical studies, and its safety, as judged from clinical experience to date, even at relatively high dose levels [Banks, 1979]. Buprenorphine's highly favorable therapeutic ratio is due to the fact that it is a partial, rather than a full, agonist at the μ receptor [Lewis et al., 1983]. Respiratory depression, a major acute toxic effect of μ agonists, is not a significant problem with buprenorphine [Jasinski et al., 1978; Lukas et al., 1984]. Because of this, overdose problems with buprenorphine are likely to be less frequent [Mello and Mendelson, 1980].

The results of two preclinical studies [Shukla et al., 1991; Witkin et al., 1991] further suggest that there is no reason to suspect an acute adverse drug interaction between cocaine and buprenorphine for subjects being administered buprenorphine. In fact, in rats pretreated with buprenorphine, a protective effect was seen in respect to cocaine-induced lethality as assessed by LD_{50} values.

Adverse Reactions. Buprenorphine has been given to about 1900 opiate abusers or opiate-dependent persons, at doses ranging from 1 to 32 mg by subcutaneous or sublingual routes without any significant adverse reactions. At least 190 subjects have been maintained uneventfully on buprenorphine for one year or more, during which about 120 were on doses of 8 mg or higher. An opiate dependent subject in a clinical study maintained on 1 mg of buprenorphine accidentally was given a 32-mg dose. The subject reported insomnia the night after being overdosed and vomiting three times, the first of which was 16 hr post dosing. She also described a pressure headache—". . . like I did four to five bags of heroin." This subject stayed in the study and resumed dosing on the fourth day following the incident. Two nondependent opiate abusers exposed to single doses of 16 and 32 mg of the drug reported the usual symptoms often seen in nondependent abusers, namely, dysphoric mood, frequent nausea and vomiting following oral intake, and constipation. However, no significant change from baseline in oxygen saturation was noted at these doses.

To the best of our knowledge, only four deaths have occurred to subjects enrolled in a clinical study of buprenorphine, and none of them was thought to be attributable to the study medication. One subject had been only on methadone. The second succumbed to an acute drug overdose: Buprenorphine and cocaine were found in his blood, and buprenorphine, cocaine, and morphine were found in his urine. The third death was attributed to coronary thrombosis in a 47-year-old male with an extensive history of hypertension who had been treated with a beta blocker, an angiotensin-converting enzyme inhibitor, and a diuretic for 7 years. The fourth subject apparently died of sepsis, secondary to an HIV-related infection.

Acceptability

What may be buprenorphine's greatest asset is that, in addition to the qualities above, addicts find buprenorphine treatment acceptable. It produces limited, but significant, opiate-like subjective effects. The drug has a smooth onset of action (perhaps because of its slow association with the μ opiate receptor) and tolerable transition from street drugs; it is long-acting, so physical dependence is low and therefore withdrawal is very mild [Jasinski et al., 1978; Mello and Mendelson, 1980; Mello et al., 1982; Lukas et al., 1984; Johnson et al., 1989]. These characteristics have contributed greatly to buprenorphine's good name on the street. To date it appears to have avoided the stigma associated with methadone maintenance therapy.

Resnick (personal communication of March 18, 1992) has reported that at the New York University Buprenorphine Treatment Program "buprenorphine treatment, in a general medical facility, is acceptable and highly efficacious for many heroin addicts who refuse other treatment modalities. Further, buprenorphine is effective over a wide range of doses (1.5–8 mg/day) in reducing craving for heroin and its use." It would thus appear that buprenorphine may be useful in treating heroin addicts who refuse treatment with methadone.

NIDA'S ACTION PLAN

NIDA's View: Toward A Buprenorphine New Drug Application

One of the clinical trials already completed and partially analyzed, the "J/J/F 090" Maintenance Study [Johnson et al., 1991] (see Fudala and Johnson, this volume), is considered to be a key efficacy study that NIDA plans to submit to the FDA as part of the buprenorphine New Drug Application (NDA). The study was performed in NIDA's Addiction Research Center in Baltimore with 162 street opiate addicts. This randomized, double-blind, dose-comparison study had two components: The first 17 weeks was a maintenance phase, and the last 8 weeks was a gradual withdrawal phase with doses decreasing to zero. The study design compared 8 mg of sublingual buprenorphine with 20 mg and 60 mg of methadone.

The results of this study showed that during the maintenance phase buprenorphine was consistently superior to 20 mg of methadone and equal to 60 mg of methadone across the primary outcome measures, namely: (1) patient retention time in treatment, (2) illicit opiate use as assessed by urine samples negative for opioids, and (3) failure to maintain abstinence as defined by two consecutive Monday urine specimens positive for opiates. The authors concluded that buprenorphine is effective for the treatment of opiate dependence and a valuable alternative to methadone [Johnson et al., 1991].

NIDA will also submit to its NDA persuasive data from a clinical study performed at Yale by Schottenfeld et al. that was funded by and reported to the Institute in October 1993. Four and 12 mg of buprenorphine were compared with 20 and 65 mg of methadone. In a randomized, double blind design, 120 opiate- and cocaine-dependent subjects were dosed daily on sublingual buprenorphine liquid or oral methadone for 26 weeks. The results showed that 12 mg of buprenorphine was superior to both 4 mg of buprenorphine and 20 mg of methadone for retention in treatment and opiate-free urines. The proportion of cocaine-free urines did not differ significantly among the two methadone and buprenorphine 12-mg groups; however, these three treatments yielded significantly more cocaine-free urines than did the buprenorphine 4-mg group.

A recently completed maintenance study that NIDA funded was conducted in Los Angeles by Ling and associates with 225 street opiate addicts. This study is similar in design to the Johnson et al. study discussed above. Ling's study was carried out as a 1-year maintenance design, the first 6 months of which were used to evaluate efficacy. Buprenorphine, at a dose of 8 mg, was compared with doses of 30 and 80 mg of methadone. According to Ling (personal communication of Feb. 4, 1994), "Eight milligrams of buprenorphine is similar to 30 milligrams of methadone in terms of patient retention, opiate craving, and illicit opiate use. However, the optimum dose of buprenorphine still needs to be determined."

NIDA also conducted a dose comparison and challenge study with buprenorphine at the Washington, D.C. Veterans Administration Medical Center Hospital, under the direction of Stephen Deutsch. This double-blind, random assignment study of buprenorphine and methadone was designed to examine the relationship of plasma concentrations of the two drugs to subjective responses produced by an opiate challenge. Forty eight subjects were enrolled; the study is now completed. Following the opiate challenge, subjects were rapidly withdrawn from opiates, providing an opportunity to systematically compare the characteristics of methadone and buprenorphine withdrawal. At the time of this writing, data are still being analyzed.

NIDA's Medications Development Division has recently completed a large multicenter study, #999a, in 12 sites to evaluate the use of 1, 4, 8, and 16 mg of buprenorphine in the treatment of dependent opiate addicts. Seven hundred thirty three subjects were dosed in the 16-week randomized, double-blind protocol; 375 subjects have completed a minimum of 16 weeks.

An "Extension" protocol, #999a Ext, was filed to allow for successful completers of the 16-week efficacy portion to continue on buprenorphine for up to 1 year for two reasons: (1) to build the safety data base, a requisite for marketing the drug; and (2) to gain more experience with the drug for labeling purposes. The Extension, which remained double-blind, allowed for flexible dosing. Dosages were increased (doubled), decreased (halved), or remained unchanged at the investigators' discretion. There was no restriction as to the number of adjustments, although 32 mg was the highest dose allowed and 1 mg the lowest. Also, clinics had the option to dose five, six, or seven days a week, with double doses given on Friday or Saturday for those clinics who dosed five or six days per week, respectively.

Some of the clinics that dosed five days a week (double doses given on Fridays) reported that subjects began experiencing withdrawal symptoms by Sunday morning. In order to minimize subjects resorting to illicit street "therapies" and terminations from the study, we amended the protocol to allow five-day-a-week sites to dose selective subjects on the sixth day. Interestingly, a number of five-day-a-week subjects refused a dose increase in an attempt to alleviate their Sunday distress; they felt that the higher dose on weekdays would cause them too much sedation. At least 172 subjects have completed one year of buprenorphine treatment in this study.

The major efficacy measures used in this study were: (1) illicit opiate use as determined by urine testing conducted on a Monday, Wednesday, Friday schedule; (2) days of retention in treatment; (3) opiate and cocaine craving scores; and (4) global rating scores of the patient's status at the time of rating and in comparison to previous ratings, including baseline. These rating scales are being completed by both the clinical staff and the patient to reflect both objective and subjective measures.

This large study was monitored by Quintiles, Inc., under a NIDA contract. The study chairman and clinical coordinator was Walter Ling, Los Angeles; the study was administered under an Interagency Agreement with the Veterans Administration Cooperative Studies Program, Perry Point, Maryland. NIDA's Medications Development Division, the study sponsor, managed and assumed oversight for all aspects of the study.

The Johnson et al. and NIDA multicenter studies described above are pivotal for the NDA, and will serve as the base for establishing the safety and efficacy of buprenorphine for the treatment of heroin dependence. It is hoped that the NDA will be filed in early 1996.

In addition to these major studies, NIDA has funded over the past 12–15 years a number of other, smaller studies that may also serve to document buprenorphine's safety in the opiate-abusing population.

Other Agenda Items. Many, if not most, of the drug abusers in the United States today are using combinations of illicit drugs. A subset of addicts identified in NIDA Treatment Research Units and elsewhere inject themselves with a heroin/cocaine combination. Consequently, NIDA is anxious to determine whether buprenorphine will be useful in the treatment of these cocaine–heroin users. Although not a major efficacy variable in NIDA's multicenter study, we are assaying for cocaine metabolites in the urine of all subjects.

NIDA made a commitment to Congress to include a representative number of opiate-addicted women in its clinical trials of buprenorphine. To give effect to this committment, NIDA's Medications Development Division requested from each of the #999a principal investigators in the current multicenter trial a signed statement that she or he was committed to enroll one third of the subjects as women. The final female enrollment was 32 percent.

What's Needed Now?

NIDA, like other NIH institutes, has always found it essential to have a private sector company manufacture and market medications. Because established pharmaceutical companies with both the resources *and* the interest in developing and marketing buprenorphine in the United States for the treatment of opiate dependence were not readily available, NIDA initially set out to develop this medication on its own. NIDA perceived the development of buprenorphine as an unmet public health need that justified the effort. The rationale behind this was NIDA's belief that buprenorphine would represent a critically important addition to the existing limited alternatives for the treatment of opiate dependence. Subsequent to NIDA's initiation of buprenorphine development, Reckitt & Colman Pharmaceuticals, Inc., the drug's original manufacturer, approached NIDA with a proposal to share the development costs and seek approval and marketing of buprenorphine through a Cooperative Research and Development Agreement.

Combination Product: Buprenorphine + Naloxone. Buprenorphine has a relatively low abuse potential as compared to a μ agonist. In terms of *analgesic equivalency,* the Buprenex (injectable buprenorphine) package insert states that 0.3 mg of injectable buprenorphine is approximately equivalent to 10 mg of morphine. Or, eight mg of intravenous buprenorphine is equivalent to more than 25 single doses of 10 mg of morphine. To prevent diversion and to control the drug's potential for misuse on the street, NIDA currently is developing a combination formulation of buprenorphine plus the antagonist naloxone, potentially to allow for dosing outside of the typical clinic system.

NIDA's goal is to utilize buprenorphine mono as a therapeutic intervention to serve as a transition from illicit street opiates to the buprenorphine-naloxone combination product. This transition is necessary because naloxone might precipitate withdrawal if heavily dependent opiate persons are exposed to the antagonist in the combination product prior to going to buprenorphine. Another use for buprenorphine mono might be for methadone patients who wish to withdraw from their methadone medication. Limited clinical experience indicates that it is considerably easier and more comfortable for the patient to be gradually withdrawn from buprenorphine than from methadone.

NIDA is working closely with FDA on the buprenorphine project, and is fortunate to receive the guidance and support of that agency. We look forward to a successful relationship with Reckitt & Colman in the co-development of buprenorphine products in order to make this valuable medication available for the treatment of opiate dependence.

ACKNOWLEDGMENTS

The authors wish to recognize Drs. Frank Vocci, Rolley "Ed" Johnson, and Kenzie Preston for reviewing this manuscript.

REFERENCES

Altman JL, Meyer RE, Mirin SM, McNamee HB, McDougle M (1976): Opiate antagonists and the modification of heroin self-administration behavior in man: An experimental study. Int J Addict 11:485–499.

Ball JC, Ross A (1991): "The Effectiveness of Methadone Maintenance Treatment. Patients, Programs, Services and Outcomes." New York: Springer-Verlag.

Banks CD (1979): Overdosage of buprenorphine: Case report. NZ Med J 89:255–256.

Charney DS, Riordan CE, Kleber HD, Murburg M (1982): Clonidine and naltrexone: A safe, effective and rapid treatment of abrupt withdrawal from methadone therapy. Arch Gen Psychiatry 39:1327–1332.

D'Aunno T, Vaughn TE (1992): Variations in methadone treatment practices. J Am Med Assoc 267:253–258.

Federal Register (March 2, 1989): "National Institute on Drug Abuse; Methadone in Maintenance and Detoxification; Joint Revision of Conditions for Use." 21 CFR Part 291. 54:8954–8979.

Fudala PJ, Jaffe JH, Dax EM, Johnson RE (1990): Use of buprenorphine in the treatment of opiate addiction. II. Physiologic and behavioral effects of daily and alternate-day administration and abrupt withdrawal. Clin Pharmacol Ther 47:525–534.

Gay GR, Inaba DS (1976): Treating acute heroin and methadone toxicity. Anesth Analg 55:607–610.

Gerstein DR, Harwood HJ (eds) (1990): "Treating Drug Problems." Washington DC: National Academy Press.

Hambrook JM, Rance MJ (1976): The interaction of buprenorphine with the opiate receptor: Lipophilicity as a determining factor in drug-receptor kinetics. In Kosterlitz HW (ed): "Opiates and Endogenous Opioid Peptides." Amsterdam: Elsevier, pp 295–301.

Jaffe JH, Martin WR (1985): Opioid analgesics and antagonists. In Gilman AS, Goodman LS, Rall TW, Murad F (eds): "The Pharmacological Basis of Therapeutics." New York: Macmillan, pp 491–531.

Jasinski DR, Pevnick JS, Griffith JD (1978): Human pharmacology and abuse potential of the analgesic buprenorphine. Arch Gen Psychiatry 35:501–516.

Jasinski DR, Johnson RE, Kocher TR (1985): Clonidine in morphine withdrawal. Arch Gen Psychiatry 42:1063–1066.

Johnson RE, Cone EJ, Henningfield JE, Fudala PE (1989): Use of buprenorphine in the treatment of opiate addiction. I. Physiologic and behavioral effects during a rapid dose induction. Clin Pharmacol Ther 46:335–343.

Johnson RE, Fudala PJ, Jaffe JH (1991): Outpatient comparison of buprenorphine and methadone maintenance. I. Effects on opiate use and self-reported adverse effects and withdrawal symptomatology. NIDA Res Monogr 105:585–586.

Judson BA, Carney TM, Goldstein A (1981): Naltrexone treatment of heroin addiction: Efficacy and safety in a double blind dosage comparison. Drug Alcohol Depend 7:325–346.

Kreek MJ (1978): Medical complications in methadone patients. Ann NY Acad Sci 311:110–132.

Kleber HD, Riordan CE, Rounsaville BJ, Kosten TR, Charney D, Gaspari J, Hogan I, O'Connor C (1985): Clonidine in outpatient detoxification from methadone maintenance. Arch Gen Psychiatry 42:391–398.

Lewis JW, Rance MJ, Sanger DJ (1983): The pharmacology and abuse potential of buprenorphine, a new antagonist analgesic. In Mello NK (ed): "Advances in Substance Abuse: Behavioral and Biological Research," Vol. 3. Greenwich, CT: JAI Press, pp 103–154.

Ling W, Wesson DR (1990): Drugs of abuse—Opiates. West J Med 152:565–572.

Lukas SE, Jasinski DR, Johnson RE (1984): Electroencephalographic and behavioral correlates of buprenorphine administration. Clin Pharmacol Ther 36:127–132.

Martin WR, Eades CG, Thompson JA, Huppler RE, Gilbert PE (1976): The effects of morphine- and nalorphine-like drugs in the nondependent and morphine-dependent chronic spinal dog. J Pharmacol Exp Ther 197:517–532.

Mello NK, Mendelson JH (1980): Buprenorphine suppresses heroin use by heroin addicts. Science 207:657–659.

Mello NK, Mendelson JH, Kuehnle JC (1982): Buprenorphine effects on human heroin self-administration: An operant analysis. J Pharmacol Exp Ther 223:30–39.

Metzger DS, Cornish J, Woody GE, McLellan AT, Druley P, O'Brien CP (1990): Naltrexone in federal probationers. NIDA Res Monogr 95:465–466.

National Commission on Aids (July 1991): "A report: The Twin Epidemics of Substance Abuse and HIV."

Schuster CR (1989): Methadone maintenance: An adequate dose is vital in checking the spread of AIDS. NIDA Notes 4(2):3, 33.

Senay E (1988): Methadone and public policy. Br J Addict 83:257–263.

Shukla VK, Goldfrank LR, Turndorf H, Bansinath M (1991): Antagonism of acute cocaine toxicity by buprenorphine. Life Sci 49:1887–1893.

Smith DE, Wesson DR, Tusel DJ (1989): "Treating Opiate Dependency." Center City, MN: Hazelden Foundation, pp 45–59.

Vining E, Kosten TR, Kleber HD (1988): Clinical utility of rapid clonidine–naltrexone detoxification by opioid abusers. Br J Addict 83:567–575.

Witkin JM, Johnson RE, Jaffe JH, Goldberg SR, Grayson NA, Rice KC, Katz JL (1991): The partial opioid agonist, buprenorphine, protects against lethal effects of cocaine. Drug Alcohol Depend 27:177–184.

INDEX

Absorption
  animal studies, 115–118
  human studies
    methodological considerations, 125–127
    protein binding, 129
    single-dose, 127–129
Abstinence syndrome, 90–92; see also Withdrawal
Abuse potential, 71–100, 213, 234–235, 315; see also Reinforcement
  vs. cocaine, 267–273
ACTH secretion, 276–277
ADME (absorption, distribution, metabolism, and excretion) studies; see also individual properties
  animal studies, 115–125
  difficulty of, 113–115
  human studies, 125–134
Agonist activity. See Clinical pharmacology; Kinetics; Preclinical pharmacology
Alcohol, 244
Alfentanil, reinforcement studies, 77–82
Analgesic properties
  in acute pain, 155
  after chronic opioid administration, 155–157
  antinociception: recent studies, 35–37
  chemistry of, 7–9
  in chronic pain, 153–157
  clinical efficacy, 152–157
  duration, 158–159
  and intensity of noxious stimulus, 37–38
  receptor kinetics and, 157–160
  and route of administration in humans, 137–145
  vs. morphine, 37

Analysis
  by gas chromatography, 107–108
  by gas chromatography–mass spectrometry, 105–107
  by immunoassay
    enzyme immunoassay, 110
    radioimmunoassay, 109–110
  by liquid chromatography, 108
Animal studies
  abstinence syndrome, 90–92
  ADME, 115–125
  cocaine abuse, 249–257
  cocaine and luteinizing hormone (LH)
  dose response, 38–41
  general pharmacology, 35–37
  noxious stimulus intensity, 37–38
  operant behavior of schedule-controlled responding; see also under Preclinical pharmacology, 49–66
  reinforcement
    in monkeys, 74–75
    and self-administration in baboons, 75–76
  withdrawal in rats, 41–42
Antagonist activity. See Clinical pharmacology; Kinetics; Preclinical pharmacology
Antinociceptive activity. See Analgesic properties
Assay methods. See Analysis

Behavioral studies; see also under Preclinical pharmacology
  in cocaine abuse, 267–273
  operant behavior of schedule-controlled responding, 49–66
p-Benzoquinone, 3
Binding studies, 19–29; see also Receptor binding

321

Biphasic dose response curves, 38–41
(University of) Bristol (UK), 3–16, 151–160, 175–184
Buprenorphine
    medicinal chemistry; see also Medicinal chemistry, 3–13
    National Institute on Drug Abuse (NIH) perpectives on, 313–319
    pharmacology 19–101; see also Clinical Pharmacology; Kinetics; Preclinical pharmacology
Butorphonol, 291

Catapres. See Clonidine
Chromatography
    gas, 107–108
    gas–mass spectrometry, 105–107
Chronic administration
    Chronic administration
        in maintenance regimens, 216–228, 260–261
        in opioid abuse, 195–197, 207
        in pain, 125, 133, 141, 153–154
Churchill Hospital. See Oxford University
Clinical efficacy. See Clinical pharmacology
Clinical pharmacology
    in analgesic applications
        efficacy, 152–157, 255–256
        receptor kinetics and, 157–160
    in epidural and intrathecal use, 165–172
    NIDA perspective on, 313–314
    in opioid abuse
        acute effects, 191–195, 206–207
        chronic effects, 195–197, 207
        in dependent subjects, 201–204, 213–239
        interaction with opioids, 203–207
        in nondependent subjects, 191–201
        withdrawal and dependency potential, 197–201
    in psychiatric disorders
        case report (schizophrenia), 179–183
        depression, 175–177
        personality disorders, 183–184
        schizophrenia, 177–178, 179–183
        studies concerning, 178–179
Clonidine, 289, 312–313
Cocaine
    and ACTH secretion, 276–277
    buprenorphine and
        comparative abuse potential, 269–273
        primate studies, 249
    drug interactions, 273–277
    dual dependence
        with heroin, 317–318
        with opiates, 257–265
    interactions with buprenorphine, 273–277
    and luteinizing hormone (LH), 250–251
    reinforcement studies, 77–82
Codeine, 189
Cornell University Medical College, 113–134
Coupland, Nick, 175–184
Cowan, Alan, 31–47
Cross-addiction. See Dual dependence and specific drugs
N-Cyclopropylmethylnororvinols, 9–11

Dependence
    in clinical analgesia, 160
    dual
        alcohol/marijuana, 244
        cocaine/opiates, 257–265
        discriminating stimulus effects and, 81–90
        heroin/cocaine, 317–318
        opiate, 213–239; see also Opiate abuse; Opiate dependence
    physical
        buprenorphine effects on other drug dependencies, 93–95
        on buprenorphine with chronic treatment, 90–92
        reinforcing effects and, 73–81
Depression, buprenorphine in treatment of, 175–177
Detoxification and naltrexone induction, 289–301; see also Naltrexone induction
Discriminative stimulus studies
    effects of buprenorphine on other drugs, 88–90
    generalization of buprenorphine studies, 81–88
Distribution, animal studies, 118–121; see also ADME studies
Dopamine receptors, 250–251, 272
Dose reduction studies, 228–233
Dose-response curves, 38–41, 139, 143, 143, 167, 180–182, 254–255, 270, 271
Dual dependence
    alcohol/marijuana, 244
    cocaine/opiates, 257–265
    maintenance regimen in, 260–261

INDEX

safety and effectiveness of
buprenorphine, 257–261
sleep pattern effects in, 258–259
conclusions and implications, 263–265
heroin/cocaine, 317–318
human inpatient studies, 257–261
human outpatient studies, 261–263
Dykstra, Linda A., 49–66

EEG sleep pattern effects, 258–259
Efficacy, 152–157
duration of, 253–256
Enzyme immunoassay, 110
Epidural administration, 165–172
Euro/DPC Ltd. (Wales), 105–112, 137–145
Excretion: animal studies, 123–124; see also ADME studies
chronic dosing, 125
lactation and, 124–125

FDA drug application status, 315–317
Fentanyl, 166–167
antagonism to, 155
Food presentation behavior studies, 51–60
Fudala, Paul J., 213–239

Gas chromatography, 107–108
Gas chromatography–mass spectrometry, 105–107
Glue, Paul, 175–184
Groves, Simon, 175–184

Hallucinations, 172
Harvard Medical School, 49–66, 241–279
Heng Xu, 19–29
Heroin abuse
buprenorphine in
human studies, 245–248
primate studies, 248–249
dual dependence with cocaine, 317–318
historical and general considerations, 241–242
self-administration study procedures in, 243–245
transfer from, 228–232
Human studies
ADME, 125–133
analgesic properties and route of administration, 137–145
detoxification and naltrexone induction, 291–301
interactions with opioids, 208
methodological considerations, 125–127

of opiate abuse/opiate dependency, 191–204, 213–239; see also Opiate abuse; Opiate dependency

Immunoassay
enzyme, 110
radio-, 109–110
Intramuscular administration, 140
Intrathecal administration, 169–170
Intravenous administration, 137–140
Inturrisi, Charles E., 113–134

Jasinski, Donald R., 189–209
J/J/F 090 Maintenance Study, 213–236, 315–317
Johns Hopkins University School of Medicine, 189–209, 213–239
Johnson, Rolley E., 213–239

Kappa receptors
antagonist activity, 58–59, 64
clinical agonism/antagonism, 151–160
and cocaine abuse, 275
and withdrawal, 208–209
Kinetics, 137–146; see also Medicinal chemistry; Pharmacokinetics
in chronic pain, 141
in humans
clinical implications, 142–145
intramuscular route, 140
intravenous route, 137–140
metabolism, 142
oral route, 141
spinal/intrathecal route, 141–142, 165–172
sublingual route, 140–141
in renal failure, 139
Kosten, Thomas R., 291–301

LAAM (levo-alpha-acetylmethadol), 311–312
Lactation, 124–125
Levo-alpha-acetylmethadol (LAAM), 311–312
Lewis, John W., 3–16, 151–160
Liquid chromatography, 108
Literature review: receptor binding, 19–29

Maintenance regimens, 216–228, 260–261
Marijuana, 243–244
Massachusetts General Hospital (Boston), 165–172
McLean Hospital–Harvard Medical School (Belmont, MA), 49–66, 241–279

McQuay, H.J., 137–145
Medicinal chemistry; *see also* Clinical pharmacology; Preclinical pharmacology
　chemical reactivity of orvinols, 11–14
　conclusions, 13
　in epidural and intrathecal use, 165–172
　in humans, 137–145
　in opiate/cocaine dependence, 266
　orvinol (N-CPM) derivatives, 9–11
　structure–activity relationships, 6–9
　synthesis and chemistry, 3–6
Mello, Nancy K., 241–279
Mendelson, Jack H., 241–279
Meperidine, 166–167
Metabolism; *see also* ADME studies
　animal studies, 121–123
　human studies, 129–133, 142
　　chronic dosing, 133
Methadone, 166–167; *see also* Naltrexone induction
　efficacy
　　vs. buprenorphine, 226–228, 299
　　vs. naltrexone, 289–291
　LAAM (levo-alpha-acetylmethadol) derivative, 311–312
　NIDA perspective on, 310–311
　substitution for, 207–209, 289–302
(University of) Michigan Medical School, 71–100
Monkeys. *See* Animal studies
Moore, R. Andrew, 105–112, 137–145
Morphine; *see also* Opiate abuse; Opiate dependency
　analgesic properties vs. buprenorphine, 37
　behavior studies vs. buprenorphine, 49–68
　comparative response in epidural administration, 166–167
　equivalents of buprenorphine ceiling dosage, 154–155
　substitution for, 207–209
Mu receptors
　binding properties, 72
　buprenorphine and, 54–57, 58, 64, 151–160
　morphine and, 52–54, 61–63
　antagonist activity
　　buprenorphine, 58, 64
　　morphine, 63

Nalbupine, 291
Naloxone
　with buprenorphine, 318
　vs. naltrexone, 291–294
Naltrexone induction; *see also* Methadone
　discussion, 299–303
　general considerations in detoxification, 289
　results, 295–299
　study methods, 291–294
　vs. naloxone, 290–291
National Institute on Drug Abuse (NIDA), 19–29, 189–209, 309–319
　and buprenorphine, 313–319
　and clonidine therapy, 312–313
　and combined treatments, 313
　and LAAM therapy, 311–312
　and methadone therapy, 310–311
　and naltrexone therapy, 312
　and substance abuse epidemiology, 309–310
Needle use/needle sharing, 262, 309–310
Negus, S. Stevens, 49–66, 71–100
Neurotoxicity, 168
Ni, Qi, 19–29
NIDA. *See* National Institute on Drug Abuse (NIDA)
(University of) North Carolina, 49–66
Noxious stimulus intensity studies, 37–38
Nutt, David, 175–184

Operant behavioral studies, 49–66; *see also under* Preclinical pharmacology
Opiate dependence
　concurrent with cocaine
　　conclusions and implications, 263–265
　　human inpatient studies, 257–261
　　human outpatient studies, 261–263
Opiate abuse; *see also* Opiate dependence
　buprenorphine in
　　in dependent users; *see also* Opiate dependence, 201–204, 213–239
　　interaction with opioids, 204–209
　　in nondependent users, 189–200
　historical considerations, 190–191
　laboratory studies, 189–209
Opiate dependence
　detoxification and naltrexone induction, 289–301
　efficacy studies
　　dose induction procedures, 214–216

of dose reduction, 228–233
historical and general considerations, 213–214
of maintenance regimens, 216–228
safety, 232–233
indications for treatment, 233–234
Opioid receptors
interactions with buprenorphine, 32–35
and physical dependency, 93
Opioids; *see also* Opiate abuse; Opiate dependence; Opioid receptors
chronic administration and transfer, 155–157, 159–160
Oral administration, 141
ORLAAM (LAAM; levo-alpha-acetylmethadol), 311–312
Orvinol derivatives, chemistry of, 9–11
Outpatient studies, 242, 261–263
detoxification and naltrexone induction, 292–293, 295–296, 299–301
Oxford University (UK), 137–145

Pain, chronic, 141, 153–157; *see also* Analgesic properties
Parenteral administration, 140
(University of) Pennsylvania School of Medicine, 213–239
Pentazocine, 291
Personality disorders, buprenorphine in treatment of, 183–184
Perspectives on substance abuse therapies, 309–319
Pharmacology
clinical, 151–184; *see also* Clinical pharmacology
preclinical, 19–101; *see also* Preclinical pharmacology
Pharmacodynamics. *See* Medicinal chemistry
Physical dependence, 90–96; *see also* Abuse potential; Dependence; Discriminating stimulus effects; Reinforcement
Polydrug use. *See* Dual dependence
Precipitated abstinence, 155–156; *see also* Withdrawal
Preclinical pharmacology
cocaine interactions, 272
discriminative stimulus effects
effects on other drugs, 88–90
generalizability to other drugs, 81–88
general
abrupt withdrawal in rats, 41–42

biphasic dose-response curves, 38–41
historical considerations, 32
intensity of noxious stimulus, 37–38
opioid receptor interactions, 32–35
perspective, 42–43
properties of buprenorphine, 31–32
recent data, 35–37
kinetics in humans, 137–145
operant behavior of scheduled-control responding
agonist effects, 50–58, 61–63
antagonist effects, 58–61, 64
food presentation studies, 51–58
kappa receptor antagonism, 58–60, 64
mu receptor agonism/antagonism, 54–57, 58, 64
shock termination/titration studies, 61–65
physical dependence
on buprenorphine with chronic treatment, 90–91
effects of buprenorphine on other drugs, 92–95
future research directions, 95–96
receptor binding: review of literature, 19–26
reinforcing effects
electrical brain stimulation studies, 77
placement-conditioning study, 76–77
on reinforcing effects of other drugs, 77–81
self-administration studies, 73–77
Preston, Kenzie L., 189–209
Primate studies. *See* Animal studies
Progressive-ratio reinforcement, 77
Protein binding, 129
Pruritus, 168, 171
Psychiatric disorders
case report (schizophrenia), 179–183
depression, 175–177
personality disorders, 183–184
schizophrenia, 177–178, 179–183
studies, 178–179
Psychopharmacology. *See* Psychiatric disorders

Radcliffe Infirmary. *See* Oxford University
Radioimmunoassay, 109–110
Rats. *See* Animal studies

Receptor binding
  affinity, 50
  agonist/antagonist properties, 50–51, 272–274
  in clinical analgesia, 157–160
  efficacy, 50
  general principles of, 19
  opioid receptor interactions, 32–35
  and physical dependency, 93
  review of literature
    competitive binding, 21–23
    conclusions, 23–26
    direct binding, 20–21
  study design and, 50
  vs. methadone, 300–301
Reckitt & Colman Products (UK), 113–134
Reinforcement, 209
  in alfentanil and cocaine, 77–82
  buprenorphine effect on other drugs, 77–82
  electrical brain stimulation studies, 77
  placement-conditioning studies, 76–77
  progressive-ratio-schedule, 77
  and self-administration in baboons, 75–76
  vs. cocaine, 77–82, 269–273
Renal failure, 139
Respiratory depression, 157, 158–159, 168, 171, 311
Rosen, Marc, 291–301
Rosow, Carl E., 165–172
Rothman, Richard B., 19–29
Routes of administration
  epidural, 165–172
  in humans, 137–145
  intrathecal, 169–170
  and reversal, 159–160
  spinal, 165–172
  sublingual, 159–160
  toxicity and, 170–172

Safety. See Toxicity
Schizophrenia, 177–178, 179–183
Schuster, Charles R., 309–319
Segal, Doralie L., 309–319
Self-administration studies, 73–77
  in humans, 243–257; see also Cocaine abuse; Dual dependence; Heroin abuse; Opiate abuse; Opiate dependence
  in primates, 249–256
Shock termination/titration studies, 60–65
Side effects; see also Toxicity
  of buprenorphine, 232–233, 246–247, 258–259

of clonidine, 312–313
  NIDA perspective on, 314–315
Sleep pattern effects, 258–259
Spinal administration, 141–142, 165–172
Study design, and receptor binding results, 50
Sublingual administration, 140–141, 159–160
Substance abuse; see also specific substances
  cocaine and heroin, 241–279
  detoxification and naltrexone induction, 289–302
  NIDA/NIH perspectives on, 309–319
  opioids
    clinical studies, 213–239
    laboratory studies, 180–209
Synthesis of buprenorphine, 3–6

Temple University (Philadelphia), 31–47
Tolerance, 172
  to buprenorphine, 56–57, 63–64
  to morphine, 53–54, 63–64
Toxicity, 232–233, 246–247, 260–261, 311
  NIDA perspective on, 314–315
  in spinal administration, 168, 170–172
Trexan. See Naltrexone

Urinary retention, 168, 172

Walter, Donald S., 113–134
Weight gain/weight loss, and withdrawal, 41
Withdrawal, 155–156
  abstinence syndrome, 90–92
  in detoxification and naltrexone induction, 29–298, 294–295, 299–300
  and dose induction, 214
  and dose reduction regimens, 228–232
  kappa receptor antagonism and, 208–209
  naltrexone vs. naloxone, 291
  rat studies, 41–42
  and weight gain, 41
meWoods, James H., 71–100

Xu, Heng, 19–29

Yale University School of Medicine, 291–301